# Development and Application of Green or Sustainable Strategies in Analytical Chemistry

# Development and Application of Green or Sustainable Strategies in Analytical Chemistry

Editor

**Attilio Naccarato**

MDPI • Basel • Beijing • Wuhan • Barcelona • Belgrade • Manchester • Tokyo • Cluj • Tianjin

*Editor*
Attilio Naccarato
University of Calabria
Italy

*Editorial Office*
MDPI
St. Alban-Anlage 66
4052 Basel, Switzerland

This is a reprint of articles from the Special Issue published online in the open access journal *Separations* (ISSN 2297-8739) (available at: https://www.mdpi.com/journal/separations/special_issues/Development_Chemistry).

For citation purposes, cite each article independently as indicated on the article page online and as indicated below:

LastName, A.A.; LastName, B.B.; LastName, C.C. Article Title. *Journal Name* **Year**, *Volume Number*, Page Range.

**ISBN 978-3-0365-6636-8 (Hbk)**
**ISBN 978-3-0365-6637-5 (PDF)**

© 2023 by the authors. Articles in this book are Open Access and distributed under the Creative Commons Attribution (CC BY) license, which allows users to download, copy and build upon published articles, as long as the author and publisher are properly credited, which ensures maximum dissemination and a wider impact of our publications.

The book as a whole is distributed by MDPI under the terms and conditions of the Creative Commons license CC BY-NC-ND.

# Contents

**About the Editor** .................................................. vii

**Attilio Naccarato**
Development and Application of Green or Sustainable Strategies in Analytical Chemistry
Reprinted from: *Separations* 2023, 10, 32, doi:10.3390/separations10010032 .............. 1

**Theodoros Chatzimitakos, Phoebe Anagnostou, Ioanna Constantinou, Kalliroi Dakidi and Constantine Stalikas**
Magnetic Ionic Liquids in Sample Preparation: Recent Advances and Future Trends
Reprinted from: *Separations* 2021, 8, 153, doi:10.3390/separations8090153 .............. 5

**Danielle P. Arcon and Francisco C. Franco, Jr.**
A Simple Microextraction Method for Toxic Industrial Dyes Using a Fatty-Acid Solvent Mixture
Reprinted from: *Separations* 2021, 8, 135, doi:10.3390/separations8090135 .............. 33

**Zhaojin Zhang, Yinan Li, Jing Gao, Alula Yohannes, Hang Song and Shun Yao**
Removal of Pyridine, Quinoline, and Aniline from Oil by Extraction with Aqueous Solution of (Hydroxy)quinolinium and Benzothiazolium Ionic Liquids in Various Ways
Reprinted from: *Separations* 2021, 8, 216, doi:10.3390/separations8110216 .............. 45

**Ismat H. Ali, Mutasem Z. Bani-Fwaz, Adel A. El-Zahhar, Riadh Marzouki, Mosbah Jemmali and Sara M. Ebraheem**
Gum Arabic-Magnetite Nanocomposite as an Eco-Friendly Adsorbent for Removal of Lead(II) Ions from Aqueous Solutions: Equilibrium, Kinetic and Thermodynamic Studies
Reprinted from: *Separations* 2021, 8, 224, doi:10.3390/separations8110224 .............. 63

**Jiawei Huang, Zhaoping Zhong, Yueyang Xu and Yuanqiang Xu**
Study on the Preparation of Magnetic Mn–Co–Fe Spinel and Its Mercury Removal Performance
Reprinted from: *Separations* 2021, 8, 225, doi:10.3390/separations8110225 .............. 81

**Matthias Harder, Rania Bakry, Felix Lackner, Paul Mayer, Christoph Kappacher, Christoph Grießer, et al.**
The Crosslinker Matters: Vinylimidazole-Based Anion Exchange Polymer for Dispersive Solid-Phase Extraction of Phenolic Acids
Reprinted from: *Separations* 2022, 9, 72, doi:10.3390/separations9030072 .............. 99

**Tobias Schlappack, Nina Weidacher, Christian W. Huck, Günther K. Bonn and Matthias Rainer**
Effective Solid Phase Extraction of Toxic Pyrrolizidine Alkaloids from Honey with Reusable Organosilyl-Sulfonated Halloysite Nanotubes
Reprinted from: *Separations* 2022, 9, 270, doi:10.3390/separations9100270 .............. 117

**Ahmed I. Foudah, Faiyaz Shakeel, Mohammed H. Alqarni, Tariq M. Aljarba, Sultan Alshehri and Prawez Alam**
Simultaneous Detection of Chlorzoxazone and Paracetamol Using a Greener Reverse-Phase HPTLC-UV Method
Reprinted from: *Separations* 2022, 9, 300, doi:10.3390/separations9100300 .............. 133

**Mohammed H. Alqarni, Faiyaz Shakeel, Ahmed I. Foudah, Tariq M. Aljarba, Aftab Alam, Sultan Alshehri and Prawez Alam**
Comparison of Validation Parameters for the Determination of Vitamin D3 in Commercial Pharmaceutical Products Using Traditional and Greener HPTLC Methods
Reprinted from: *Separations* 2022, 9, 301, doi:10.3390/separations9100301 .............. 145

**Mara Calleja-Gómez, Juan Manuel Castagnini, Ester Carbó, Emilia Ferrer, Houda Berrada and Francisco J. Barba**
Evaluation of Pulsed Electric Field-Assisted Extraction on the Microstructure and Recovery of Nutrients and Bioactive Compounds from Mushroom (*Agaricus bisporus*)
Reprinted from: *Separations* **2022**, *9*, 302, doi:10.3390/separations9100302 . . . . . . . . . . . . . . . **157**

**Catia Balducci, Marina Cerasa, Pasquale Avino, Paolo Ceci, Alessandro Bacaloni and Martina Garofalo**
Analytical Determination of Allergenic Fragrances in Indoor Air
Reprinted from: *Separations* **2022**, *9*, 99, doi:10.3390/separations9040099 . . . . . . . . . . . . . . . **175**

**Hongmei Zhou, Junhui Zhang, Aihong Duan, Bangjin Wang, Shengming Xie and Liming Yuan**
Facile Preparation of Phenyboronic-Acid-Functionalized $Fe_3O_4$ Magnetic Nanoparticles for the Selective Adsorption of Ortho-Dihydroxy-Containing Compounds
Reprinted from: *Separations* **2023**, *10*, 4, doi:10.3390/separations10010004 . . . . . . . . . . . . . . . **187**

# About the Editor

**Attilio Naccarato**

Dr. Attilio Naccarato is an Assistant Professor at the University of Calabria. His main research activity include the method development for the analysis of pollutants in environmental, food, and clinical matrices using eco-friendly analytical strategies such as microextraction approaches, mass spectrometry-based techniques, and multivariate optimization. He also serves as an Associate Editor on the editorial board of Chemical Papers (Springer), Frontiers in Analytical Science, and Separations (MDPI).

*Editorial*

# Development and Application of Green or Sustainable Strategies in Analytical Chemistry

Attilio Naccarato

Department of Chemistry and Chemical Technologies, University of Calabria, Via P. Bucci Cubo 12D, I-87030 Arcavacata di Rende, CS, Italy; attilio.naccarato@unical.it

Analytical chemistry is bound to face growing challenges in the near future, especially for the quantification of trace analytes in complex matrices. Although the development of increasingly sensitive and specific instrumental techniques has achieved remarkable results, sample preparation is still a fundamental step, often limiting the whole workflow.

In the context spawned by recent international environmental policies that are responsive to the rapport of human activities with the surrounding environment, chemistry cannot hesitate to give its contribution. Almost pioneeringly, in analytical chemistry, we have been talking for some time about "green analytical chemistry", its guiding principles, and the development of eco-friendly analytical approaches.

Since then, awareness of eco-sustainability in analytical chemistry has evolved significantly, also targeting the establishment of referable and comparable metric systems to assess the environmental cost of analysis and, more specifically, that of sample preparation [1,2]. The analytical benefits of effective sample-prep are well known, and among the approaches that provide the greatest environmental benefits is the wide portfolio of microextraction techniques, which include not only frontier solutions but also commercially available and affordable strategies [3–5]. However, the new and still-open challenge is to advance not only in eco-compatibility but mainly in eco-sustainability, thus rooting the future of analytical chemistry in new paradigms which are founded on the use of materials that are recycled, reusable, or from natural sources, therefore being sustainable [6].

It must be clear how the environmental footprint left behind must now be assessed comprehensively, with a broad look that encompass not only the laboratory practices but it is important to embrace the entire analytical process, from sampling to final determination. Hence, for example, it is vital to develop strategies that simplify sampling and counteract the need for costly and time-consuming transportation to analytical facilities [2,7]. While this urgency may sound somewhat unimportant to the most affectionate laboratory folks, it actually shows its weight in the context of international monitoring networks where sampling sites are multiple and often far from the few analytical centers [8–13].

Equally important is the oversight of energy demand to conduct the analysis. Using very sophisticated (and often energy-intensive) instruments is not always a real need, but rather an academic or research pursuit, which also requires high levels of skill training and operating costs. From gravimetric and volumetric techniques to sophisticated high-resolution mass spectrometry, a wide set of analytical instruments are encountered, and cost-effectiveness (including environmental) must be the guideline for a proper choice of strategy [14].

These evaluations are not unrelated to a proper knowledge of the analytical system under consideration and the potential of the used method, both aspects for which traditional optimization and evaluation approaches do not permit a full understanding. Although already used by some scholars and analysts [15–18], it is still desirable to increasingly involve chemometric and data analysis techniques in real case studies and method development. Only the release of optimization and response surfaces makes it possible to thoroughly get to know a method of analysis and its behavior, somewhat as one does with a person when

**Citation:** Naccarato, A. Development and Application of Green or Sustainable Strategies in Analytical Chemistry. *Separations* **2023**, *10*, 32. https://doi.org/10.3390/separations10010032

Received: 2 December 2022
Accepted: 21 December 2022
Published: 5 January 2023

**Copyright:** © 2023 by the author. Licensee MDPI, Basel, Switzerland. This article is an open access article distributed under the terms and conditions of the Creative Commons Attribution (CC BY) license (https://creativecommons.org/licenses/by/4.0/).

one gets to know his or her personality and understands what are the most sensitive points that determine his or her emotional reaction.

Considering the above, this Special Issue aimed to collect studies that show the progress in analytical chemistry based on the arguments previously raised and discussed, with a particular reference to eco-compatibility and eco-sustainability. The collected articles range from studies aimed at the development of new materials to address the removal or extraction of target compounds from environmental and food matrices, to the use of less sophisticated techniques for analysis in the pharmaceutical industry.

The Special Issue includes a review by Chatzimitakos et al. [19], which provides a literature update on the current trends of magnetic ionic liquids (MILs) in different modes of sample preparation, along with the current limitations and the prospects of the field. MILs combine the advantageous properties of ionic liquids along with the magnetic properties, creating an unsurpassed combination, and their use in different extraction approaches including dispersive liquid–liquid microextraction, and matrix solid-phase dispersion were surveyed.

Pulsed electric field (PEF), as a sustainable innovative technology for the recovery of nutrients and bioactive compounds from *A. bisporus*, was explored in the contribution by Barba's group [20]. PEF facilitates the sustainable and economic isolation of compounds by using water as a solvent, thus reducing the use of organic solvents. In addition, it is a technology that reduces the temperature and time required for the extraction of the different compounds, thus preserving the thermolabile components. The application of PEF technology under optimal conditions to mushrooms increases the extraction of carbohydrates, proteins, antioxidant compounds, and minerals such as P, Mg, Fe, and Se compared to conventional methodology.

A low-volume fatty-acid mixture-based solvent was used by Francisco C. Franco and co-workers [21] for a simple and efficient microextraction method for the removal of dyes in aqueous solutions. The fatty-acid mixture presented a green and economic procedure for the extraction of toxic dyes in wastewater treatment. The experimental results reveal that even at a microvolume solvent availability, a fatty-acid mixture performs efficiently even towards hydrophobic contaminants 200 times its volume.

Ali et al. [22] reported the synthesis of a gum arabic–magnetite composite (GA/MNPs), which was characterized and assessed by several spectroscopic and analytical methods as an adsorbent for Pb(II) ions from synthetic wastewater. The GA/MNP composite is a partially bio-based material, and it demonstrated unique properties which permit a removal efficiency of 99.3% at the optimum conditions.

Additionally, an element known for its potentially toxic effects on humans and ecosystems is mercury, which was the target of the study carried out by Huang and co-workers [23]. Here, a manganese-doped manganese–cobalt–iron spinel adsorbent was prepared by the sol–gel self-combustion method and characterized by XRD, SEM, and VSM. Its use was explored providing theoretical guidance and a research basis for the development of efficient and recyclable spinel ferrite adsorbents for the trapping of gaseous elemental mercury ($Hg^0$).

The Special Issue included two contributions from Rainer's group reporting the studies on reusable materials to improve the extraction of target compounds. More precisely, in one study, they synthesized a novel [$C_6$-bis-VIM] [Br] crosslinked anion exchange polymer and subsequently develop an efficient extraction procedure for phenolic acids from aqueous samples [24]. The reusability of the sorbent used for dispersive- solid-phase extraction was investigated, and it resulted in being an efficient and sustainable anion exchange material with maximum recoveries ranging between 84.1 and 92.5%.

In the second study, the use of modified halloysite nanotubes was explored in the selective solid-phase extraction of toxic pyrrolizidine alkaloids as alternative candidates to polymeric resins [25]. Satisfactory results were obtained in an aqueous pyrrolizidine alkaloid mixture containing four of the six main structures of the pyrrolizidine alkaloid group as well as in spiked honey samples. Furthermore, halloysite nanotubes can once again be presented as an economical and environmentally friendly resource due to their massive natural occurrence and resulting low cost.

The use of less demanding and sophisticated instrumentation was addressed in the two contributions by Alam and co-workers, who employed the high-performance thin-layer chromatography for vitamin D3 estimation in commercial pharmaceutical products and the quantification of chlorzoxazone and paracetamol in commercial capsules and tablets [26,27]. The greenness of the proposed HPTLC-UV methods was assessed quantitatively by the "Analytical GREENness" (AGREE)metric.

I hope that you will enjoy reading the collection of papers included in this Special Issue.

**Conflicts of Interest:** The authors declare no conflict of interest.

# References

1. Sajid, M.; Płotka-Wasylka, J. Green Analytical Chemistry Metrics: A Review. *Talanta* **2022**, *238*, 123046. [CrossRef] [PubMed]
2. Wojnowski, W.; Tobiszewski, M.; Pena-Pereira, F.; Psillakis, E. AGREEprep—Analytical Greenness Metric for Sample Preparation. *TrAC Trends Anal. Chem.* **2022**, *149*, 116553. [CrossRef]
3. Naccarato, A.; Gionfriddo, E.; Elliani, R.; Pawliszyn, J.; Sindona, G.; Tagarelli, A. Investigating the Robustness and Extraction Performance of a Matrix-Compatible Solid-Phase Microextraction Coating in Human Urine and Its Application to Assess 2-6-Ring Polycyclic Aromatic Hydrocarbons Using GC-MS/MS. *J. Sep. Sci.* **2018**, *41*, 929–939. [CrossRef] [PubMed]
4. Amico, D.; Tassone, A.; Pirrone, N.; Sprovieri, F.; Naccarato, A. Recent Applications and Novel Strategies for Mercury Determination in Environmental Samples Using Microextraction-Based Approaches: A Review. *J. Hazard. Mater.* **2022**, *433*, 128823. [CrossRef] [PubMed]
5. Naccarato, A.; Tagarelli, A. Recent Applications and Newly Developed Strategies of Solid-Phase Microextraction in Contaminant Analysis: Through the Environment to Humans. *Separations* **2019**, *6*, 54. [CrossRef]
6. Mafra, G.; García-Valverde, M.T.; Millán-Santiago, J.; Carasek, E.; Lucena, R.; Cárdenas, S. Returning to Nature for the Design of Sorptive Phases in Solid-Phase Microextraction. *Separations* **2019**, *7*, 2. [CrossRef]
7. Naccarato, A.; Moretti, S.; Sindona, G.; Tagarelli, A. Identification and Assay of Underivatized Urinary Acylcarnitines by Paper Spray Tandem Mass Spectrometry. *Anal. Bioanal. Chem.* **2013**, *405*, 8267–8276. [CrossRef]
8. Martino, M.; Tassone, A.; Angiuli, L.; Naccarato, A.; Dambruoso, P.R.; Mazzone, F.; Trizio, L.; Leonardi, C.; Petracchini, F.; Sprovieri, F.; et al. First Atmospheric Mercury Measurements at a Coastal Site in the Apulia Region: Seasonal Variability and Source Analysis. *Environ. Sci. Pollut. Res.* **2022**, *29*, 68460–68475. [CrossRef]
9. Chianese, E.; Tirimberio, G.; Dinoi, A.; Cesari, D.; Contini, D.; Bonasoni, P.; Marinoni, A.; Andreoli, V.; Mannarino, V.; Moretti, S.; et al. Particulate Matter Ionic and Elemental Composition during the Winter Season: A Comparative Study among Rural, Urban and Remote Sites in Southern Italy. *Atmosphere* **2022**, *13*, 356. [CrossRef]
10. Naccarato, A.; Tassone, A.; Martino, M.; Moretti, S.; Macagnano, A.; Zampetti, E.; Papa, P.; Avossa, J.; Pirrone, N.; Nerentorp, M.; et al. A Field Intercomparison of Three Passive Air Samplers for Gaseous Mercury in Ambient Air. *Atmos. Meas. Tech.* **2021**, *14*, 3657–3672. [CrossRef]
11. Moretti, S.; Tassone, A.; Andreoli, V.; Carbone, F.; Pirrone, N.; Sprovieri, F.; Naccarato, A. Analytical Study on the Primary and Secondary Organic Carbon and Elemental Carbon in the Particulate Matter at the High-Altitude Monte Curcio GAW Station, Italy. *Environ. Sci. Pollut. Res.* **2021**, *28*, 60221–60234. [CrossRef] [PubMed]
12. Moretti, S.; Salmatonidis, A.; Querol, X.; Tassone, A.; Andreoli, V.; Bencardino, M.; Pirrone, N.; Sprovieri, F.; Naccarato, A. Contribution of Volcanic and Fumarolic Emission to the Aerosol in Marine Atmosphere in the Central Mediterranean Sea: Results from Med-Oceanor 2017 Cruise Campaign. *Atmosphere* **2020**, *11*, 149. [CrossRef]
13. Bencardino, M.; Andreoli, V.; D'Amore, F.; de Simone, F.; Mannarino, V.; Castagna, J.; Moretti, S.; Naccarato, A.; Sprovieri, F.; Pirrone, N. Carbonaceous Aerosols Collected at the Observatory of Monte Curcio in the Southern Mediterranean Basin. *Atmosphere* **2019**, *10*, 592. [CrossRef]
14. Tassone, A.; Moretti, S.; Martino, M.; Pirrone, N.; Sprovieri, F.; Naccarato, A. Modification of the EPA Method 1631E for the Quantification of Total Mercury in Natural Waters. *MethodsX* **2020**, *7*, 100987. [CrossRef] [PubMed]
15. Naccarato, A.; Tassone, A.; Martino, M.; Elliani, R.; Sprovieri, F.; Pirrone, N.; Tagarelli, A. An Innovative Green Protocol for the Quantification of Benzothiazoles, Benzotriazoles and Benzosulfonamides in PM10 Using Microwave-Assisted Extraction Coupled with Solid-Phase Microextraction Gas Chromatography Tandem-Mass Spectrometry. *Environ. Pollut.* **2021**, *285*, 117487. [CrossRef] [PubMed]
16. Elliani, R.; Naccarato, A.; Malacaria, L.; Tagarelli, A. A Rapid Method for the Quantification of Urinary Phthalate Monoesters: A New Strategy for the Assessment of the Exposure to Phthalate Ester by Solid-Phase Microextraction with Gas Chromatography and Tandem Mass Spectrometry. *J. Sep. Sci.* **2020**, *43*, 3061–3073. [CrossRef] [PubMed]
17. Naccarato, A.; Gionfriddo, E.; Elliani, R.; Sindona, G.; Tagarelli, A. A Fast and Simple Solid Phase Microextraction Coupled with Gas Chromatography-Triple Quadrupole Mass Spectrometry Method for the Assay of Urinary Markers of Glutaric Acidemias. *J. Chromatogr. A* **2014**, *1372*, 253–259. [CrossRef]

18. Naccarato, A.; Elliani, R.; Sindona, G.; Tagarelli, A. Multivariate Optimization of a Microextraction by Packed Sorbent-Programmed Temperature Vaporization-Gas Chromatography–Tandem Mass Spectrometry Method for Organophosphate Flame Retardant Analysis in Environmental Aqueous Matrices. *Anal. Bioanal. Chem.* **2017**, *409*, 7105–7120. [CrossRef]
19. Chatzimitakos, T.; Anagnostou, P.; Constantinou, I.; Dakidi, K.; Stalikas, C. Magnetic Ionic Liquids in Sample Preparation: Recent Advances and Future Trends. *Separations* **2021**, *8*, 153. [CrossRef]
20. Calleja-Gómez, M.; Castagnini, J.M.; Carbó, E.; Ferrer, E.; Berrada, H.; Barba, F.J. Evaluation of Pulsed Electric Field-Assisted Extraction on the Microstructure and Recovery of Nutrients and Bioactive Compounds from Mushroom (*Agaricus bisporus*). *Separations* **2022**, *9*, 302. [CrossRef]
21. Arcon, D.; Franco, F. A Simple Microextraction Method for Toxic Industrial Dyes Using a Fatty-Acid Solvent Mixture. *Separations* **2021**, *8*, 135. [CrossRef]
22. Ali, I.H.; Bani-Fwaz, M.Z.; El-Zahhar, A.A.; Marzouki, R.; Jemmali, M.; Ebraheem, S.M. Gum Arabic-Magnetite Nanocomposite as an Eco-Friendly Adsorbent for Removal of Lead(II) Ions from Aqueous Solutions: Equilibrium, Kinetic and Thermodynamic Studies. *Separations* **2021**, *8*, 224. [CrossRef]
23. Huang, J.; Zhong, Z.; Xu, Y.; Xu, Y. Study on the Preparation of Magnetic Mn–Co–Fe Spinel and Its Mercury Removal Performance. *Separations* **2021**, *8*, 225. [CrossRef]
24. Harder, M.; Bakry, R.; Lackner, F.; Mayer, P.; Kappacher, C.; Grießer, C.; Neuner, S.; Huck, C.W.; Bonn, G.K.; Rainer, M. The Crosslinker Matters: Vinylimidazole-Based Anion Exchange Polymer for Dispersive Solid-Phase Extraction of Phenolic Acids. *Separations* **2022**, *9*, 72. [CrossRef]
25. Schlappack, T.; Weidacher, N.; Huck, C.W.; Bonn, G.K.; Rainer, M. Effective Solid Phase Extraction of Toxic Pyrrolizidine Alkaloids from Honey with Reusable Organosilyl-Sulfonated Halloysite Nanotubes. *Separations* **2022**, *9*, 270. [CrossRef]
26. Foudah, A.I.; Shakeel, F.; Alqarni, M.H.; Aljarba, T.M.; Alshehri, S.; Alam, P. Simultaneous Detection of Chlorzoxazone and Paracetamol Using a Greener Reverse-Phase HPTLC-UV Method. *Separations* **2022**, *9*, 300. [CrossRef]
27. Alqarni, M.H.; Shakeel, F.; Foudah, A.I.; Aljarba, T.M.; Alam, A.; Alshehri, S.; Alam, P. Comparison of Validation Parameters for the Determination of Vitamin D3 in Commercial Pharmaceutical Products Using Traditional and Greener HPTLC Methods. *Separations* **2022**, *9*, 301. [CrossRef]

**Disclaimer/Publisher's Note:** The statements, opinions and data contained in all publications are solely those of the individual author(s) and contributor(s) and not of MDPI and/or the editor(s). MDPI and/or the editor(s) disclaim responsibility for any injury to people or property resulting from any ideas, methods, instructions or products referred to in the content.

*Review*

# Magnetic Ionic Liquids in Sample Preparation: Recent Advances and Future Trends

**Theodoros Chatzimitakos \*, Phoebe Anagnostou, Ioanna Constantinou, Kalliroi Dakidi and Constantine Stalikas**

Laboratory of Analytical Chemistry, Department of Chemistry, University of Ioannina, 45110 Ioannina, Greece; foebes@gmail.com (P.A.); pch01378@uoi.gr (I.C.); pch01372@uoi.gr (K.D.); cstalika@uoi.gr (C.S.)
\* Correspondence: chatzimitakos@outlook.com; Tel.: +30-265-1008-725

**Abstract:** In the last decades, a myriad of materials has been synthesized and utilized for the development of sample preparation procedures. The use of their magnetic analogues has gained significant attention and many procedures have been developed using magnetic materials. In this context, the benefits of a new class of magnetic ionic liquids (MILs), as non-conventional solvents, have been reaped in sample preparation procedures. MILs combine the advantageous properties of ionic liquids along with the magnetic properties, creating an unsurpassed combination. Owing to their unique nature and inherent benefits, the number of published reports on sample preparation with MILs is increasing. This fact, along with the many different types of extraction procedures that are developed, suggests that this is a promising field of research. Advances in the field are achieved both by developing new MILs with better properties (showing either stronger response to external magnetic fields or tunable extractive properties) and by developing and/or combining methods, resulting in advanced ones. In this advancing field of research, a good understanding of the existing literature is needed. This review aims to provide a literature update on the current trends of MILs in different modes of sample preparation, along with the current limitations and the prospects of the field. The use of MILs in dispersive liquid–liquid microextraction, single drop microextraction, matrix solid-phase dispersion, etc., is discussed herein among others.

**Keywords:** magnetic ionic liquid; sample preparation; dispersive liquid–liquid microextraction; single drop microextraction; GC; HPLC

**Citation:** Chatzimitakos, T.; Anagnostou, P.; Constantinou, I.; Dakidi, K.; Stalikas, C. Magnetic Ionic Liquids in Sample Preparation: Recent Advances and Future Trends. *Separations* **2021**, *8*, 153. https://doi.org/10.3390/separations8090153

Academic Editors: Attilio Naccarato and Jared L. Anderson

Received: 12 August 2021
Accepted: 7 September 2021
Published: 13 September 2021

**Publisher's Note:** MDPI stays neutral with regard to jurisdictional claims in published maps and institutional affiliations.

**Copyright:** © 2021 by the authors. Licensee MDPI, Basel, Switzerland. This article is an open access article distributed under the terms and conditions of the Creative Commons Attribution (CC BY) license (https://creativecommons.org/licenses/by/4.0/).

## 1. Introduction

The advancements in the production of new materials in liquid and solid form have significantly impacted the field of sample preparation in analytical chemistry [1]. In the past few years, a wide variety of them with exceptional properties have been utilized in the preconcentration and separation of analytes from different matrices [2,3]. The ionic liquids (ILs) are salts composed of organic cations and organic or inorganic anions [4]. They have melting points at or below 100 °C and hence, they exist as liquids in a wide temperature range. They have low melting points, negligible vapor pressures, outstanding chemical and thermal stabilities, and good affinities for both organic and inorganic analytes [4,5]. In addition, their viscosity, miscibility with organic or inorganic phases, and selectivity for several applications may be tuned based on the needs. Because of these unique physicochemical properties, they have emerged as an excellent class of alternative extraction media. Despite these advantages, poor phase separation, particularly in solvent-based extractions, remains a major challenge in analytical sample preparation [4–7].

Magnetic materials possess a prominent position in this research field as they have attracted considerable attention [3]. Recent developments have indicated that the synthesis of the single component magnetic ILs (MILs) where magnetic metals are not incorporated externally but as a part of the IL can lead to better phase separations by way of an external magnet [8]. MILs are liquids that combine the unique properties of ionic liquids with magnetism, thus facilitating their easy retrieval during the extraction process. MILs are easy to disperse in solutions and can magnetically be separated from the other phases, negating the need for centrifugation, commonly employed in classical extraction procedures, thus, reducing the time and energy required for it. The careful tailoring of their components allows their successful application in the extraction of both hydrophilic and hydrophobic analytes from several media [3,4,8].

Much effort has been put into developing MILs based on transition metals that are less subjected to hydrolysis and possess improved magnetic properties. Therefore, MILs containing transition elements such as manganese, cobalt, gadolinium, and dysprosium have also been reported while metal-free MILs with a paramagnetic component based on organic radicals have also been synthesized [8–13].

So far, the use of MILs in sample preparation has greatly advanced, and many reports are being published with innovative concepts. Since this topic or research has great potential, our aim is to provide a literature update on the current trends of MILs in sample preparation under different conditions of operation, along with the current limitations and the prospects of the field.

## 2. Dispersive Liquid–Liquid Phase Microextraction

Dispersive liquid–liquid phase microextraction (DLLME) is one of the most popular solvent-based microextraction techniques. MILs have been investigated in DLLME-based approaches. Because of their favorable physicochemical properties, they can be an environmentally benign alternative to toxic organic solvents that are commonly employed. Furthermore, MILs can increase the sample throughput as they are easily harvested with the aid of a simple magnet, thereby eliminating the need for a centrifugation step.

### 2.1. Direct Use of MILs

2.1.1. Procedures for Organic Compounds Determination in Environmental Samples

A MIL can be used as an extraction phase for DLMME. In the study of Silva et al., such a sample preparation procedure was developed and combined with high-performance liquid chromatography with diode array detection (HPLC-DAD) for the determination of organic contaminants in river water samples [14]. The organic contaminants included pharmaceuticals (estriol, estrone, carbamazepine, diazepam, ketoprofen, ibuprofen, 17 α-ethynyl estradiol), plastic, and personal care additives (bisphenol A, triclocarban, methylparaben, ethylparaben), and pesticides (aldicarb, methyl parathion, metolachlor, and diuron). In comparison with other studies, this method does not require a centrifugation step and uses smaller volumes of solution and short time sample preparation. Although this method was published lately (compared with other procedures discussed later on), its design is rather simple.

A unique approach of MIL aqueous two-phase system (MIL-ATPs) coupled with HPLC was pursued by Yao et al. [15], for the first time. The magnetic ionic liquid was synthesized based on guanidinium cation to effectively separate chloramphenicol in a water environment. The method did not require any organic solvent and the extraction equilibrium was almost achieved when an aqueous two-phase phenomenon was applied. In these systems, different solutes can be separated from each other and separately distributed into two immiscible aqueous phases. A simple external magnetic field can be used to assist in phase separation. First, the MIL was dissolved in a cap containing only a water sample, then the anhydrous potassium phosphate was dissolved in the solution to avoid the change of ATPs temperature during salt addition. Finally, the magnetic collection was activated to separate the extraction phase and an amount of ionic liquid phase was directly injected into HPLC for quantitative analysis. Good linear range, low limit of detection and quantitation, good recovery were some of the advantages of this method. It is noteworthy that the use of different salts has a varying effect on phase separation. This information can be of interest for future studies to tune the performance of the method.

In another report of two-phase system development, three novel chiral MIL ([$C_{2-4}$MIM-Tempo][L-Pro])(4-Hydroxy-2,2,6,6-tetramethyl piperidine 1-oxyl free radical (4-OH-tempo))

(L-Pro:L-Pro) were used to construct an ATP system with inorganic salts ($K_2HPO_4$, $K_2CO_3$, $K_3PO_4$, $Na_2SO_4$, $Na_2CO_3$) for the extractive resolution of racemic phenylalanine (DL) [16]. Firstly, the MIL was added into a centrifuge tube and dissolved in water. Then, a certain amount of DL-Phe and $Cu(OAc)_2$ was added, and the mixture was shaken to form a clear solution. A volume of anhydrous dipotassium phosphate was added, so the solution turned cloudy, and the chiral MIL-ATPs was formed. Two aqueous phases formed and the enantiomer concentration of amino acid in the top phase and the bottom phase were determined by chiral HPLC-MS. This system combined the advantages of organic solvent-free, magnetic phase separation, and rapid extraction. It is worth mentioning that the MIL could be recycled at least six times with good resolution ability. This type of methods that focuses on the separation of racemic mixtures are scanty and sparse, and thus, more attention should be paid to such applications, since there is a demand for optically pure compounds.

A group of polycyclic aromatic hydrocarbons (PAHs) was extracted by three MILs and quantified with HPLC and fluorescence detection (HPLC-FD) [12]. The three MILs used were benzyltrioctylammonium bromotrichloroferrate (III) (MIL A), methoxybenzyltrioctylammonium bromotrichloroferrate (III) (MIL B), and 1,12-di(3-benzylbenzimidazolium) dodecane bis[(trifluoromethyl)sulfonyl)] imide bromotrichloroferrate (III) (MIL C). The analysis was applied to real aqueous samples, including tap water, wastewater, and a tea infusion and the analytes were benzo(a)anthracene, chrysene, benzo(a)pyrene, benzo(b)fluoranthene and benzo(k)fluoranthene. The results showed that the MIL A had the best performance for the extraction of PAHs (MIL B showed adequate extraction efficiency but was lower than that obtained with MIL A, whereas MIL C extracted the PAHs poorly). The method was able to achieve sufficient quantitation of high molecular weight PAHs, high sensitivity with low limits of detection (LODs), and sufficient reproducibility and efficiency. At the same time, a low volume of MIL and low consumption of organic solvent were required. This study corroborated the fact that MILs with different hydrophobicity have different extraction potential for the analytes, with better performance usually recorded for more hydrophobic MILs. However, the higher the hydrophobicity of a MIL the more difficult is to achieve a more homogeneous dispersion. Thus, the procedures can be improved by adding dispersive solvents or altering the pH of the solution.

In the study of Deng et al., a new hydrophobic magnetic room temperature ionic liquid {(trihexyltetradecylphosphonium tetrachloroferrate (III) ($[3C_6PC_{14}][FeCl_4]$)} was synthesized [17]. The purpose of synthesizing this MIL was to investigate its possible use as a separation solvent for the phenolic compounds from soil samples, because of its paramagnetic characteristics as a response to an external magnet. The distribution ratios of the phenolic compounds were influenced by the pH of the aqueous phase, the nature of the ILs, and the chemical structure of the phenols themselves. Furthermore, the MIL technique showed a much higher extraction capacity than traditional nonfunctionalized RTILs. Finally, the MIL method might be efficient in real samples such as industrial, river and lake water samples, or it can be tested in recovery and recycling the MIL (probably by using centrifugation techniques or strong magnets).

Three hydrophobic MILs were synthesized and employed as extraction solvents in DLLME coupled to HPLC employing ultraviolet radiation (UV) detection [18]. The three MILs [tetrachloromanganate(II) [$MnCl_4^{2-}$]anion, aliquat tetrachloromanganate (II) ([Aliquat$^+$]$_2$[$MnCl_4^{2-}$]), methyltrioctylammonium [$MnCl_4^{2-}$]([$N_{1,8,8,8}^+$]$_2$[$MnCl_4^{2-}$]), and trihexyltetradecylphosphonium [$MnCl_4^{2-}$] ([$P_{6,6,6,14}^+$]$_2$[$MnCl_4^{2-}$])], used in the extraction of pharmaceuticals, phenolics, insecticides, and polycyclic aromatic hydrocarbons. The MIL was mixed with the disperser solvent, and then the mixture was pipetted with the sample solution, followed by shaking. The MIL containing the analytes was retrieved using a magnet rod. Finally, the MIL was dissolved in acetonitrile, and that solution was injected into an HPLC. The [$P_{6,6,6,14}^+$]$_2$[$MnCl_4^{2-}$] was the MIL that exhibited the best extraction results for most analytes and the cleanest chromatographic background. Low LODs were obtained for all target analytes, and acceptable recoveries for water and lake samples were accomplished. In this study, an important topic was taken under consideration as most MILs based on the iron tetrachlorate anion are prone to hydrolysis. This can limit the potential of MILs for use in aqueous samples. Moreover, this anion exhibits strong UV absorbance which also limits the potential use of HPLC-UV systems for separation and analysis. Therefore, other magnetic anions, such as the manganese tetrachloride anion should be examined.

Another informative case is the investigation of three MILs based on the cation $[P_{6,6,6,14}]^+$, combined with tetrachloroferrate(III), ferricyanide, and dysprosium thiocyanate [19]. These MILs were evaluated as extraction solvents of three different microextraction strategies for the efficient extraction and preconcentration of four estrogens, estrone, estradiol, estriol, and ethinylestradiol, from environmental water. In the first and second one, the MIL was suspended to the aqueous solution, and stirring was performed with an external magnet using an orbital shaker and a stir bar, respectively. In the third one, MIL was first immobilized on a stir bar and then, the aqueous solution was added, with MIL remaining on the rod due to its high viscosity and strong paramagnetism. In the latter case, highest extraction recoveries of analytes were achieved with the use of $[P_{6,6,6,14}][FeCl_4]$, $[P_{6,6,6,14}]_3[Fe(CN)_6]$, and $[P_{6,6,6,14}]_5[Dy(SCN)_8]$ MILs. Of these three MILs, the $[P_{6,6,6,14}]_3[Fe(CN)_6]$ was found to be the most suitable due to its reduced cost, hydrophobicity, and easier synthesis. It is important that the design of this MIL, along with the advantages it offers, has barely been examined for its application in the analytical field. Compared with other reported techniques, due to the low consumption of solvents and the use of ionic liquids as an organic phase, it responds well to the green chemistry principles. Compared with solid–phase–based microextraction and other DLLME-based techniques, it provides similar or lower extraction times.

Along these lines, Chatzimitakos and co-workers proposed a stirring-assisted drop breakup microextraction combined with HPLC for the determination of selected phenols and acidic pharmaceuticals in aqueous matrices [20]. In this mode, an aqueous sample was added to a glass beaker and temperature and pH was adjusted. The addition of one drop of the MIL (16 ± 1 mg) accompanied by stirring, initially led to the decomposition of the droplets and the subsequent reunion into one which contained the analytes. The $N_{8,8,8,1}$ [$FeCl_4$] MIL was detached from the solution with a magnet and dissolved in a mixture of DDW: acetonitrile (1:3) for the injection into the HPLC-DAD system. In the authors' view, this is the first attempt to use this MIL for analytical aims through a simple, efficient, environmentally friendly, and low-cost drop-breakup microextraction for small molecules.

2.1.2. Procedures for the Determination of Organic Substances in Food Samples

The application of different MIL can also be applied in DLLME to the determination of six estrogens in samples of milk and cosmetics [21]. Six estrogens (estrone, estradiol, 17-α-hydroxyprogesterone, chloromadinone, 17-acetate, megestrol 17-acetate and medroxyprogesterone 17-acetate) were extracted by employing four MILs ($[P_{6,6,6,14}^+][FeCl_4^-]$, $[P_{6,6,6,14}^+]_2[MnCl_4^{2-}]$, $[P_{6,6,6,14}^+]_2[CoCl_4^{2-}]$ and $[P_{6,6,6,14}^+]_2[NiCl_4^{2-}]$). The MIL that gave the best results including low chromatographic background, wide linear range, low detection limit, and good recovery, was the $[CoCl_4^{2-}]$-based MIL. In comparison with $[P_{6,6,6,14}^+]_2[MnCl_4^{2-}]$ and $[P_{6,6,6,14}^+]_2[CoCl_4^{2-}]$, the $[P_{6,6,6,14}^+]_2[CoCl_4^{2-}]$ shows good selectivity for six analytes, and the color of MIL based on $[MnCl_4^{2-}]$ is light yellow, which is difficult to observe and separate. In this case, also, the iron-based anion was avoided, reducing the hydrolysis of the MIL, and the color of the MIL was also taken under consideration, in order to make easier the visualization of the droplets and their magnetic harvest. The method presented fast, accurate and precise results along with accuracy, precision, effectiveness, economical, and environmentally friendly.

Wang et al. developed a MIL that can extract triazine herbicides from vegetable oils with a DLLME [22]. The 1-hexyl-3-methylimidazolium tetrachloroferrate ($[C_6mim][FeCl_4]$) was used for extracting triazine herbicides from vegetable oils, two soybean oils, three maize oils, and two sunflower seed oils samples. Moreover, carbonyl iron powder (CIP) was applied to minimize magnetic separation time, thus can be magnetically attracted by the MIL to form a combination of CIP and MIL (CIP-MIL). Briefly, a dilution of vegetable oil sample with n-hexane was performed, then an ultrasound extraction with the MIL and the adding volume of the CIP. Next, the CIP-MIL was collected with a strong magnet and washed with n-hexane. A volume of deionized water and ethyl acetate was added to dissolve the MIL and extract the target analytes. After evaporation with a nitrogen stream, the final sample was redissolved with acetonitrile, followed by LC analysis (Figure 1). Compared with previously developed methods [23–27], the performance achieved by this method was found to be acceptable. The use of ultrasonication was an asset for this method, as it reduced the extraction time to 7 min. Thus, not only dispersion agents, but ultrasonication can also be employed for the dispersion of the MILs.

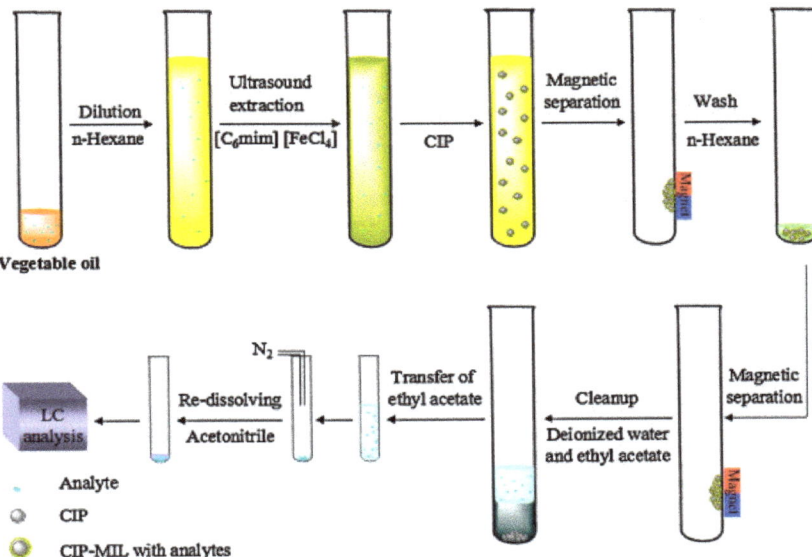

**Figure 1.** Schematic diagram of the DLLME extraction procedure of triazines from vegetable oils. Reproduced with permission from [22]. Copyright Elsevier, 2014.

Li et al. developed an ultrasonic-assisted extraction method (UAE) for sinomenine (SIN) microextraction from *Sinomenium acutum*, utilizing a range of MILs based on imidazolium cations and iron(III) anions [28]. S. acutum powder was added to a tube along with the MIL. The solution was ultrasonicated for the extraction of SIN and subsequently was centrifuged. The surfactant bis (2-ethylhexyl) sulfosuccinate sodium salt (AOT) was used for the reversed micellar extraction of SIN. The AOT was dissolved in isooctane and an amount of water was added into the tube. The resulting transparent AOT/isooctane reversed micellar system was added to a tube together with the aqueous MIL phase. The mixture was shaken and then placed in a separating funnel. The AOT/isooctane micellar system was separated from the aqueous phase and absolute ethanol was added to destroy the micellar system. After stirring, the mixture was transferred to a separation funnel for the reversed extraction of SIN into the ethanol phase. Compared with other methods the heat reflux extraction using MIL, ethanol-heat reflux extraction, $H_2O$-heat reflux extraction, Ethanol-UAE, $H_2O$-UAE), MIL-UAE reveals the maximum extraction yield of 10.57 mg $g^{-1}$, indicating the excellent extraction ability of the MIL and the rapidity of the method. Moreover, due to multi-interactions (ionic/charge-charge and hydrogen bonding,) between SIN and the MIL, the solubility of the SIN in the solution is enhanced in the present MIL-UAE method. However, this method is still complex and contains many laborious steps, thus it needs to be further simplified in order to be used more widely.

Feng et al. reported an extraction method for polyphenols in tea leaves using magnetic ionic liquids [29]. Pulverized tea leaves were added into a conical flask along with $C_3mimFeCl_4$ MIL and the extraction was ultrasonic-assisted. The pH was adjusted to 3–3.5 and the solution was left to stand. After the removal of tea residue with filtration, NaOH was added, and the forming tea polyphenols-MIL complex was precipitated. The mixture was centrifuged and filtered. The liquid phase was diluted and analyzed with HPLC to determine its contents. The filter cake was dissolved and diluted with methanol to determine its contents. In this, a CIP was added, and the MIL was combined with the CIP. The solution was left to stand and then a magnet was used for the separation of the phases. The residual solution was diluted and analyzed with HPLC. The extraction efficiency of tea polyphenols was much higher than that achieved by the traditional solvents such as water and alcohol-water. The MIL that was used in this study has high selectivity, so it can be removed from target analytes easily. In conclusion, MILs can be a useful solvent for the extraction and determination of natural products. More attention should be paid to the iron-containing MILs, due to their increased absorbance in the UV region, which can hinder the identification and quantification of polyphenols.

A more enhanced version of MIL-DLLME came to the fore by Chatzimitakos and co-workers, who succeeded in identifying triazines (TZs) and sulfonamides (SAs) [9]. The peculiarity of this new method was the combination of a water-insoluble solid support with $[P_{6,6,6,14}^+][Dy(III)(hfacac)_4^-]$ MIL and the separation of the analytes in a one-pot, pH-modified procedure. By mixing the solid support with the MIL, difficulties related to the weighing and the uniform dispersion of the MIL were avoided. To do so, materials such as quartz silica microparticles, insoluble silica, and soluble inorganic salts were tested, with quartz silica eventually prevailing. In brief, the liquid sample was mixed with trisodium citrate and a quantity of a $[P_{6,6,6,14}^+][Dy(III)(hfacac)_4^-]$-quartz silica mixture under stirring, which led to the formation of tiny droplets. Magnetic isolation (using a neodymium cylinder magnet), and dissolution in acetonitrile solution, was followed by the HPLC analysis. A key point of the process was the pH adjustment. It was observed that by primarily adjusting the pH to 9.0 and then lowering it to 3.0, efficient extraction was achieved first of the TZs and then of the SAs. Apart from the beneficial use of the solid supporting material and the achievement of the process in one-pot, it is worth noting that both classes of analytes were simultaneously separated, with advantageous recoveries. This study is one of the few that tried to address the issue of poor MIL dispersibility, a common problem for hydrophobic MILs and paved the way for the use of solid support materials, whose sole purpose is to control the dispersibility of the MIL. The solid support materials can be reused since they do not take part in the extraction process.

Mousavi and co-workers synthesized a magnetic ionic liquid that was used in the DLLME of ultra-trace amounts of parabens in wine, beer, and water samples [30]. $[N_{1,8,8,8}^+][FeCl_4^-]$, a magnetic room temperature IL, was synthesized through an easy reaction between methyltrioctylammonium chloride ($[N_{1,8,8,8}^+][Cl^-]$) and iron(III) chloride hexahydrate ($FeCl_3\ 6H_2O$). What stands out in the present study is the possibility of micelle formation due to the surface-active material of MIL. It was investigated, for the first time, as an important factor in phase separation, affecting the recovery of the MIL. For this purpose, 20 mL of extraction solvent ($[N_{1,8,8,8}^+][FeCl_4^-]$) was diluted in 250 mL of disperser solvent (acetone). The solution was placed into a capped glass containing the sample solution with NaCl (25%, $w/v$). After sonication for 1 min, a cloudy solution was formed. From this, the supernatant was removed, and the analyte-rich MIL phase was retrieved by a magnet. The analysis was further continued via HPLC-UV. According to the results, great extraction efficiencies and impressive recoveries were caused by the extensive surface contact between the $[N_{1,8,8,8}^+][FeCl_4^-]$ droplets and the sample, which forms stronger intermolecular forces (e.g., $p$-$p$, $n$-$p$) than those between water and parabens.

### 2.1.3. Procedures for the Determination of Organic Substances in Biological Samples

MILs can also be used as extraction phases in DLLME for the analysis of hormones (estriol, 17-β-estradiol, 17-α-ethynylestradiol, and estrone) in human urine samples [11]. Merib et al. showed that the MILs trihexyltetradecylphosphonium tetrachloromanganate (II) ($[P_{6,6,6,14}^+]_2[MnCl_4^{2-}]$) and aliquat tetrachloromanganate (II) ($[Aliquat^+]_2[MnCl_4^{2-}]$), followed by separation/detection with HPLC, could be of a biological interest too. Briefly, a mixture of a disperser solvent (methanol) and the MIL ($[P_{6,6,6,14}^+]_2[MnCl_4^{2-}]$) was added to urine samples, and a manual shaking step was applied to facilitate the formation of microdroplets in the solution. Right after that, the MIL was reserved with a rod magnet and the extraction phase containing the enriched analytes was desorbed in an portion of acetonitrile, before injection into the HPLC-DAD system. This extraction step was performed in a short time (90 s) without the need for a centrifugation step. In addition, chromatographic separations were conducted within 10 min allowing for high-throughput analysis and LODs, comparable to previously reported data using other microextraction techniques requiring longer extraction times. Analyzing biological samples in such an efficient and environmentally friendly way is needed in analytical chemistry.

The work of Will et al. expands the potentials of MIL with the presentation of a simultaneous determination of different compounds of human urine samples [31]. The $[P_{6,6,6,14}^+][Cl^-]$ MIL was used for the extraction of eight compounds including pesticides (carbofuran, atrazine, simazine, diuron, and metolachlor), estrogenic hormones (ethinylestradiol and estrone), and a pharmaceutical compound (diclofenac). Dispersive liquid–liquid microextraction based on the MIL was used as the sample preparation procedure (MIL-DLLME). The MIL showed good selectivity for the low-polarity compounds evaluated in the study, allowing for the simultaneous determination of different classes of compounds. A MIL-DLLME-based procedure was fully optimized, and a rapid extraction methodology was performed (11 min of extraction). In addition, this method can be easily automated, which can considerably increase the throughput features of the determination. The MIL-based approach constitutes a formidable tool to avoid chlorinated organic solvents, which have frequently been used in DLLME. Moreover, the combination of automated analyses with wide applicability is highly promising for an analytical method to be employed in routine analysis. Abdelaziz and co-workers succeeded in the determination of four antihypertensive drugs of the sartan class through a gadolinium-based MIL which was used as an extraction solvent in DLLME [32]. For this reason, three hydrophobic MILs based on the trihexyl(tetradecyl)phosphonium ($P_{6,6,6,14}$) cation containing different paramagnetic metal-halide anions ($FeCl_4^-$, $MnCl_4^{2-}$, and $GdCl_6^{3-}$) were synthesized according to previous reports. Although all three revealed compatibility with common reversed-phase HPLC solvents and low miscibility with aqueous samples during extraction, some complications affected the extraction efficiency. The $[P_{6,6,6,14}^+]_3[GdCl_6^{3-}]$ was found to be the most beneficial. It combines advantages such as higher magnetic susceptibility, better adhesion to sartans, lower toxicity, satisfactory viscosity, and it is less likely to undergo hydrolysis in aqueous samples. A comparison between the proposed method and SPE and IL-DLLME, confirms its advantageous position. $[P_{6,6,6,14}^+]_3[GdCl_6^{3-}]$ based DLLME provides faster and better automation. Although LOQ values were lower for the SPE methods, much lower quantitation limits should be achieved if the proposed one is coupled with highly sensitive MS/MS detection and/or applied to larger sample volumes.

In a more recent study, two room temperature MILs were synthesized to preconcentrate, determine, and separate the carbamazepine drug in urine and wastewater samples [33]. To achieve this, dispersive micro-solid phase extraction was used in conjunction with HPLC. Iron and cobalt-containing MILs $[OA]FeCl_4$ and $[OA]CoCl_3$ ((Z)-octadec-9-en-1-aminium tetrachloroferrate (III) and (Z)-octadec-9-en-1-aminium trichlorocobaltate (II)) were tested as sorbents with the satisfactory results setting the stage for future applications for the extraction of pharmaceutical trace contaminants in the water samples. For the extraction procedure, 10 mL of 300 $\mu g \cdot L^{-1}$ NaCl and 40 mL of MIL were mixed in a glass beaker on a magnetic stirrer. Owing to the magnetic properties, the MIL adsorbent was obtained through a magnet, and analysis was continued by HPLC-DAD. The extraction was based on electrostatic interactions, between the MIL and the analyte, whereas the acetonitrile used for the desorption, exhibited stronger dipole-dipole interactions with the analyte and removed it from the MIL. Reaping such mechanisms of interaction for the extraction of the analytes can sometimes be advantageous, since hydrophobic interactions need to use more hydrophobic MILs. Thus, other type of problems may arise, whereas electrostatic interactions are weaker and are more sensitive to temperature variations.

2.1.4. Procedures for Metal Species Determination in Food Samples

Fiorentini et al. investigated a MIL-DLLME method for the determination of trace Cadmium in honey samples [34]. The honey sample was added in a tube along with ammonium diethyldithiophosphate (DDTP) and HCl. The DDTP was added to form a hydrophobic chelate with Cd(II), highly stable at pH values below three, thus the efficiency of the extraction increases. HCl was added to acidify the solution. The mixture was stirred and left to rest to ensure the formation of the Cd-DDTP complex. Subsequently, the trihexyl(tetradecyl)phosphonium, $[P_{6,6,6,14}]FeCl_4$, MIL phase and acetonitrile, as a dispersant, were added. After stirring the mixture, a magnetic bar was used to collect the MIL phase. An amount of $HNO_3$ was utilized to back-extract the analyte from the MIL material since the direct determination of Cd in the MIL is not attainable due to the interference of Fe in this type of solvent. Finally, the preconcentrated sample solution was injected into the graphite furnace of an electrothermal atomic absorption spectrometry (ETAAS). The LOD was $0.4\ ng·L^{-1}$ Cd and extraction efficiency was 93%. This work probably constitutes the first report of $[P_{6,6,6,14}]FeCl_4$ MIL application along with the DLLME method for the determination of trace Cd in honey samples.

Another research on the MIL-DLLME technique developed also by Fiorentini et al. concerns the preconcentration and microextraction of the highly toxic Arsenic in honey samples [35]. $[P_{6,6,6,14}]FeCl_4$ MIL was utilized as extractant material and the analysis was performed using ETAAS. Before the microextraction procedure of As, aqueous solutions of 1% $(w/v)$ honey were prepared. The pH was adjusted and then methanol was added as a dispersant material along with 2-(5-bromo-2-pyridylazo)-5-diethylamino-phen as a complexing agent. The solution was left to stand, and chloroform was added, followed by a vortex. Initially, for the determination of As, the supernatant from the above-mentioned clean-up procedure was acidified with HCl, and KI was added to reduce As(V) to As(III). After that, the mixture was left to rest to ensure the As reduction and DDTP was added followed by acetonitrile addition. The DDTP was added to form a hydrophobic chelate with As(III), highly stable under acidic conditions, thus the efficiency of the extraction increases. Acetonitrile works as a dispersant. The solution was left for a second time to rest in order to ensure the formation of the As(III)-DDTP complex. The extraction phase was added, and the sample was stirred. The MIL phase was retrieved via an external magnetic rod and an adequate amount of sample was injected into the ETAAS for trace As determination. The LOD of the analysis procedure was $12\ ng·L^{-1}$ As and the extraction efficiency was 99%. The advantage of this study is the tremendous analytical recovery of 95.2–102% despite the matrix complexity along with the utilization of the $[P_{6,6,6,14}]FeCl_4$ MIL as an excellent extractant.

Fiorentini et al. have also introduced the MIL, $[P_{6,6,6,14}]FeCl_4$, for the determination of chromium in honey samples, followed by ETAAS [36]. In this work, the sample solution was acidified with HCl, and Fe(II) was added to the sample to prevent the oxidation of Cr(III). Then, $[P_{6,6,6,14}]FeCl_4$ was added and the sample was stirred for 10 min. The MIL extracts the analyte during the stirring process. The MIL phase is collected via a magnetic rod and diluted with $CHCl_3$. Thereafter, the sample was injected into the ETAAS for Cr determination. The obtained efficiency of the extraction was 98% and the LOD was $5\ ng·L^{-1}$ Cr. The dominant result derived from this study is the avoidance of centrifugation due to the facile separation of the extraction phase by an external magnetic field, using a Fe-containing MIL.

Recently, Fiorentini et al. reported the application of trihexyl(tetradecyl)phosphonium tetrachloromanganate (II), $[P_{6,6,6,14}]_2MnCl_4$, MIL along with the DLLME method for the determination of trace Pb in bee products (honey, mead, honey beer, and honey vinegar) for the first time [37]. The sample was acidified with HCl, then 1,5-diphenylcarbazide was added for the complex formation, NaCl to adjust the ionic strength and acetonitrile as a dispersant. The MIL phase was added, and the mixture was vortexed. Finally, the MIL phase was separated from the solution by a magnetic rod and an adequate amount of this containing the preconcentrated analyte was injected into the ETAAS instrument. A LOD of 3 $ng·L^{-1}$ Pb and an extraction efficiency of 97% were obtained. This approach has an excellent advantage. The MIL used for the Pb detection does not contain Fe, thus the interference effects are avoided. All the above methods from the same research team, point towards the need for careful selection of the metal moiety in the anion of the MIL, since it can interfere with the detection of metal ions.

Wang et al. reported in 2016 a novel magnetic ionic liquid-based up-and-down-shaker-assisted DLLME for the speciation and determination of Selenium from five rice samples (white rice, brown rice, parboiled rice, glutinous rice, and rice flour) [38]. An aqueous solution containing Se(IV) and Se(VI) was inserted into a centrifuge tube. Further, 2,3-diaminonaphthalene, utilized as a chelating agent, was added to the tube along with diluted HCl for the pH adjustment. The MIL 1-butyl-3-methylimidazolium tetrachloroferrate, $[C_4mim][FeCl_4]$, was then injected into the solution and the tube was shaken by an up-and-down shaker. Subsequently, the Se(IV)-2,3-diaminonaphthalene complex was separated from the aqueous phase by an external magnetic field. Due to the high viscosity of the MIL phase that complicates the sample injection into the analysis instrument, an amount of $HNO_3$ in ethanol (1:1, $v/v$) was added to reduce the viscosity of the MIL. The analysis was performed using a graphite furnace atomic absorption spectrometer. After the sample preparation, the aforementioned steps were followed to determine the total Se concentration and the Se(IV) concentration. The total inorganic Se concentration was determined as well, following the same experimental course. Thus the Se(VI) concentration can be calculated by subtracting the Se(IV) concentration from the total inorganic Se concentration. This technique presents good accuracy, relative standard deviation (RSD) lower than 3%, LOD of 0.018 $μg·L^{-1}$ and repeatability of <3.0% for Se(IV).

A novel ultrasound-assisted surfactant-enhanced emulsification microextraction combined with micro-solid phase extraction using the MIL butyl-3-methylimidazolium tetrachloroferrate ($[C_4mim][FeCl_4]$) was reported by Yao et al. for the determination of cadmium and lead in six edible vegetable oils (olive oil, soybean oil, maize oil, sunflower seed oil, and two peanut oils) [39]. For the preconcentration and detection of Cd and Pb, the sample was inserted into a tube. Non-ionic surfactant Triton X-100 was added as an emulsifier reagent along with the MIL and the mixture was ultrasonicated. Then, an amount of $Fe_3O_4$ nanoparticles were added and the solution was stirred. The $Fe_3O_4$ nanoparticles were utilized to enhance the efficiency of the MIL phase separation by absorption. The MIL-nanoparticle phase was collected after applying an external magnetic field. $HNO_3$ was added to the tube for ultrasonicated dissolution of the MIL. The $Fe_3O_4$ nanoparticles containing the target analytes were collected with a magnetic bar and injected into the GFAAS. The ultrasound procedure applied in this work accelerates the homogenous dispersed solution and the mass transfer between the analyte and the extractant. Compared with other reports in the literature, this method presents similar or lower LODs (0.002 for Cd and 0.02 for Pb), and among the advantages are simplicity, rapidity, and satisfactory sensitivity. The main advantage, however, is the avoidance of the centrifugation step that reduces the analysis time, leading to an excellent alternative technique to oil samples microextraction.

A very interesting study was reported by Wang et al. in 2018 for the determination of arsenite and arsenate in five types of leafy vegetable samples (leaf lettuce, bok choy, spinach, celery, and coriander) applying a novel effervescence tablet-assisted magnetic ionic liquids-based microextraction (ETA-MILs-ME) [40]. For the ETA-MILs-ME procedure, the sample was added into a tube along with ammonium molybdate, and ascorbic acid and the mixture was acidified (Figure 2). An effervescent tablet was added to the solution and bubbles rose rapidly from the bottom of the tube. During this step, the MIL 1-butyl-3-methylimidazolium tetrachloroferrate ([C$_4$mim][FeCl$_4$]) dispersed homogeneously in the aqueous phase and the formed As(V)-ammonium molybdate complex was extracted by the MIL. The MIL phase was separated from the aqueous solution with a magnetic rod. Finally, it was diluted with ethanol and injected into the GFAAS. For the total As the sample was weighed into a PTFE reactor. HNO$_3$ and H$_2$O$_2$ were added, and the mixture was digested. Then, As(III) was oxidized to As(V) and the fore-mentioned ETA-MILs-ME procedure was followed for the total As calculation. For the detection of As(V), the sample was weighed into a tube and nitric acid was added to extract inorganic arsenic. The solution was sonicated and centrifuged to obtain the extract. The supernatant was filtered with a cellulose acetate membrane and the ETA-MILs-ME methodology was performed. For the calculation of the total inorganic As the formerly obtained supernatant was filtered and an amount of KMnO$_4$ solution was added to oxidize As(III) to As(V) with the aid of sonication. The sample was subjected to the ETA-MILs-ME procedure for the total inorganic measurement. As(III) was calculated by subtracting the As(V) from the total inorganic As concentration. This work presents a simple, rapid, and efficient method with a LOD of 0.007 µg·L$^{-1}$ and 97.9–105.8% recovery for As(V) in a variety of leafy vegetable samples. The use of effervescence is another way to address the issue of poor MIL dispersibility, similar to the use of solid support materials, as mentioned above. As this is one of the main restrictions of (hydrophobic) MILs usage, more alternative options like the ones mentioned above, should be examined.

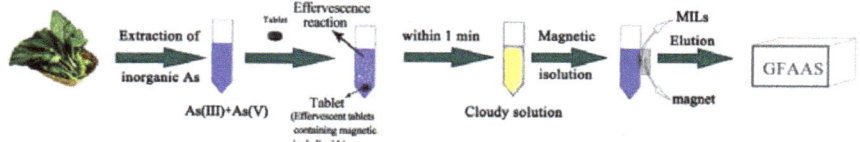

**Figure 2.** The scheme of the proposed ETA-MILs-ME. Reproduced with permission from [40]. Copyright Elsevier, 2018.

### 2.1.5. Procedures for Metal Species Determination in Environmental Samples

In 2016, Wang et al. proposed a novel magnetic ionic liquid-based air-assisted liquid–liquid microextraction (MIL-AALLME) for the trace Arsenite and Arsenate species determination in five environmental water (river, pond, tap water), sediment, and soil samples [41]. This technique combines MIL and air-assisted liquid–liquid microextraction. The 1-butyl-3-methylimidazolium tetrachloroferrate ([C$_4$mim][FeCl$_4$]) MIL was used as the extractant phase. For the detection of the total inorganic As in water samples, Na$_2$S$_2$O$_3$ and KI were added to the sample to reduce arsenate to arsenite. The solution was left to rest to ensure the reduction. In the mixture containing ammonium pyrrolidine dithiocarbamate as a chelating reagent, diluted nitric acid/ammonia was added for the pH adjustment. Subsequently, the MIL was inserted into the tube and the solution was rapidly withdrawn and rapidly injected into the tube for 10 times via a syringe to accelerate the MIL dispersion in the sample. The MIL phase containing the As(III)-ammonium pyrrolidine dithiocarbamate complex was collected by an external magnet rod and HNO$_3$ in ethanol (1:1, $v/v$) was added to reduce the viscosity, thus the aliquot sample could be injected into the graphite furnace atomic absorption spectrometry (GFAAS) instrument. The total inorganic As concentration and the As(III) concentration were determined following the above-mentioned steps. As(V) concentration is calculated by subtracting the As(III) concentration from the total inorganic As concentration. For the detection of the total As in solid and sediment samples As(V) was reduced to As(III) and then the MIL-AALLME process was performed. An amount of the solid and sediment samples was weighed and transferred to a polytetrafluoroethylene

(PTFE) reactor. $HNO_3$ in ethanol (1:1, $v/v$), $H_2O_2$ and HF were injected into the PTFE reactor and the samples were digested by a microwave digestion system. The PTFE reactor was cooled and then KI and $Na_2S_2O_3$ were added to the residual solution for the reduction of As(V) to As(III) with microwave assistance. Finally, the MIL-AALLME process was performed for the calculation of the total As concentration. For the determination of As(V) and As(III), the samples were weighed and added into a tube along with the extraction solution (phosphoric acid and ascorbic acid). The mixture was sonicated and centrifuged for obtaining the extracted As(V) and As(III). In order to obtain As(III) concentration, an aliquot of the filtered supernatant was diluted and subjected to the MIL-AALLME procedure. KI and $Na_2S_2O_3$ solutions were added to another aliquot of filtered supernatant for the reduction of As(V) and As(III) with microwave assistant. The total inorganic arsenic concentration was evaluated by the MIL-AALLME procedure. The As(V) concentration was calculated by subtracting the As(III) concentration from the total inorganic arsenic. Compared with different approaches reported in the literature for trace inorganic As species determination in natural samples, this method has low sample consumption and short extraction time, it shows an acceptable LOD (0.029 $\mu g \cdot mL^{-1}$) in the range of and the linear dynamic range is acceptable (0.04–10.0 $\mu g \cdot L^{-1}$) compared with other reported techniques.

In a recently reported work, Oviedo et al. developed a MIL-DLLME method to determine inorganic Antimony species in natural water samples (tap, dam, mineral, wetland, underground, rain, and river water) [42]. For the determination of Sb(III) the sample was mixed with HCl and DDTP solution as a chelating reagent. The solution was left to stand to ensure the formation of the Sb(III)-DDTP complex. Then, NaCl was added to adjust the ionic strength and the acetonitrile in order to disperse the extraction phase. The MIL hexyl(tetradecyl)phosphonium tetrachloroferrate ($[P_{6,6,6,14}]FeCl_4$) was used as an extractant material. The sample was vortexed, and the extraction phase was separated from the aqueous phase with a magnetic rod and subsequently was diluted with chloroform. An aliquot sample was injected into the graphite furnace of ETAAS. For the total inorganic Sb determination, KI was inserted in the sample and then the solution was acidified with HCl. The mixture was left to stand and the above-mentioned MIL-DLLME technique was performed to calculate the total inorganic Sb. The Sb(V) concentration was calculated by subtracting the Sb(III) concentration from the total inorganic Sb concentration. This method constitutes the first application of the MIL for extraction and preconcentration of Sb in natural water samples. The extraction efficiency for Sb(III) was 98.0% and for Sb(V) was 92.6%, LOD for Sb(III) was 0.02 $\mu g \cdot L^{-1}$ and the linear range was 0.08–20 $\mu g \cdot L^{-1}$.

Aguirre et al. have proposed an analytical MIL-DLLME approach for the detection of cadmium in three fuel samples (engine oil, gasoline, and diesel) [43]. For the microextraction and detection of Cd, sample was mixed with an amount of the MIL bis(1-ethyl-3-methylimidazolium) tetrathiocyanatocobaltate (II), $[Emim]_2[Co(SCN)_4]$, and the tube was vortexed (Figure 3). The MIL phase was collected with the aid of a magnetic rod and an aliquot was transferred into a $HNO_3$ solution for the back-extraction of the target analyte to the aqueous phase. This is the first study that demonstrates MIL-DLLME and back-extraction procedures in combination with the ETAAS instrument for Cd determination in engine oil, gasoline, and diesel fuels. The obtained LOD value (0.084 $\mu g \cdot kg^{-1}$) reveals that the MIL-DLLME technique can enhance the sensitivity of the ETAAS analysis. The extraction of Cd from the difficult-to-handle fuel samples via a compatible aqueous phase is an advantage that makes this approach a promising method for facile and sensitive determination of Cd in comparable types of samples. As can be seen so far, most of the methods developed for metal species determination are more complex, compared with those developed for organic compounds. This is an issue that needs further attention, since future methods for metal species should be simpler, so that they can be used in routine analysis.

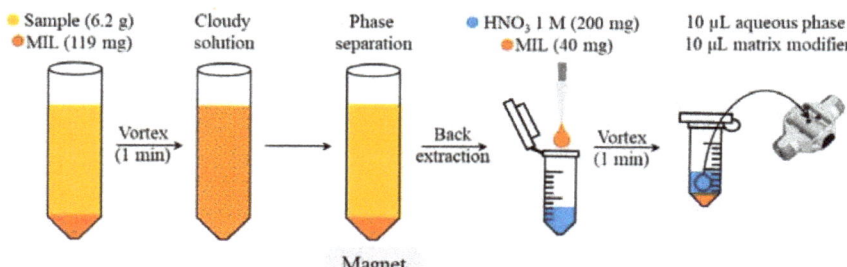

**Figure 3.** Scheme of the analytical procedure for Cd quantification. Reproduced with permission from [43]. Copyright Elsevier, 2020.

*2.2. In-Situ Formation of MILs*

The in-situ DLLME approach is an adjustment of the classical DLLME method that uses ILs as extraction solvents [44]. A limited number of reports have exploited this alternative formation of MILs in micro-extraction approaches, which until now include the determination of harmful analytes in water and food samples, and most recently the extraction of DNA [45]. In these cases, the magnetism of MILs enables magnetic isolation of the extraction solvent with a strong magnet, thus replacing the centrifugation and filtration steps [44]. For this application, mixing of a hydrophilic IL with a metathesis reagent is required, leading to an anion exchange and generation of a hydrophobic IL. Through this reaction, multiple hydrophobic IL micro-droplets are formed that are able to interact with the analytes [46]. Most of the time, the MILs designed contain paramagnetic anions. However, some restrictions in their use in DLLME have been observed, such as hydrolysis in water at room temperature and incompatibility with HPLC [44].

To overcome the problems associated with these types of MILs, Anderson and co-workers reported a new generation containing paramagnetic cations [44]. The in-situ MIL-DLLME was compared with the conventional MIL-DLLME for the verification of the successful extraction of polar and non-polar pollutants in aqueous samples. In combination with HPLC the determination of UV filters, polycyclic aromatic hydrocarbons (PAHs), alkylphenols, a plasticizer, and a preservative was achieved. Therefore, five different MILs consisting of cations containing Ni(II) centers coordinated with four ligands of N-alkylimidazole and chloride anions with different alkyl substituents (R), were synthesized and tested as extraction solvents in in-situ DLLME. These can go through a metathesis reaction with the bis[(trifluoromethyl)sulfonyl]imide ($[NTf_2^-]$) anion. Among them, $[Ni(C_4IM)_4^{2+}]_2[Cl^-]$ and $[Ni(BeIM)_4^{2+}]_2[Cl^-]$ showed the highest extraction efficiencies. A mixture of the aqueous MIL solution in the chlorinated $[Cl^-]$ form and the extraction solvent, was added to a glass vial containing an aqueous solution of the analytes. A certain amount of ion-exchange reagent was added to achieve the formation of MIL:$[Li^+][NTf_2^-]$ in a ratio of 1:1, 1:2, or 1:3. Assisted by vortex, the metathesis reaction was accelerated and thus the configuration of the magnetic drops containing the analytes. These were magnetically isolated and diluted in acetonitrile solution to reduce viscosity. To illustrate the superiority of the designed method, a comparison was made with MIL-DLLME. It was found that the extraction efficiencies for the in-situ MIL-DLLME, were higher and ranged from 46.8–88.6% and 65.4–97.0% for the $[Ni(C_4IM)_4^{2+}]_2[Cl^-]$ and the $[Ni(BeIM)_4^{2+}]_2[Cl^-]$ MILs, respectively.

In view of this new category of MILs, four novel organic ones were synthesized by Yao and Du [47]. Through a similar approach, an in-situ MIL-DLLME coupled with HPLC was established to simultaneously separate, preconcentrate, and determine trace amounts of sulfonamides in milk samples for the first time. Along the same lines with the previous report, the MIL was dissolved in a vial containing the milk sample with the analytes. Then, the addition of the ion-exchange reagent resulted in the immediate formation of a light red turbid solution. The hydrophobic MIL drops were directly isolated via a neodymium magnetic bar (30 s) and immersed into methanol for subsequent HPLC analysis. A comparison between this developed method and others that have been reported has shed light on its multiple benefits. Through its application, the extraction time is reduced, as steps such as centrifugation and agitation are avoided. In addition, a small amount of an inorganic salt is generated and no organic toxic solvents are required. Taking everything into account, this is the first time to realize all these advantages in an analytical method.

In a recently published work, a unique analytical method, known as in-situ parallel-dispersive droplet extraction evolved for the first time [45]. Combined with HPLC-DAD, micropollutants in aqueous environmental samples were successfully detected. To begin with, for the formation of the hydrophobic MIL, a cation precursor (CP), and an anion exchange reagent (AER) were required. Three different hydrophilic MILs (CPs) $[Ni(C_4IM)_4^{2+}]_2[Cl^-]$, $[Ni(BeIM)_4^{2+}]_2[Cl^-]$, $[Co(C_4IM)_4^{2+}]_2[Cl^-]$ were evaluated, with the latter prevailing. $[Li^+][NTf_2^-]$ was used as AER. One of the main advantages of the method is its automation. This is achieved using a 96-well plate system on which NdFeB were adjusted, thus increasing the throughput of the process. By adding 1.25 mL of sample and 100 µL of an aqueous solution of $[Co(C_4IM)_4^{2+}]_2[Cl^-]$ at a concentration of 40 g·L$^{-1}$ to the plates vials, a hydrophilic compound was formed. After 5 min of vigorous agitation, the addition of 40 µL of an aqueous solution of $LiNTf_2$ converts it to a hydrophobic, maintaining agitation for 75 min. Eventually, the MIL microdroplets attached to the rod magnets were diluted in ACN and the solution was injected in the HPLC-DAD. The advantages of this method are not limited to its automation. Compared with others used for the determination of analytes in water samples, less extraction time is required (0.78 min) and a low amount of toxic organic solvents and sample is consumed (MILs are synthesized in aqueous media, 20 µL of ACN and 1.25 mL of sample are needed). In addition, due to the increase of the surface area of the MIL, higher extraction efficiency is achieved. On the other hand, it is worth noting that LODs were slightly higher than those mentioned by other studies, and concerns associated with the solubility of MIL in aqueous samples were noticed.

The same concept was introduced for the in-situ DLLME of DNA samples. To give an illustration of that, the research carried out by Bowers and co-workers proved the successful extraction of long and short double-stranded DNA through the formation of hydrophobic MILs droplets [46]. As previously described, the aqueous solution of the MIL in the chloride form was mixed with the DNA sample and the dispersive solvent (dimethylformamide). Then, with the addition of the ion exchange reagent ($[Li^+][NTf_2^-]$), the hydrophobic MIL droplets settled at the bottom of the vial. An aliquot of the upper aqueous phase was used for further analysis. In this study, up to ten different MILs consisting of N-substituted imidazole ligands (with butyl-, benzyl-, or octyl-groups as substituents) coordinated to different metal centers ($Ni^{2+}$, $Mn^{2+}$, or $Co^{2+}$) as cations, and chloride anions were synthesized. Co-based MILs provided the highest EFs (>85%) while Ni-based MILs showed the greatest selectivity in extracting the different sized duplex DNA fragments. It is worth noting that the subsequent analysis was performed by both HPLC-DAD and fluorescence spectroscopy with the latter technique to be more suitable for the faster detection of DNA. This is due to various chromatographic complexities such, as high solvent consumption and frequent column cleaning. Despite this, the preparation method was compared both with other IL- and MIL-based extraction methods and the conventional MIL-DLLME. The results showed the superiority of the method, based on its simplicity and highest extraction efficiencies. It is noteworthy that it is an affordable alternative for the extraction of DNA instead of commercially available DNA extraction kits.

Apart from the in-situ DLLME, an in-situ derivatization combined with MIL-based fast DLLME has been reported for the determination of biogenic amines (Bas) in food samples [48]. Once more, the experimental course is based on a similar philosophy as before. In a vial containing 5 mL of the sample (at set pH), the derivatizing agent, dansyl-chloride, (DNS-Cl) was added and the mix was left in the incubator for 15 min at 60 °C. The derivatization process was then completed, and the DNS-Cl was removed. An amount of MIL and methanol was added and the mixture was agitated via vortex. Next, the analytes were removed using a neodymium magnet, and redissolved in acetonitrile. The resulting solution was filtered and subjected to the HPLC system. This stage of derivatization is necessary, as BAs are not easily detected by HPLC-UV due to the lack of chromophore groups in their structure. It is worth mentioning that although cobalt(II)-based MILs have not been extensively studied yet, the authors synthesized the MIL trihexyltetradecylphosphonium tetra-chlorocobalt (II) $[P_{6,6,6,14}^{+}]_2[CoCl_4^{2-}]$ as it completely dissolves in the mobile phase, exhibits low absorbance eliminating background and allowing sensitive analysis of the analyte. This method seems to be a suitable and fast way, with high sensitivity for the determination of amines in real samples such as wine and fish. In fact, by comparing it to other methods such as SPE and UPLC/Q-TOFMS, it has been shown to provide lower LOD values and improved analytical performance as it can pre-concentrate and extract at the same time. From the above-mentioned applications, it can be seen that the method of in-situ formation of MILs addresses properly some issues that exist with DLLME procedures. Therefore, such procedures are more promising and worth further research to advance this topic.

*2.3. Single Drop Microextraction Procedures*

Single-drop microextraction is a straightforward, environmentally friendly technique that has been used, mainly, for the extraction and subsequent determination of low-molecular-weight compounds after coupling to various chromatographic techniques. The extraction is performed using a few microliters of an organic solvent either immersed into an aqueous sample or exposed to the headspace of the matrix with the aid of a micro syringe. MILs can provide a viable alternative to organic solvents in this technique because of their low volatility and high viscosity/hydrophobicity, overcoming the problem of droplet instability.

A new method named: weighing paper-assisted magnetic ionic liquid headspace single-drop microextraction, using microwave distillation followed by gas chromatography–mass spectrometry (WP-MIL-HS-SDME), was developed to determine a total of 39 volatile compounds in 16 lavender samples from three different harvest years with principal component analysis [49]. An amount (9 µL) of the magnetic ionic liquid 1-octyl-3-methylimidazolium tetrachloroferrate ($[C_8mimFeCl_4]$) could stably be divided on the weighing paper for long time extraction. The lavender samples were placed into a headspace vial. The $[C_8mim]FeCl_4$ was distributed into the weighing paper, which was adhered to the sealing cap containing PTFE-silicone septum, suspended in the headspace of the lavender sample. The bottle was sealed, and headspace extraction was achieved by microwave irradiation. After that, the weighing paper, with the analyte, was transferred directly into a polyethylene (PE) tube containing a back-extractant (cyclohexane). Next, the PE tube was vortexed, centrifuged, and then was forced to magnetic separation. Finally, a sample of back-extractant was injected into the GC-MS instrument for analysis. The increase of the extractant volume resulted in a significant increase in the extraction efficiency of the WP-MIL-HS-SDME than MIL-HS-SDME method, while the disadvantages of traditional HS-SDME, such as the microdroplet in the needlepoint being easy to drop and high operational requirements got through. Although this method is a rather unique approach, it is complex enough and many steps are needed for the extraction. These two drawbacks override the benefits.

The team of Jiwoo et al. investigated with HS-SDME and a DLLME methods, two tetrachloromanganate ($[MnCl_4^{2-}]$)-based MIL as extraction solvents for the determination of twelve aromatic compounds, including four polyaromatic hydrocarbons from lake water samples [50]. The optimized HS-SDME method was compared to the DLLME method employing the same two MILs as extraction compounds. The method of DLLME with the two MIL(($[P_{6,6,6,14}^{+}]_2[MnCl_4^{2-}]$) and ($[Aliquat^+]_2[MnCl_4^{2-}]$) showed much faster extraction and higher enrichment for analytes with low vapor pressure. On the other hand,

HS-SDME showed advantages in extracting analytes possessing relatively high vapor pressure. Both methods provided low LODs and high precision for the target analytes as well as acceptable relative recoveries from the samples, suggesting that the examined MILs can be exercised in microextraction techniques. The disadvantage of the HS-SDME method compared with DLLME was that in DLLME the total sampling time required was less than 5 min, which demonstrates its potential as a high throughput sampling technique. However, the HS-SDME method can easily be employed in cases of complex matrices. Generally, the use of [$MnCl_4^{2-}$]-based MILs provides advantages, such as convenient ways of extractions, low UV absorbance which permits the direct coupling to HPLC for chromatographic analysis, and high extraction selectivity.

A novel technique was presented in the manuscript of Trujillo-Rodríguez et al., which was developed by using a vacuum headspace single-drop microextraction method based on the use of magnetic ionic liquids (vacuum MILHS-SDME) (Figure 4) [51]. This method provided a successful approach for the determination of a group of short-chain free fatty acids (FFAs) (from $C_3$ to n-$C_7$), responsible for the aroma of milk and other dairy products. The use of MIL ([$P_{6,6,6,14}^+$][$Mn(hfacac)^{3-}$]) demonstrated advantages at decreased pressure conditions, with analytes reaching equilibrium faster than regular atmospheric pressure MIL-HS-SDME showed an improvement in the extraction efficiency for all analytes, at any extraction time. Furthermore, the method does not require derivatization of the free fatty acids to their methyl ester analogues and combined with vacuum headspace single-drop microextraction, analytes are determined in an automated approach using GC-MS without any interferences from the MIL solvent.

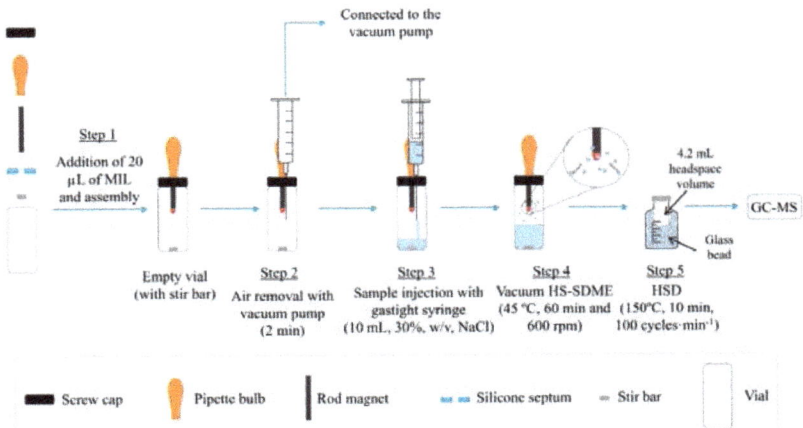

**Figure 4.** Schematic of the vacuum MIL-HS-SDME procedure under optimum conditions. Reproduced with permission from [51]. Copyright Elsevier, 2017.

In another study, a microextraction technique using a MIL was coupled with voltammetric determination of ascorbic acid (AA) in samples of vitamin C effervescent compounds and orange juice [52]. The MIL ([$Aliquat^+$]$_2$[$MnCl_4^{2-}$]) was used as the extracting solvent and was exposed directly on the surface of the working electrode. The MIL had a double purpose: as a cleanup (no interfering species) and as an electrode modifier (with $TiO_2$ nanoparticles). With this method, no dilution step was needed and in comparison to other voltammetry techniques, lower LOD, as well as increased sensitivity was achieved because a preconcentration step was performed before the electrochemical measurement. For the first time, the assets of the ionic characteristics of the IL as well as of the $Mn^{2+}$ ion for the modification of the electrode and enhancement of the electron transfer of AA were taken advantage of.

Fernández et al. used 1-ethyl-3- methylimidazolium tetraisothiocyanatocobaltate(II) ([Emim]$_2$[Co(NCS)$_4$]) as a MIL to extract nine chlorobenzenes (i.e., 1,2-dichlorobenzene,1,3-dichlorobenzene, 1,4-dichlorobenzene, 1,2,3-trichlorobenzene, 1,2,4-trichlorobenzene, 1,3,5-trichlorobenzene, 1,2,3,4-tetrachlorobenzene, 1,2,4,5-tetrachlorobenzene, and pentachlorobenzene) for analyses from water samples (tap water, pond water, and wastewater) [53]. In this study, with the method of magnetic headspace single-drop microextraction, the MIL was located on one end of a small neodymium magnet, and the extracts were determined by GC-MS. This approach showed lower LOD values than in the IL-based HS-SDME method with shorter extraction times and lower IL supply.

## 2.4. Matrix Solid-Phase Dispersion Procedures

For the most part, solvent-based extraction methods are the leading for the extraction of analytes from fatty solid samples. However, lately, matrix solid-phase dispersion (MSPD) has become an upward trend, mainly due to the low consumption of organic solvents. Likewise, DLLME, has attracted much attention considering its relatively high extraction efficiency. Wang and co-workers combined these two methods with a MIL followed by UFLC-UV to determine six triazine herbicides in oilseeds (Figure 5) [54]. Two types of MIL [C$_4$mim][FeCl$_4$] and [C$_6$mim][FeCl$_4$] were tested to observe the influence of the structure. Admittedly, the polarity of [C$_4$mim][FeCl$_4$] with the target analytes, which was slightly different compared to [C$_6$mim][FeCl$_4$], offered higher recoveries. Moreover, this addition was found to be highly advantageous, since the MSPD-MIL-DLLME method achieved better precision and lower LODs compared with results by QuEChERS coupled with UFLC. The MIL succeeds in replacing the centrifugation step in QuEChERS with magnetic separation and thus simplifies the method. Another key fact to mention is that LODs and LOQs obtained by the developed method are similar to or lower than those reported in methods for the determination of target analytes in solid, fatty matrices. Although the combination of the two techniques is innovative, the fact that both are employed increases the overall time of analysis.

In a more recent work, MIL [P$_{6,6,6,1,4}^+$][Co(II)(hfacac)$_3^-$] was directly used for the first time in a matrix solid-phase extraction procedure [55]. Developed by Chatzimitakos and co-workers, this method provided a valuable approach for the determination of multi-class pesticides residues in raw vegetables. Taking advantage of its magnetic properties, MIL was readily harvested after the extraction step by simply using a magnet. In addition, its hydrophobic and viscous nature made separation and retrieval feasible and assisted in mixing with the matrix. Described in more detail, chopped vegetables were pulverized and one drop of MIL was added to the sample, forming tiny droplets. Then, a saturated sodium chloride solution was added resulting in coalescing and creation of larger droplets which were separated more easily from the bulk phase. The homogenized mixture was ultrasonicated. Due to the dark red color of the chosen MIL, discrimination and harvesting from the solution were easily done. Droplets coupled with target analytes were transferred via magnet to 1 mL of acetonitrile in order to be dissolved (Figure 6). After the evaporation with a gentle nitrogen stream, the sample was ready to be injected into the HPLC-DAD system for further separation and detection of pesticides. In order to find the most efficient conditions, authors examined the effect of vegetable matrix and dispersion material as well as the selection of the most suitable MIL. For this purpose, MIL that differs in the metal and the number of anionic ligands were tested with the results showing that neither plays a decisive role in the extraction, on the contrary, the extraction is based, mainly, on the cation. Continuing in this line, the selection was based on the visual discrimination from the rest of the system, with dark red [P$_{6,6,6,1,4}^+$][Co(II)(hfacac)$_3^-$] preferred. As regards the vegetable matrix, low recoveries occurred with matrices of low water content. The best combination was found to be vegetables with "soft" texture and a high water-content matrix (e.g., potatoes). Lastly, silica, quartz silica, sodium chloride, and sodium sulfate were tried as dispersion materials in the developed method. Based on the results, solid dispersing materials, as well as the addition of both salts were found to be ineffective. However, when only one salt was added, the reproducibility of the procedure was improved (without the salt solution, RSD of five measurements was 9.5% and with the salt solution was 6.0%). Such direct procedures are highly welcome, since they can be carried out by analysts with less expertise and they have reduced cost, compared with other procedures.

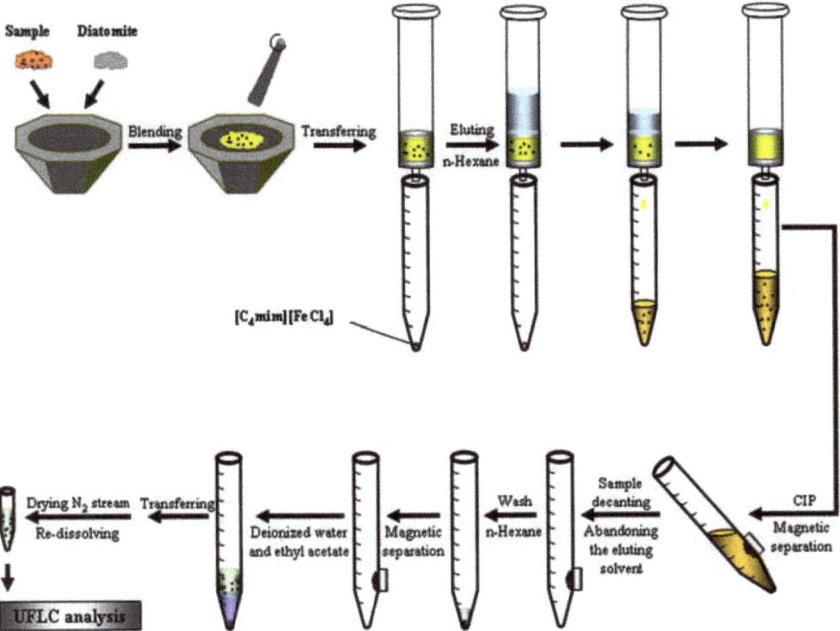

**Figure 5.** Schematic diagram for the MSPD-DLLME extraction procedure. Reproduced with permission from [54]. Copyright Elsevier, 2015.

*2.5. Stir-Bar Dispersive Procedures*

A novel hybrid microextraction method called stir bar dispersive liquid microextraction (SBDLME) opens new insights into the microextraction field, due to the facile retrieval of the target analytes [56]. This approach combines the advantages of stir bar sorptive extraction and DLLME. The first report of this approach was by Peng et al., 2012. His team developed a SBDLME method for the determination of three fungicide residues in real water samples (tap water, rainwater, and lake water), utilizing the MILs 1-butyl-3-methylimidazolium hexafluorophosphate ([$C_4$mim][$PF_6$]), and 1-hexyl-3-methylimidazolium hexafluorophosphate ([$C_6$mim][$PF_6$]) as extractant. The ionic liquid and magnetic stir bar were held within a sealed PCR tube pierced with many micro-holes on the wall. The ionic liquid magnetic bar was then placed in the aqueous solution for extraction. Meanwhile, the magnetic stirrer was switched on and the ionic liquid magnetic bar dispersed freely in the sample. When the magnetic stirrer was switched off the MIL was retrieved on the rod magnet The obtained ionic liquid extract was too viscous to be injected directly into the HPLC system, thus it was firstly diluted with acetonitrile. LODs were varying from 1.4 to 3.4 $\mu g \cdot L^{-1}$, RSD values ranged from 2.9 to 6.0%, and recoveries of carbamate pesticides at spiking levels of 5 and 50 $\mu g \cdot L^{-1}$ were in the range of 85-98.0%, 80-98% respectively, leading to a simple, practical and efficient method for the determination of trace level of carbamates in environmental samples.

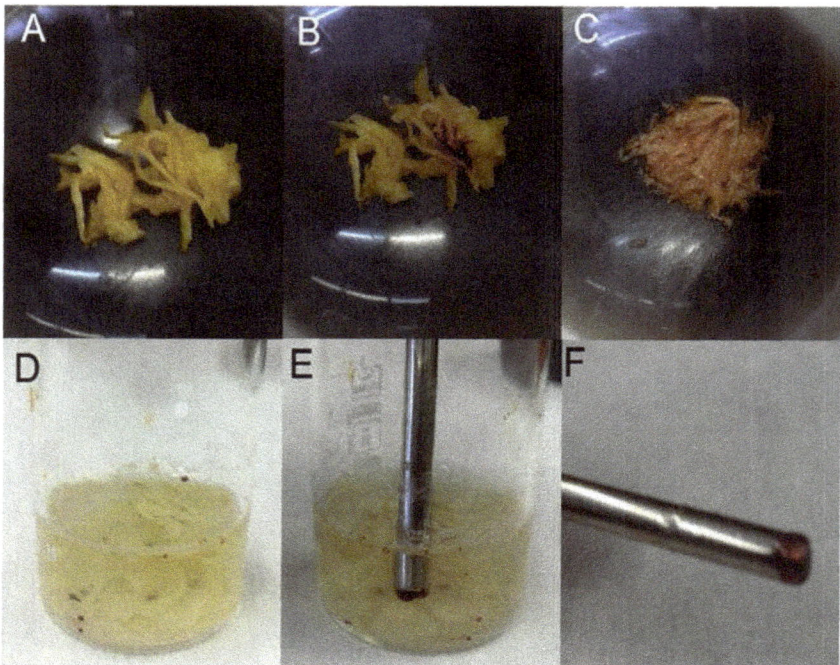

**Figure 6.** Representative pictures of the various steps of the developed procedure: (**A**) chopped potato in a mortar, (**B**) chopped potato with a drop of MIL before mixing, (**C**) chopped potato and MIL after mixing, (**D**) mixture of potato and MIL in saturated sodium chloride solution after ultrasonication, (**E**) harvesting the MIL droplets with a magnetic rod and (**F**) the collected MIL droplets on the magnetic rod. Reproduced with permission from [55]. Copyright Elsevier, 2018.

Juan et al. proposed a SBDLME procedure for the determination of (ultra)trace amounts of 10 PAHs in three natural water samples (river, tap, and rainwater) [57]. Regarding this, they utilized the $[P_{6,6,6,14}{}^+][Ni(II)(hfacac)_3{}^-]$ MIL as an extraction material that magnetically coats a neodymium magnetic stir bar. When the stirring rate is low, the MIL remains on the stir bar. At a high stirring rate, the MIL is dispersed in the solution and achieves the extraction of the analytes. As long as the stirring rate decreases, once again the MIL attaches to the magnetic bar. The MIL-coated stir bar is subsequently thermally desorbed into a GC system coupled to a MS detector. The determination of trace amounts of PAHs in water samples at the low ng·$L^{-1}$ level indicates the sensitivity of the SBDLME method in addition to the simplicity and efficiency of this approach. Moreover, compared to previous methods for the determination of PAHs, this method requires little sample manipulation, reduces the analysis time and it does not require solvent evaporation and external magnetic field.

In another study employed from the aforementioned research group, eight lipophilic organic UV filters from environmental water samples (river, sea, and swimming pool water samples) were determined by the above method using the $[P_{6,6,6,14}{}^+][Co(II)(hfacac)_3{}^-]$ and $[P_{6,6,6,14}{}^+][Ni(II)(hfacac)_3{}^-]$ MILs [58]. This work contributes to the development of expedient and sensitive methods for the determination of trace compounds in aqueous media and enables the use of tailor-made solvents (i.e., MILs).

In another interesting study, Trujillo-Rodríguez et al. attempted to develop an in-situ stir bar dispersive liquid–liquid microextraction technique for the first time [10]. Using three different MILs containing $Ni^{2+}$ or $Co^{2+}$ metal centers coordinated with Nbutylimidazole or N-octylimidazole ligands as extraction solvents ($[Ni(C_4IM)_4^{2+}]2[Cl^-]$, $[Ni(C_8IM)_4^{2+}]2[Cl^-]$ and $[Co(C_8IM)_4^{2+}]2[Cl^-]$ MILs), they determined seven organic pollutants on tap and mineral water. The microextraction method was combined with headspace gas chromatography mass spectrometry. The procedure of microextraction consists of the addition of the sample solution and the NdFeB stir bar into the extraction vial, followed by the insertion of MIL and dispersive solvent to the vial under stirring. The ion-exchange reagent ($[Li^+][NTf_2^-]$) was then added to achieve a 1:2 MIL:$[Li^+][NTf_2^-]$ molar ratio and the stirring rate was increased. During stirring, the MIL (in the $[NTf_2^-]$-form generated from the metathesis reaction) was dispersed within the extraction vial. When the stirring was stopped, the hydrophobic MIL immediately settled to the bottom of the vial and was collected onto the rod magnet. Finally, the MIL-coated stir bar was transferred to a HS vial. LODs down to 10 $\mu g \cdot L^{-1}$, adequate reproducibility, and relative recoveries between 72.5% and 102% were obtained. This method containing an in-situ metathesis reaction during the microextraction procedure revealed that long alkyl chains substituents in the ligand can increase the thermal stability of the MILs, indicated the important role of ligand in the MIL to the optimum extraction conditions required for the method.

*2.6. Other Procedures*

Definite proof of the beneficial combination of nanomaterials with magnetic ionic liquids is the research carried out by Zhu and co-workers [59]. Two endocrine-disrupting chemicals, bisphenol A and 4-nonylphenol were determined in vegetable oils by DLLME followed by HPLC-MS/MS. Although MILs can facilitate separation, in the present work, the magnetism of the ionic liquid was found to be insufficient. This limitation was resolved by the addition of $Fe_3O_4$ nanoparticles. Briefly, a diluted sample of vegetable oil (0.5 g in 10 mL of n-hexane) was mixed with the diluted sample of MIL (80 µL in 400 µL acetone) with the aid of stirring in vortex. This was followed by the addition of $Fe_3O_4$ nanoparticles (20 mg) and centrifugation to remove the supernatant n-hexane layer. The magnetic ionic liquid layer that interests us, was extracted through p-xylene (2 × 300 µL), which was isolated by using a magnet. The sample was prepared for further analysis by HPLC-MS/MS, after evaporation under nitrogen atmosphere, filtration through a membrane (0.22 µm), and dissolution in methanol (0.5 mL). After testing various MILs ($[C_4mim][FeCl_4]$, $[C_8mim][FeCl_4]$) they concluded that $[C_6mim][FeCl_4]$ was the most suitable, providing a higher EF for BPA and 4-NP. The proposed method could find application in other analyses of different pollutants in vegetable oil.

Hongmei et al. prepared a novel magnetic ionic liquid-gold nanoparticles/porous silicon (MIL/Au NPs/PSi) active substrate through the immersion method for the detection of Arsenic in pure water samples with three easy steps [60]. First, the preparation of PSi, second the reduction of Au-NPs on PSi with the assistance of microwave, and last one the deposition of MIL layer on the composite. To prove its high-sensitive detection of Arsenite, a set of arsenic solutions of different concentrations was prepared, and then the MIL/Au NPs/PSi substrate was immersed in 1.5 mL of each one for 1h. Afterward, it was left to dry at room temperature and the surface-enhanced Raman spectroscopy (SERS) spectrum was obtained. In the same way, the procedure was repeated with the Au NPs/PSi and PIL/Au NPs/PSi substrates, respectively. To do so, the magnetic ionic liquid 1-methyl-3-hexyl imidazole ferric tetrachloride ($[C_6mim]FeCl_4$) and the poly ionic liquid poly (1-hexyl-3-vinyl imidazolium tetrafluoroborate ($[Hmim]BF_4$) were used for the separate modifications of the Au NPs/PSi SERS substrate. Under the same Raman test conditions, the functionalized substrates (MIL-Au NPs/PSi and PIL-Au NPs/PSi) were found to have a stronger SERS response than Au NPs/PSi SERS on their own. This occurs due to the EM (electromagnetic mechanism) and CM (chemical mechanism) enhancement that is caused by Au NPs, and the IL enriched analytes. Furthermore, much stronger signal intensity was achieved with the addition of MIL, via a specific binding between NPs and arsenite. As reported, this is the first time that this direct modification took place, and it is characterized as a simple and cost-effective method to load a MIL film on Au NPs/PSi and improve the stability and sensitivity of the detection (the detection limit is as low as 0.5 ppb, and the RSD only 1.6%). Considering the convenience of designing and grafting functional groups in MIL, it is expected to have a successful use for the qualitative and quantitative detection of trace components in complex systems and gain practical value in environmental or biological sample detection.

In another study, the magnetic 1-allyl-3-octylimidazolium tetrachloroferrate ionic liquid was synthesized and combined with a molecularly imprinted polymer for the extraction of phenolic acids in apple samples [61]. Tashakkori et al. developed a magnetic sorbent for SPE instead of using the method itself, to minimize the extraction time. The MIP was prepared by suspension polymerization using the MIL as a functional monomer and chlorogenic acid as a template molecule. The selection of the resulting magnetic imprinted polymer as a sorbent provided not only a wide linear range (1-1000 $\mu g \cdot L^{-1}$ for chlorogenic acid) with a small amount of it (2 mg) but also lower LOD values (0.31 $\mu g \cdot L^{-1}$ for chlorogenic acid) compared with other SPE methods. Analytical parameters, such as the type and volume of elution solution, pH of sample solution, sorbent amount, extraction, and the desorption time were optimized, before the analysis by HPLC. This environmentally friendly way achieves separation and pre-concentration of phenolic acids, with high reproducibility and reliability. A summary of all the sample preparation procedures discussed herein along with their analytical figures of merit is presented in Table 1.

Table 1. Summary of developed analytical procedures based on MILs; LOD: limit of detection.

| Ionic Liquid | Extraction Technique | Matrix | Target Analytes | LOD (µg·kg$^{-1}$ or µg·L$^{-1}$) | Recoveries (%) | Analytical Instrumental System | Reference |
|---|---|---|---|---|---|---|---|
| [P$_{6,6,6,14}$$^+$][Dy(III)(hfacac)$^{4-}$] | MIL-DLLME | water samples | sulfonamides and triazines | 0.011–0.029 and 0.013–0.030 | 90–101 and 89–98 | HPLC-DAD | [9] |
| [Ni(C$_8$IM)$_2$$^{2+}$]$_2$[NTf$_2$$^-$] | SBSDME | tap and mineral water | organic pollutants | <10 | 72.5–102 | HS-GC-MS | [10] |
| trihexyltetradecylphosphonium tetrachloromanganate (II) ([P$_{6,6,6,14}$$^+$]$_2$[MnCl$_4$$^{2-}$]) | DLLME | human urine | estriol, 17-β-estradiol, 17-α-ethynylestradiol, and estrone | 2 | 67.5–115.6 | HPLC | [11] |
| benzyltrioctylammonium bromotrichloroferrate (III) | magnet-based microextraction | aqueous sample | benzo(a)anthracene (BaA), chrysene (Chy), benzo(a)pyrene (BaPy), benzo(b)fluoranthene (BbF) and benzo(k)fluoranthene (BkF) | 0.005–0.02 | 91.5–119 | HPLC-FID | [12] |
| [P$_{6,6,6,14}$$^+$]$_2$[MnCl$_4$$^{2-}$] | DLLME | river water | estrone, estradiol, 17-α-hydroxyprogesterone, chloromadinone 17-acetate, megestrol 17-acetate and medroxyprogesterone 17-acetate | 1.5–15.1 | 56–123 | HPLC-DAD | [14] |
| [P$_{6,6,6,14}$$^+$]$_2$[CoCl$_4$$^{2-}$] | DLLME | milk and cosmetics | estrone, estradiol, 17-α-hydroxyprogesterone, chloromadinone 17-acetate, megestrol 17-acetate and medroxyprogesterone 17-acetate | 5–15 | 98.5–109.3 and 96.3–111.4 | HPLC | [21] |
| [TMGl][TEMPO OSO$_3$] | MILATPs | environmental waters | chloramphenicol | 0.14 | 94.6–99.72 | HPLC | [15] |
| [C$_4$MIM-Tempo][L-Pro] | aqueous two-phase (ATPs) system | - | phenylalanine (D-L) | - | - | HPLC | [16] |
| trihexyltetradecylphosphonium [MnCl$_4$$^{2-}$] ([P$_{6,6,6,14}$$^+$]$_2$[MnCl$_4$$^{2-}$]) | DLLME | tap water, wastewater, and a tea infusion | pharmaceutical drugs, phenolics, insecticides, and polycyclicaromatic hydrocarbons | 0.25–1.00 | 53.8–114.7 (spiking 5 µg·L$^{-1}$ for phenanthrene) 106.7–150 (spiking 37.5 µg·L$^{-1}$ for phenanthrene) | HPLC, UV | [18] |
| 1-hexyl-3-methylimidazolium tetrachlo-roferrate ([C$_6$mim][FeCl$_4$]) | liquid–liquid microextraction technique (DLLME) | vegetable oils, two soybean oils, three maize oils and two sunflower seed oils | triazine herbicides | 1.31–1.49 | 81.8–114.2 | HPLC | [22] |

Table 1. Cont.

| Ionic Liquid | Extraction Technique | Matrix | Target Analytes | LOD (µg·kg$^{-1}$ or µg·L$^{-1}$) | Recoveries (%) | Analytical Instrumental System | Reference |
|---|---|---|---|---|---|---|---|
| [P$_{6,6,6,14}$$^+$][Cl$^-$] | Dispersive liquid-liquid microextraction | human urine | carbofuran, atrazine, simazine, diuron, metalochlor, ethinylestradiol, estrone, diclofenac | - | 75-130 | HPLC | [31] |
| 1-ethoxyl-3-methyl-imidazoliumtetrachloroferrate [C$_2$OHmim][FeCl$_4$] | UAE | sinomenium acutum | sinomenine (SIN) | - | 81.3 | HPLC | [28] |
| C$_3$MIMFeCl$_4$ | | tea leaves | polyphenols | - | 99.8 | HPLC-UV-Vis | [29] |
| [P$_{6,6,6,14}$$^+$]$_3$[GdCl$_6$$^{3-}$] | MIL-DLLME | river and tap water | four antihypertensive drugs | - | 82.5-101.48 | HPLC-UV | [32] |
| [P$_{6,6,6,14}$]$_3$[Fe(CN)$_6$] | IL-on SBME | environmental water | four estrogens | 0.2-0.5 | 88.5-99.6 and 88.4-99.9 | HPLC-UV | [19] |
| methyltrioctylammonium tetrachloroferrate (N$_{8,8,8,1}$][FeCl$_4$]) | SADBME | aqueous matrices | phenols and acidic pharmaceuticals | 1.05-33.0 | 89-94 | HPLC-DAD | [20] |
| methyltriocty lammonium tetrachloroferrate ([N$_{1,8,8,8}$$^+$][FeCl$_4$]) | MIL-DLLME | water, beer and beverage samples | parabens | 300-500 | 95-103 | HPLC-UV | [30] |
| [OA][FeCl] | D-µSPE | human urine and wastewater samples | carbamazepine drug | 0.51 | 85.5-98 | HPLC-DAD | [33] |
| trihexyl(tetradecyl)phosphonium tetrachloroferrate(III) ([P$_{6,6,6,14}$][FeCl$_4$]) | MIL-DLLME | honey | Cd | 0.0004 | 95.5-102 | ETAAS | [34] |
| trihexyl(tetradecyl)phosphonium tetrachloroferrate (III) ([P$_{6,6,6,14}$][FeCl$_4$]) | DLLME | honey | As | 0.012 | 95.2-102 | ETAAS | [35] |
| exyl(tetradecyl)phosphonium tetrachloroferrate ([P$_{6,6,6,14}$][FeCl$_4$]) | DLLME | honey | Cr | 0.005 | 94.0-101 | ETAAS | [36] |
| trihexyl(tetradecyl)phosphonium tetrachloromanganate (II) ([P$_{6,6,6,14}$]$_2$[MnCl$_4$]) | DLLME | honey, mead, honey vinegar and honey beer | Pb | 0.003 | 94.8-101 | ETAAS | [37] |
| 1-butyl-3-methylimidazolium tetrachloroferrate [C$_4$mim][FeCl$_4$] | MIL-UDSA-DLLME | rice | Se | 18 | 94.9-104.8 | GFAAS | [38] |
| 1-butyl-3-methylimidazolium tetrachloroferrate ([C$_4$mim][FeCl$_4$]) | AALLME | environmental water, sediment and soil samples | As | 29 | 93.0-108.5 | GFAAS | [41] |

Table 1. Cont.

| Ionic Liquid | Extraction Technique | Matrix | Target Analytes | LOD (µg·kg$^{-1}$ or µg·L$^{-1}$) | Recoveries (%) | Analytical Instrumental System | Reference |
|---|---|---|---|---|---|---|---|
| trihexyl(tetradecyl)phosphonium tetrachloroferrate ([P$_{6,6,6,14}$][FeCl$_4$]) | DLLME | n tap, dam, mineral, wetland, underground, rain and river water samples | Sb | 0.02 | 94.0–100 | ETAAS | [42] |
| bis(1-ethyl-3-methylimidazolium) tetrathiocyanatocobaltate (II) [Emim]$_2$[Co(SCN)$_4$] | DLLME | engine oil, gasoline and diesel | Cd | 0.084 | 95–110 | ETAAS | [43] |
| butyl-3-methylimidazolium tetra-chloroferrate ([C$_4$mim][FeCl$_4$]) | UASEME | vegetable oil | Cd, Pb | 0.002, 0.02 | 95.0–105.8 | GFAAS | [39] |
| 1-butyl-3-methylimidazolium tetrachloroferrate ([C$_4$mim][FeCl$_4$]) | ETA-MILs-ME | vegetable samples | As | 7 | 97.9–105.8 | GFAAS | [40] |
| ([Ni(C$_n$IM)$_4^{2+}$]$_2$[Cl$^-$] and [Ni(BeIM)$_4^{2+}$]$_2$[Cl$^-$]) | in situ MIL-DLLME | aqueous samples | polar and non-polar pollutants | 0.13–5.2 and 0.012–1.6 | 67.7–120 and 86.5–96.6 | HPLC-DAD | [44] |
| [Co(C$_4$IM)$^{+2}$$_4$]$_2$[NTf$_2$] | in situ Pa-DDE/MIL | aqueous environmental samples | organic micropollutants | 7.5 | 53.9–129.1 | HPLC-DAD | [45] |
| [P$_{6,6,6,14}^+$][Ni(II)(hfacac)$^{3-}$] | in situ MIL-DLLME | | long and short double-stranded DNA | - | - | fluorescence emission spectroscopy | [46] |
| [C$_4$MIM-TEMPO]Cl | in-situ MIL-DLLME | milk samples | sulfonamides | 0.534–0.891 | 95–105 | HPLC-UV | [47] |
| [P$_{6,6,6,14}^+$]$_2$[CoCl$_4^{2-}$] | in situ derivatization-MIL-DLLME | wine and fish samples | six biogenic amines | 1.3–3.9 and 1.2–3.8 | 93.2–103.1 and 94.5–102.3 | LC-UV | [48] |
| 1-octyl-3-methylimidazolium tetrachloroferrate ([C$_8$MIM]FeCl$_4$) | weighing paper-assisted magnetic ionic liquid headspace single-drop microextraction (WP-MIL-HS-SDME) | 16 lavender samples | 39 volatile compounds | - | - | GC-MS | [49] |
| ([P6,6,6,14$^+$]$_2$[MnCl$_4^{2-}$]) and ([Aliquat$^+$]$_2$[MnCl$_4^{2-}$]) | HS-SDME, DLLME | lake water samples | twelve aromatic compounds and four polyaromatic hydrocarbons | 0.04–1.0 and 0.05–1.0 | 70.2–109.6 and 68.7–104.5 | HPLC | [50] |
| [P$_{6,6,6,14}^+$][Mn(hfacac)$^{3-}$] | vacuum MIL-HS-SDME | milk samples | free fatty acids (FFAs) | 14.5–70.3 | 79.5–111 | GC-MS | [51] |
| aliquat tetrachloromanganate (II) [Aliquat$^+$]$_2$[MnCl$_4^{2-}$] | single drop microextraction | aqueous samples | ascorbic acid | 0.012–18.7 | 101.0–104.1 | voltammetric determination | [52] |

Table 1. Cont.

| Ionic Liquid | Extraction Technique | Matrix | Target Analytes | LOD (µg·kg$^{-1}$ or µg·L$^{-1}$) | Recoveries (%) | Analytical Instrumental System | Reference |
|---|---|---|---|---|---|---|---|
| 1-ethyl-3-methylimidazolium tetraisothiocyanatocobaltate(II) ([Emim]$_2$[Co(NCS)$_4$]) | magnetic headspace single-drop microextraction (Mag-HS-SDME) | water samples | 1,2-dichlorobenzene, 1,3-dichlorobenzene, 1,4-dichlorobenzene, 1,2,3-trichlorobenzene, 1,2,4-trichlorobenzene, 1,3,5-trichlorobenzene, 1,2,3,4-tetrachlorobenzene, 1,2,4,5-tetrachlorobenzene, and pentachlorobenzene | 0.003–0.152 | 82–114 | GC-MS | [53] |
| 1-butyl-3-methylimidazolium tetrachloroferrate ([C$_4$MIM][FeCl$_4$]) | MSPD-MIL-DLLME | oilseeds | triazine herbicides | 1.20–2.72 | 82.9–113.7 | UFLC-UV | [54] |
| [P$_{6,6,6,14}$$^+$][Co(II)(hfacac)$_3$$^-$] | MSPD | raw vegetables | ten pesticides | - | 65–85 | HPLC-DAD | [55] |
| [C$_6$MIM][PF$_6$] | ILMB-ME | water samples | carbamate pesticides | 1.4–3.4 | 85–98.0 (spiking 5 µg·L$^{-1}$), 80–98 (spiking 50 µg·L$^{-1}$) | HPLC-DAD | [56] |
| [P$_{6,6,6,14}$$^+$][Ni(II)(hfacac)$_3$$^-$] | SBDLME | natural water samples | polycyclic aromatic hydrocarbons (PAHs) | 0.0005–0.0087 | 84–115 | GC-MS | [57] |
| [P$_{6,6,6,14}$$^+$][Ni(hfacac)$_3$$^-$] | SBDLME | environmental water samples | lipophilic organic UV filters | 0.0099–0.027 | 87–113 (river water), 91–117 (sea-water), 89–115 (swimming pool water) | TD-GC-MS | [58] |
| 1-hexyl-3-methylimidazolium tetrachloroferrate ([C$_6$MIM][FeCl$_4$]) | DLLME | vegetable oil | bisphenol A and 4-nonylphenol | 0.1 and 0.06 | 70.4–112.3 | HPLC-MS/MS | [59] |
| 1-methyl-3-hexyl imidazole ferric tetrachloride ([C$_6$MIM][FeCl$_4$]) | immersion method | water samples | arsenic | 0.500 | - | - | [60] |
| 1-allyl-3-octylimidazolium tetrachloroferrate | SPE | apple samples | phenolic acids | 0.31–1.72 | 81–100 | HPLC-DAD | [61] |
| trihexyltetradecylphosphonium tetrachloroferrate (III) ([3C$_6$PC$_{14}$][FeCl$_4$]) | magnetic room temperature ionic liquid | soil samples | phenol (Ph), 4-nitrophenol (4-NP), 2-chlorophenol (2-CP), 4-chlorophenol (4-CP), 2,4-dichlorophenol (2,4-DCP), 3,5-dichlorophenol (3,5-DCP), pentachlorophenol (penta-CP), and 2-benzyl-4-chlorophenol (2-Bn-4-CP) | - | - | UV-Vis-NIR spectrometer, HPLC | [17] |

## 3. Conclusions and Future Perspectives

The development of novel sample preparation procedures is a never-ending field of research, as the need for more accurate and easier analytical methods increases. To this end, much effort has been put into employing MILs in sample preparation and great advancements have been made up to now. The results presented in this review article corroborate the significance of designing various kinds of magnetic ILs in order to be successfully applied to sample preparation. The MILs not only exhibit high extraction capacity activities and moderate selectivity, but also showed facile recovery and recyclability. These features render them suitable choices for designing environmentally benign sample preparation procedures in (bio)chemical analysis. However, efforts should be made towards heightening the magnetic susceptibility so that a more efficient retrieval of the MIL is allowed for more complex matrixes, under a lower magnetic field. Moreover, properties of the MILs such as the hydrophobicity and the viscosity should be more easily tuned, to broaden their applicability. Tuning the hydrophobicity of the MILs will assist in the extraction of different analytes. On the other hand, the high viscosity of the MILs generally causes many problems regarding the use of exact amounts of MILs and their handling and generation of droplets during dispersion. This can be addressed either by using supporting materials or by in-situ formation of MILs. However, these solutions are difficult to be employed in complex matrixes. Finally, the direct injection of MILs in HPLC systems, currently has many limitations, such as high back pressure, immiscibility with the mobile phase and absorbance in the UV region. Thus, the synthesis of new MILs with other organic groups that can alter the physicochemical characteristics of the MILs can address the problems, thus, avoiding time consuming steps, such as back extraction. Despite the current limitations, this topic of research is developing rapidly, and many advancements are expected to occur in the near future.

**Author Contributions:** Conceptualization, T.C. and C.S.; writing—original draft preparation, T.C., P.A., I.C. and K.D.; writing—review and editing, T.C. and C.S. All authors have read and agreed to the published version of the manuscript.

**Funding:** This research received no external funding.

**Conflicts of Interest:** The authors declare no conflict of interest.

## References

1. Reyes-Garcés, N.; Gionfriddo, E.; Gómez-Ríos, G.A.; Alam, N.; Boyacı, E.; Bojko, B.; Singh, V.; Grandy, J.; Pawliszyn, J. Advances in solid phase microextraction and perspective on future directions. *Anal. Chem.* **2018**, *90*, 302–360. [CrossRef]
2. Zhang, B.-T.; Zheng, X.; Li, H.-F.; Lin, J.-M. Application of carbon-based nanomaterials in sample preparation: A review. *Anal. Chim. Acta* **2013**, *784*, 1–17. [CrossRef]
3. Sajid, M. Magnetic ionic liquids in analytical sample preparation: A literature review. *TrAC Trends Anal. Chem.* **2019**, *113*, 210–223. [CrossRef]
4. Chatzimitakos, T.; Stalikas, C. Carbon-based nanomaterials functionalized with ionic liquids for microextraction in sample preparation. *Separations* **2017**, *4*, 14. [CrossRef]
5. Feng, J.; Loussala, H.M.; Han, S.; Ji, X.; Li, C.; Sun, M. Recent advances of ionic liquids in sample preparation. *TrAC Trends Anal. Chem.* **2020**, *125*, 115833. [CrossRef]
6. Ruiz-Aceituno, L.; Sanz, M.; Ramos, L. Use of ionic liquids in analytical sample preparation of organic compounds from food and environmental samples. *TrAC Trends Anal. Chem.* **2013**, *43*, 121–145. [CrossRef]
7. Yavir, K.; Konieczna, K.; Marcinkowski, Ł.; Kloskowski, A. Ionic liquids in the microextraction techniques: The influence of ILs structure and properties. *TrAC Trends Anal. Chem.* **2020**, *130*, 115994. [CrossRef]
8. Clark, K.D.; Emaus, M.N.; Varona, M.; Bowers, A.N.; Anderson, J.L. Ionic liquids: Solvents and sorbents in sample preparation. *J. Sep. Sci.* **2018**, *41*, 209–235. [CrossRef] [PubMed]
9. Chatzimitakos, T.G.; Pierson, S.A.; Anderson, J.L.; Stalikas, C.D. Enhanced magnetic ionic liquid-based dispersive liquid-liquid microextraction of triazines and sulfonamides through a one-pot, pH-modulated approach. *J. Chromatogr. A* **2018**, *1571*, 47–54. [CrossRef] [PubMed]
10. Trujillo-Rodríguez, M.J.; Anderson, J.L. In situ generation of hydrophobic magnetic ionic liquids in stir bar dispersive liquid-liquid microextraction coupled with headspace gas chromatography. *Talanta* **2019**, *196*, 420–428. [CrossRef]

11. Merib, J.; Spudeit, D.A.; Corazza, G.; Carasek, E.; Anderson, J.L. Magnetic ionic liquids as versatile extraction phases for the rapid determination of estrogens in human urine by dispersive liquid-liquid microextraction coupled with high-performance liquid chromatography-diode array detection. *Anal. Bioanal. Chem.* **2018**, *410*, 4689–4699. [CrossRef]
12. Trujillo-Rodríguez, M.J.; Nacham, O.; Clark, K.D.; Pino, V.; Anderson, J.L.; Ayala, J.H.; Afonso, A.M. Magnetic ionic liquids as non-conventional extraction solvents for the determination of polycyclic aromatic hydrocarbons. *Anal. Chim. Acta* **2016**, *934*, 106–113. [CrossRef] [PubMed]
13. Maximino, C.; de Brito, T.M.; da Silva Batista, A.W.; Herculano, A.M.; Morato, S.; Gouveia, A., Jr. Measuring anxiety in zebrafish: A critical review. *Behav. Brain Res.* **2010**, *214*, 157–171. [CrossRef] [PubMed]
14. Da Silva, A.C.; Mafra, G.; Spudeit, D.; Merib, J.; Carasek, E. Magnetic ionic liquids as an efficient tool for the multiresidue screening of organic contaminants in river water samples. *Sep. Sci. Plus* **2019**, *2*, 51–58. [CrossRef]
15. Yao, T.; Yao, S. Magnetic ionic liquid aqueous two-phase system coupled with high performance liquid chromatography: A rapid approach for determination of chloramphenicol in water environment. *J. Chromatogr. A* **2017**, *1481*, 12–22. [CrossRef] [PubMed]
16. Yao, T.; Li, Q.; Li, H.; Peng, L.; Liu, Y.; Du, K. Extractive resolution of racemic phenylalanine and preparation of optically pure product by chiral magnetic ionic liquid aqueous two-phase system. *Sep. Purif. Technol.* **2021**, *274*, 119024. [CrossRef]
17. Deng, N.; Li, M.; Zhao, L.; Lu, C.; de Rooy, S.L.; Warner, I.M. Highly efficient extraction of phenolic compounds by use of magnetic room temperature ionic liquids for environmental remediation. *J. Hazard. Mater.* **2011**, *192*, 1350–1357. [CrossRef]
18. Yu, H.; Merib, J.; Anderson, J.L. Faster dispersive liquid-liquid microextraction methods using magnetic ionic liquids as solvents. *J. Chromatogr. A* **2016**, *1463*, 11–19. [CrossRef]
19. Berton, P.; Siraj, N.; Das, S.; de Rooy, S.; Wuilloud, R.G.; Warner, I.M. Efficient low-cost procedure for microextraction of estrogen from environmental water using magnetic ionic liquids. *Molecules* **2020**, *26*, 32. [CrossRef]
20. Chatzimitakos, T.; Binellas, C.; Maidatsi, K.; Stalikas, C. Magnetic ionic liquid in stirring-assisted drop-breakup microextraction: Proof-of-concept extraction of phenolic endocrine disrupters and acidic pharmaceuticals. *Anal. Chim. Acta* **2016**, *910*, 53–59. [CrossRef]
21. Feng, X.; Xu, X.; Liu, Z.; Xue, S.; Zhang, L. Novel functionalized magnetic ionic liquid green separation technology coupled with high performance liquid chromatography: A rapid approach for determination of estrogens in milk and cosmetics. *Talanta* **2020**, *209*, 120542. [CrossRef] [PubMed]
22. Wang, Y.; Sun, Y.; Xu, B.; Li, X.; Jin, R.; Zhang, H.; Song, D. Magnetic ionic liquid-based dispersive liquid–liquid microextraction for the determination of triazine herbicides in vegetable oils by liquid chromatography. *J. Chromatogr. A* **2014**, *1373*, 9–16. [CrossRef] [PubMed]
23. García-Reyes, J.F.; Ferrer, C.; Thurman, E.M.; Fernández-Alba, A.R.; Ferrer, I. Analysis of herbicides in olive oil by liquid chromatography time-of-flight mass spectrometry. *J. Agric. Food Chem.* **2006**, *54*, 6493–6500. [CrossRef]
24. Ferrer, C.; Gómez, M.J.; García-Reyes, J.F.; Ferrer, I.; Thurman, E.M.; Fernández-Alba, A.R. Determination of pesticide residues in olives and olive oil by matrix solid-phase dispersion followed by gas chromatography/mass spectrometry and liquid chromatography/tandem mass spectrometry. *J. Chromatogr. A* **2005**, *1069*, 183–194. [CrossRef] [PubMed]
25. Fuentes, E.; Báez, M.E.; Quiñones, A. Suitability of microwave-assisted extraction coupled with solid-phase extraction for organophosphorus pesticide determination in olive oil. *J. Chromatogr. A* **2008**, *1207*, 38–45. [CrossRef] [PubMed]
26. Cai, M.; Chen, X.; Wei, X.; Pan, S.; Zhao, Y.; Jin, M. Dispersive solid-phase extraction followed by high-performance liquid chromatography/tandem mass spectrometry for the determination of ricinine in cooking oil. *Food Chem.* **2014**, *158*, 459–465. [CrossRef]
27. Wang, W.-X.; Yang, T.-J.; Li, Z.-G.; Jong, T.-T.; Lee, M.-R. A novel method of ultrasound-assisted dispersive liquid–liquid microextraction coupled to liquid chromatography–mass spectrometry for the determination of trace organoarsenic compounds in edible oil. *Anal. Chim. Acta* **2011**, *690*, 221–227. [CrossRef] [PubMed]
28. Li, Q.; Wu, S.; Wang, C.; Yi, Y.; Zhou, W.; Wang, H.; Li, F.; Tan, Z. Ultrasonic-assisted extraction of sinomenine from Sinomenium acutum using magnetic ionic liquids coupled with further purification by reversed micellar extraction. *Process Biochem.* **2017**, *58*, 282–288. [CrossRef]
29. Feng, X.; Zhang, W.; Zhang, T.; Yao, S. Systematic investigation for extraction and separation of polyphenols in tea leaves by magnetic ionic liquids. *J. Sci. Food Agric.* **2018**, *98*, 4550–4560. [CrossRef]
30. Mousavi, K.Z.; Yamini, Y.; Seidi, S. Dispersive liquid–liquid microextraction using magnetic room temperature ionic liquid for extraction of ultra-trace amounts of parabens. *New J. Chem.* **2018**, *42*, 9735–9743. [CrossRef]
31. Will, C.; Omena, E.; Corazza, G.; Bernardi, G.; Merib, J.; Carasek, E. Expanding the applicability of magnetic ionic liquids for multiclass determination in biological matrices based on dispersive liquid–liquid microextraction and HPLC with diode array detector analysis. *J. Sep. Sci.* **2020**, *43*, 2657–2665. [CrossRef]
32. Abdelaziz, M.A.; Mansour, F.R.; Danielson, N.D. A gadolinium-based magnetic ionic liquid for dispersive liquid–liquid microextraction. *Anal. Bioanal. Chem.* **2021**, *413*, 205–214. [CrossRef]
33. Hassan, A.A.; Tanimu, A.; Alhooshani, K. Iron and cobalt-containing magnetic ionic liquids for dispersive micro-solid phase extraction coupled with HPLC-DAD for the preconcentration and quantification of carbamazepine drug in urine and environmental water samples. *J. Mol. Liq.* **2021**, *336*, 116370. [CrossRef]
34. Fiorentini, E.F.; Escudero, L.B.; Wuilloud, R.G. Magnetic ionic liquid-based dispersive liquid-liquid microextraction technique for preconcentration and ultra-trace determination of Cd in honey. *Anal. Bioanal. Chem.* **2018**, *410*, 4715–4723. [CrossRef]

35. Fiorentini, E.F.; Canizo, B.V.; Wuilloud, R.G. Determination of As in honey samples by magnetic ionic liquid-based dispersive liquid-liquid microextraction and electrothermal atomic absorption spectrometry. *Talanta* **2019**, *198*, 146–153. [CrossRef] [PubMed]
36. Fiorentini, E.F.; Oviedo, M.N.; Wuilloud, R.G. Ultra-trace Cr preconcentration in honey samples by magnetic ionic liquid dispersive liquid-liquid microextraction and electrothermal atomic absorption spectrometry. *Spectrochim. Acta Part B At. Spectrosc.* **2020**, *169*, 105879. [CrossRef]
37. Fiorentini, E.F.; Botella, M.B.; Wuilloud, R.G. A simple preconcentration method for highly sensitive determination of Pb in bee products by magnetic ionic liquid dispersive liquid-liquid microextraction and electrothermal atomic absorption spectrometry. *J. Food Compos. Anal.* **2020**, *95*, 103661. [CrossRef]
38. Wang, X.; Chen, P.; Cao, L.; Xu, G.; Yang, S.; Fang, Y.; Wang, G.; Hong, X. Selenium speciation in rice samples by magnetic ionic liquid-based up-and-down-shaker-assisted dispersive liquid-liquid microextraction coupled to graphite furnace atomic absorption spectrometry. *Food Anal. Methods* **2017**, *10*, 1653–1660. [CrossRef]
39. Yao, L.; Liu, H.; Wang, X.; Xu, W.; Zhu, Y.; Wang, H.; Pang, L.; Lin, C. Ultrasound-assisted surfactant-enhanced emulsification microextraction using a magnetic ionic liquid coupled with micro-solid phase extraction for the determination of cadmium and lead in edible vegetable oils. *Food Chem.* **2018**, *256*, 212–218. [CrossRef]
40. Wang, X.; Xu, G.; Guo, X.; Chen, X.; Duan, J.; Gao, Z.; Zheng, B.; Shen, Q. Effervescent tablets containing magnetic ionic liquids as a non-conventional extraction and dispersive agent for speciation of arsenite and arsenate in vegetable samples. *J. Mol. Liq.* **2018**, *272*, 871–877. [CrossRef]
41. Wang, X.; Xu, G.; Chen, P.; Liu, X.; Fang, Y.; Yang, S.; Wang, G. Arsenic speciation analysis in environmental water, sediment and soil samples by magnetic ionic liquid-based air-assisted liquid–liquid microextraction. *RSC Adv.* **2016**, *6*, 110247–110254. [CrossRef]
42. Oviedo, M.N.; Fiorentini, E.F.; Lemos, A.A.; Wuilloud, R.G. Ultra-sensitive Sb speciation analysis in water samples by magnetic ionic liquid dispersive liquid–liquid microextraction and multivariate optimization. *Anal. Methods* **2021**, *13*, 1033–1042. [CrossRef] [PubMed]
43. Aguirre, M.Á.; Canals, A.; López-García, I.; Hernández-Córdoba, M. Determination of cadmium in used engine oil, gasoline and diesel by electrothermal atomic absorption spectrometry using magnetic ionic liquid-based dispersive liquid-liquid microextraction. *Talanta* **2020**, *220*, 121395. [CrossRef] [PubMed]
44. Trujillo-Rodríguez, M.J.; Anderson, J.L. In situ formation of hydrophobic magnetic ionic liquids for dispersive liquid–liquid microextraction. *J. Chromatogr. A* **2019**, *1588*, 8–16. [CrossRef]
45. Will, C.; Huelsmann, R.D.; Mafra, G.; Merib, J.; Anderson, J.L.; Carasek, E. High-throughput approach for the in situ generation of magnetic ionic liquids in parallel-dispersive droplet extraction of organic micropollutants in aqueous environmental samples. *Talanta* **2021**, *223*, 121759. [CrossRef]
46. Bowers, A.N.; Trujillo-Rodríguez, M.J.; Farooq, M.Q.; Anderson, J.L. Extraction of DNA with magnetic ionic liquids using in situ dispersive liquid–liquid microextraction. *Anal. Bioanal. Chem.* **2019**, *411*, 7375–7385. [CrossRef]
47. Yao, T.; Du, K. Simultaneous determination of sulfonamides in milk: In-situ magnetic ionic liquid dispersive liquid-liquid microextraction coupled with HPLC. *Food Chem.* **2020**, *331*, 127342. [CrossRef]
48. Cao, D.; Xu, X.; Xue, S.; Feng, X.; Zhang, L. An in situ derivatization combined with magnetic ionic liquid-based fast dispersive liquid-liquid microextraction for determination of biogenic amines in food samples. *Talanta* **2019**, *199*, 212–219. [CrossRef]
49. Chen, P.; Liu, X.; Wang, L.; Wang, C.; Fu, J. Weighing paper-assisted magnetic ionic liquid headspace single-drop microextraction using microwave distillation followed by gas chromatography-mass spectrometry for the determination of essential oil components in lavender. *J. Sep. Sci.* **2021**, *44*, 585–599. [CrossRef]
50. An, J.; Rahn, K.L.; Anderson, J.L. Headspace single drop microextraction versus dispersive liquid-liquid microextraction using magnetic ionic liquid extraction solvents. *Talanta* **2017**, *167*, 268–278. [CrossRef]
51. Trujillo-Rodríguez, M.J.; Pino, V.; Anderson, J.L. Magnetic ionic liquids as extraction solvents in vacuum headspace single-drop microextraction. *Talanta* **2017**, *172*, 86–94. [CrossRef] [PubMed]
52. Jahromi, Z.; Mostafavi, A.; Shamspur, T.; Mohamadim, M. Magnetic ionic liquid assisted single-drop microextraction of ascorbic acid before its voltammetric determination. *J. Sep. Sci.* **2017**, *40*, 4041–4049. [CrossRef] [PubMed]
53. Fernández, E.; Vidal, L.; Canals, A. Hydrophilic magnetic ionic liquid for magnetic headspace single-drop microextraction of chlorobenzenes prior to thermal desorption-gas chromatography-mass spectrometry. *Anal. Bioanal. Chem.* **2017**, *410*, 4679–4687. [CrossRef] [PubMed]
54. Wang, Y.; Sun, Y.; Xu, B.; Li, X.; Wang, X.; Zhang, H.; Song, D. Matrix solid-phase dispersion coupled with magnetic ionic liquid dispersive liquid–liquid microextraction for the determination of triazine herbicides in oilseeds. *Anal. Chim. Acta* **2015**, *888*, 67–74. [CrossRef]
55. Chatzimitakos, T.G.; Anderson, J.L.; Stalikas, C.D. Matrix solid-phase dispersion based on magnetic ionic liquids: An alternative sample preparation approach for the extraction of pesticides from vegetables. *J. Chromatogr. A* **2018**, *1581–1582*, 168–172. [CrossRef]
56. Peng, S.; Xiao, J.; Cheng, J.; Zhang, M.; Li, X.; Cheng, M. Ionic liquid magnetic bar microextraction and HPLC determination of carbamate pesticides in real water samples. *Microchim. Acta* **2012**, *179*, 193–199. [CrossRef]
57. Benedé, J.L.; Anderson, J.L.; Chisvert, A. Trace determination of volatile polycyclic aromatic hydrocarbons in natural waters by magnetic ionic liquid-based stir bar dispersive liquid microextraction. *Talanta* **2018**, *176*, 253–261. [CrossRef]

58. Chisvert, A.; Benedé, J.L.; Anderson, J.L.; Pierson, S.A.; Salvador, A. Introducing a new and rapid microextraction approach based on magnetic ionic liquids: Stir bar dispersive liquid microextraction. *Anal. Chim. Acta* **2017**, *983*, 130–140. [CrossRef]
59. Zhu, S.; Wang, L.; Su, A.; Zhang, H. Dispersive liquid-liquid microextraction of phenolic compounds from vegetable oils using a magnetic ionic liquid. *J. Sep. Sci.* **2017**, *40*, 3130–3137. [CrossRef]
60. Li, H.; Wang, Q.; Gao, N.; Fu, J.; Yue, X.; Lv, X.; Zhong, F.; Tang, J.; Wang, T. Facile synthesis of magnetic ionic liquids/gold nanoparticles/porous silicon composite SERS substrate for ultra-sensitive detection of arsenic. *Appl. Surf. Sci.* **2021**, *545*, 148992. [CrossRef]
61. Tashakkori, P.; Erdem, P.; Bozkurt, S.S. Molecularly imprinted polymer based on magnetic ionic liquid for solid phase extraction of phenolic acids. *J. Liq. Chromatogr. Relat. Technol.* **2017**, *40*, 657–666. [CrossRef]

Article

# A Simple Microextraction Method for Toxic Industrial Dyes Using a Fatty-Acid Solvent Mixture

Danielle P. Arcon and Francisco C. Franco, Jr. *

Chemistry Department, De La Salle University, 2401 Taft Avenue, Manila 0922, Philippines; danielle_arcon@dlsu.edu.ph
* Correspondence: francisco.franco@dlsu.edu.ph

**Abstract:** A mixture of dodecanoic and hexanoic fatty acids was used to perform a simple and efficient microextraction method for industrial dyes such as methylene blue (MB), methyl violet (MV), and malachite green (MG) in aqueous solution. The fatty-acid microextractants were simply mixed and heated until the mixture became homogeneous before adding it to the dye solutions. The fatty-acid solvent and its components were characterized with Fourier transform infrared spectroscopy (FTIR) and proton nuclear magnetic resonance ($^1$H NMR) measurements, while the dye concentrations were measured using UV-Vis spectroscopy. The performance of the extracting mixture was observed to vary across different dye contaminants, dosages of the extractant, concentrations of the dyes, and contact times. High extraction efficiencies of up to ~99% were obtained for MG as well as MV, and ~73% efficiency was achieved for MB. The study shows how a mixture of fatty acids can be used as a simple, efficient, green, and sustainable low-volume method for the removal of toxic industrial dyes in aqueous solutions.

**Keywords:** dyes; fatty acids; microextraction

**Citation:** Arcon, D.P.; Franco, F.C., Jr. A Simple Microextraction Method for Toxic Industrial Dyes Using a Fatty-Acid Solvent Mixture. *Separations* **2021**, *8*, 135. https://doi.org/10.3390/separations8090135

Academic Editor: Marcello Locatelli

Received: 2 August 2021
Accepted: 19 August 2021
Published: 26 August 2021

**Publisher's Note:** MDPI stays neutral with regard to jurisdictional claims in published maps and institutional affiliations.

**Copyright:** © 2021 by the authors. Licensee MDPI, Basel, Switzerland. This article is an open access article distributed under the terms and conditions of the Creative Commons Attribution (CC BY) license (https://creativecommons.org/licenses/by/4.0/).

## 1. Introduction

A vast range of micropollutants heavily permeate several industries. Though broadly relevant in the production of several commodities, the adverse impacts of micropollutants such as dyes, pesticides, and pharmaceutical residues characterize a risk to the environment and exposed organisms [1–3]. Dyes alone have been linked to detrimental threats such as cytotoxicity and mutagenic effects [4,5]. As a potent micropollutant chiefly used in the textile, food, and printing industries, the impact of dye contamination reduces water quality and interferes with the photosynthetic activity of microorganisms [6,7]. The accumulation of toxic dyes in bodies of water can cause severe environmental and health hazards. Therefore, an efficient method to treat dye effluent is vital.

Conventional water treatment methods have been the traditional practice in addressing contamination by dye effluents, but collective efforts have been drawn to circumvent their disadvantages such as difficulties in regeneration and high amounts of sludge formation [8,9]. In recent years, numerous natural and synthetic routes have been used to explore alternatives such as ionic liquids (ILs) [10], advanced oxidation processes (AOPs) [11], and a diverse set of bioremediation approaches [12,13]. However, while these alternative techniques are able to reduce the concentration of dyes in aquatic environments, the combination of efficiency, sustainability, and economic feasibility remains to be an elusive balance to be achieved from a single method.

The use of fatty-acid mixtures has been gaining attention in recent years for their use in green applications. Beyond their role as phase-change materials (PCMs) [14,15], fatty acids have been applied for a broad range of uses that encompass applications in latent heat thermal energy storage (LHTES) [16] and water remediation to reduce the presence of micropollutants in aqueous environments [17–19]. In maximizing their desirable properties, previous studies have investigated their role in the extraction of compounds such as alkylphenols [20], fluoroquinolones [21], and endocrine-disrupting compounds (EDCs) [22] that threaten the survival of aquatic organisms.

Binary mixtures of fatty acids represent a vast potential in wastewater remediation, and, to expand their role towards green initiatives, their performance is herein assessed. This study explores their performance in dye remediation using the binary combination of hexanoic acid and dodecanoic acid as a microextractant for methylene blue (MB), methyl violet (MV), and malachite green (MG) as shown in Figure 1. With many potential advantages including rapid dye absorption, facile preparations, high efficiency, and minimal volume requirement, the use of fatty-acid-based mixtures presents a green and promising solvent for microextraction methods.

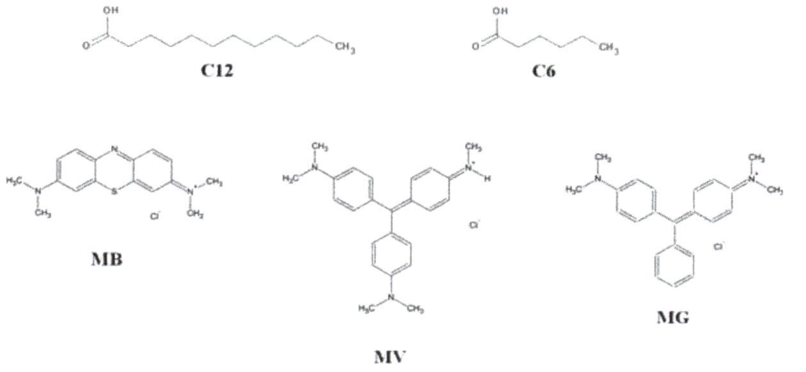

Figure 1. Chemical structures of the fatty acids and dyes used in this study.

The study aims to investigate the spectral profile of the fatty-acid mixture and to probe the influencing variables that may boost or hinder its performance as a dye extractant. FTIR measurements were conducted to explore if the characteristic peaks of the mixture components would indicate significant changes before and after mixing. $^1$H NMR measurements were carried out to establish how mixing the fatty acids may impact hydrogen bond strength and the solvent's behavior upon exposure to water which could influence its extraction capacity. With the use of a liquid–liquid microextraction method and the determination of the solvent's efficiency via UV-Vis measurements, the proposed approach of the study leveraged the benefits of simplicity and the rapid route of both solvent preparation and analyses.

## 2. Results and Discussion

### 2.1. FTIR and $^1$H NMR Measurements

Figure 2 shows the FTIR spectra of the C12:C6 fatty-acid mixture prepared. Vibrations found at ~2900 cm$^{-1}$ and ~2800 cm$^{-1}$ wavenumbers correspond to the stretching -CH$_3$ and -CH$_2$ vibrations, respectively. The 1700 cm$^{-1}$ peak is attributed to the stretching mode of C=O, while the broad peaks found between 3000 cm$^{-1}$ and 3500 cm$^{-1}$ are assigned to the hydroxyl group of the fatty acids. Comparing between the spectra of the mixture and its individual components, it was observed that they had the same characteristic peaks, which suggests that there was no chemical reaction that occurred between the hexanoic and dodecanoic fatty-acid components, consistent with previous reports [23,24].

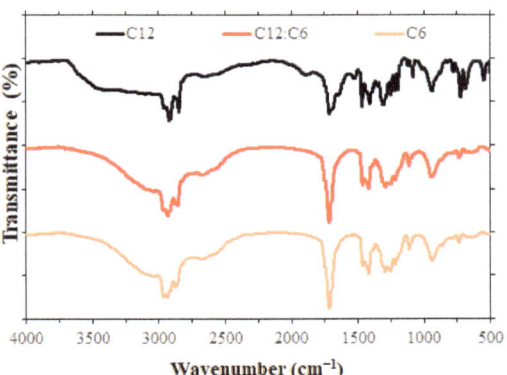

**Figure 2.** FTIR spectra of the fatty-acid mixture prepared and its individual components: C12 (black line), C12:C6 (red line), and C6 (orange line).

Figure 3 shows the $^1$H NMR spectra and $^1$H assignments of the C12 and C6 fatty acids and their C12:C6 mixture. As observed in the FTIR spectra above, the $^1$H NMR spectra also confirm that the fatty acids in the mixture did not have any chemical reaction as no additional peaks were observed after mixing the C12 and C6 fatty acids [25]. Additionally, the resonance signals did not have significant changes in chemical shifts (Supplementary Materials, Table S1) before and after mixing which shows that the hydrogen bonding strengths did not vary greatly between the individual fatty acids and when combined to form the mixtures. The results show that the preparation of the solvent in this study promoted a maximized atom economy without the formation of side products and could be a sustainable solvent [26,27].

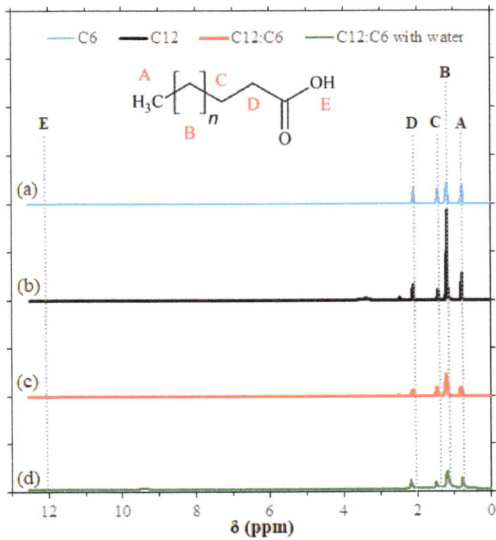

**Figure 3.** $^1$H NMR spectra of the C12:C6 fatty-acid mixture prepared (red line) and its individual components, C12 (black line), and C6 (blue line). The $^1$H NMR spectra of the C12:C6 mixture after mixing with water (green line) is also shown.

The $^1$H NMR spectra of the C12:C6 mixture after mixing with water was also determined to understand possible changes in the mixture during extractions in aqueous solution, as shown in Figure 3. Water was first introduced in a vial and was continuously mixed with the fatty-acid layer using a pipette by a series of suction and expulsion steps to form temporary emulsion droplets. The mixture was then allowed to stand until phase separation between the aqueous phase and the fatty-acid layer was achieved, and the hydrophobic layer was then subjected to $^1$H NMR measurement. From the results, the hydrogens in the aliphatic chain were not greatly affected by the introduction of water and their chemical shifts were similar. However, the resonance signals corresponding to the hydrogens of the carboxyl groups were observed to have significant upfield shifts (SI, Table S1), i.e., they were more shielded, which showed that the introduction of water disrupted the hydrogen bonding interactions of the fatty-acid components, similarly observed in previous studies upon the addition of water [28,29]. This attribute could be seen as an advantage as earlier studies also positively noted the influence of water in potentially improving the efficiency of solvents for various extraction methods and applications [18,30].

### 2.2. Liquid–Liquid Microextraction

The extraction efficiencies at various experimental conditions of the fatty-acid mixture and dye solutions were varied: (1) dye solution, (2) dye concentration, (3) C12:C6 dose, and (4) contact time, and the changes in the extraction efficiencies were presented. Concentrations were varied between 50 μmol/L and 500 μmol/L for the three dyes: methylene blue (MB), methyl violet (MV), and malachite green (MG). Varying the volume of the solvent extractant could have led to the possibility of increasing the dye uptake but could also have led towards decreasing the solvent's extraction efficiency, and in line with this, the solvent dose was tested between 30 μL and 600 μL [31].

Figure 4 presents the extraction efficiencies for the dyes after 24 h exposure at increasing C12:C6 mixture volumes and dye concentrations. It can be observed that in the dye solutions, extraction efficiencies generally increased as the volume of the C12:C6 mixture increased from 30 μL to 600 μL, and an increase in dye concentration was also followed by an increase in the solvent's efficiency. Higher extractant volumes were able to increase the possibility of dye dissolution and extraction, and as a result, close to 100% efficiency was achieved at 600 μL of the C12:C6 mixture for the MG and MV dyes. The highest extraction efficiency of 99.3% for the MG dye was achieved at 600 μL extractant volume and 500 μmol/L dye concentration. The highest efficiency for MB (73.1%) was also achieved at similar parameters, while the maximum value for MV was recorded at 98.6%. The minimum and maximum extraction efficiencies during the 24 h extraction are summarized in Table 1.

**Table 1.** Highest and lowest extraction efficiencies of the methylene blue (MB), methyl violet (MV), and malachite green (MG) dye solutions with the C12:C6 mixture at 24 h contact time.

|     | Extraction Efficiency (%) | |
| --- | --- | --- |
|     | Lowest | Highest |
| MB  | 6.7    | 73.1    |
| MV  | 83.2   | 98.6    |
| MG  | 67.5   | 99.3    |

Comparing the effects of dye concentration in the three dyes, it can be observed that the increase in dye concentration only had a minimal effect on the 24 h extraction performance of C12:C6 on MG and MV due to their high efficiencies, whereas the increase in MB dye concentration contributed a more notable increase in efficiency. The increase in the extraction efficiency with the increase in dye concentration could be attributed to the increase in the driving force due to mass transfer [32]. When the dye concentration was low, the mass transfer was slower, and the equilibrium of the dye distribution in the fatty-acid mixture and aqueous phase was reached faster. Conversely, increasing the dye concentration may have favorably accompanied high mass transfer, and the equilibrium of the dye distribution may not have been reached, suggesting that the dye extraction could still have been ongoing even after 24 h.

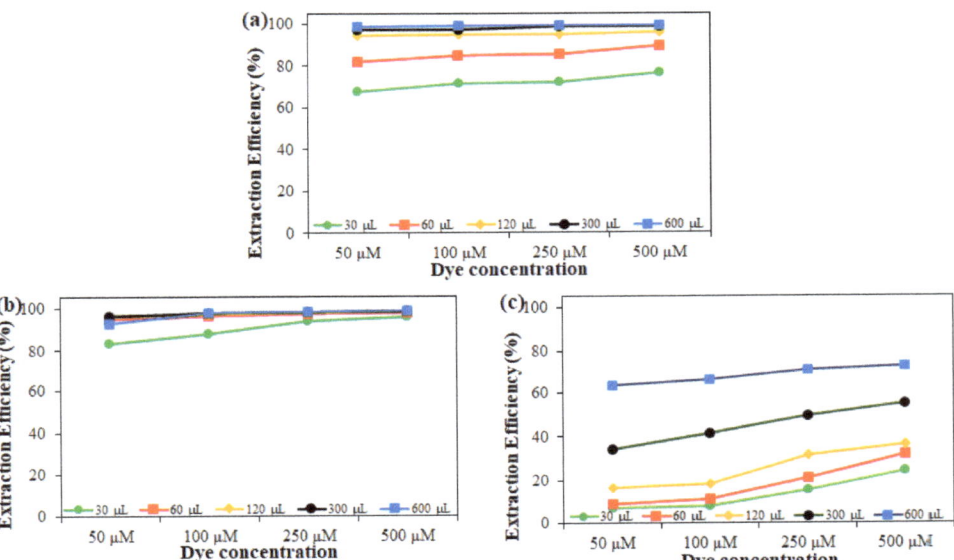

**Figure 4.** Dye extraction efficiencies after 24 h exposure of the dye solution with the C12:C6 mixture: (**a**) malachite green, (**b**) methyl violet, and (**c**) methylene blue at various dye concentrations (50 µmol/L, 100 µmol/L, 250 µmol/L, and 500 µmol/L) and extractant volume: 30 µL (green circle), 60 µL (red square), 120 µL (yellow diamond), 300 µL (black circle), and 600 µL (blue square). The error bars are the standard deviations from the three trials.

Between the three higher extractant volumes (120 µL, 300 µL, and 600 µL) used in MG extraction, the difference in efficiency was not greater than ~5% as the MG concentration increased from 50 µmol/L to 500 µmol/L. Related to this, across all concentrations for MV, the three higher volumes showed a ~3% difference in the extraction of MV for all dye concentrations used. These observations showed that the performances of C12:C6 mixtures for MG and MV were virtually the same even in large dye concentrations, especially when the extractant volume was high. However, although the C12:C6 fatty-acid mixture could efficiently extract MG and MV dyes, the same level of efficiency was not observed for MB.

With the extraction of MB, when the increments of extractant volume and dye concentration were increased, both exhibited significant effects on C12:C6 efficiency. At 50 µmol/L, there was a ~47% jump in efficiency between 120 µL and 600 µL, and at 500 µmol/L, there was a ~37% efficiency gap between a similar volume range. While the earlier examples of MG and MV showed that the performance of 600 µL could be nearly attained with a lower volume such as 120 µL, MB in contrast displayed a wide gap in efficiency between varying C12:C6 volumes. This pattern shows that the extraction for MB may be more resource demanding than MV and MG in terms of the extractant volume needed to reach the maximum efficiency for each concentration.

It was generally observed that MB appeared to have the lowest solubility in the fatty-acid-based solvent relative to MV and MG. Thus, to gain insights on the disparity of extraction efficiency values between the three dyes, the Gibbs free energy of solvation ($\Delta G_{solv}$) values were determined via quantum chemical calculations using the SMD model with water as the solvent employed in Gaussian16 [33] at the DFT/B3LYP/6-311+G(d,p) level. The results are summarized in Table 2.

Table 2. Gibbs free energy of solvation for the three dyes used using the SMD model at the DFT/B3LYP/6-311+G(d,p) level.

| Dye | $\Delta G_{solv}$ (kcal/mol) |
|---|---|
| Malachite green | −34.07 |
| Methyl violet | −34.16 |
| Methylene blue | −39.79 |

The $\Delta G_{solv}$ of MG and MV were very similar, while MB had a more negative $\Delta G_{solv}$ value which could indicate MB's higher tendency to be solvated in water than the two other dyes. Previous studies have also noted a similar pattern of MB to favor amphiphilic or more hydrophilic media instead of hydrophobic solvents [34,35]. Establishing a connection between a solvent's performance and a target compound's solubility could help in the understanding of solvent efficiency and the screening of solvent components based on polarity [36,37].

Showing very high efficiencies ~99% at a large range of concentrations for MG and MV, C12:C6 exhibited a significant capacity to extract MG and MV in high dye concentrations with nearly as much efficiency as when the dye concentration was low. The extraction capability in high-to-low concentration, as well as high-to-minimal extractant volume requirement, is desirable in industrial practice as its performance may exceed the resources it demands. Although the solvent had a lower recorded maximum value of extraction efficiency for MB than with the other two dyes, it was still at a good level at ~73% and may still be used in the right conditions. In the extraction of these dyes at a microvolume range with as low as 1:10 to 1:200 of extractant to dye solution volume ratio without further treatment, the fatty-acid-based mixture presented the attributes of being efficient, environmentally friendly, and cost effective.

To evaluate the performance of the proposed method, under optimal conditions, the figures of merit were also investigated and tabulated in Table 3. The calibration curves for the three dyes had a coefficient of determination ($R^2$) no less than 0.9965.

Table 3. Analytical performance of the proposed microextraction method.

| Parameter | Methylene Blue | Methyl Violet | Malachite Green |
|---|---|---|---|
| Coefficient of determination ($R^2$) | 0.9997 | 0.9998 | 0.9965 |
| Linear equation | y = 53170x − 0.0327 | y = 44782x − 0.0212 | y = 48699x + 0.0099 |
| Accuracy (%) | 102.45 ± 4.95 | 100.39 ± 1.57 | 92.95 ± 14.37 |
| Linearity range ($\times 10^{-6}$ mol/L) | 1–50 | 1–50 | 0.5–50 |
| Standard error of intercept | 0.0131 | 0.0074 | 0.0435 |
| Standard deviation of intercept | 0.0292 | 0.0195 | 0.1065 |
| Limit of detection (LOD, $\times 10^{-6}$) | 1.81 | 1.44 | 7.21 |
| Limit of quantification (LOQ, $\times 10^{-6}$) | 5.49 | 4.36 | 21.9 |

Contact time between the aqueous dye solution and extractant is also an important parameter in the extraction of dyes. To determine the practicality and efficiency of using the C12:C6 fatty-acid mixture at a shorter period, contact time was reduced to 5 min.

Figure 5 shows the dye extraction efficiencies after the 5 min exposure of the dye solutions with the extractant solvent, and the highest extraction efficiencies exhibited a noticeable decline compared to the 24 h exposure. For MB, the highest extraction efficiency was reduced from 73.1% to 43.6% at 600 μL extractant dose and 500 μmol/L dye concentration. On the other hand, the highest extraction efficiency of MG at 99.3% was reduced to 84.3%. The highest efficiency for MV at 98.6% also decreased, to 75.0%. The lowest and highest efficiency values obtained during the 5 min extraction are tabulated in Table 4.

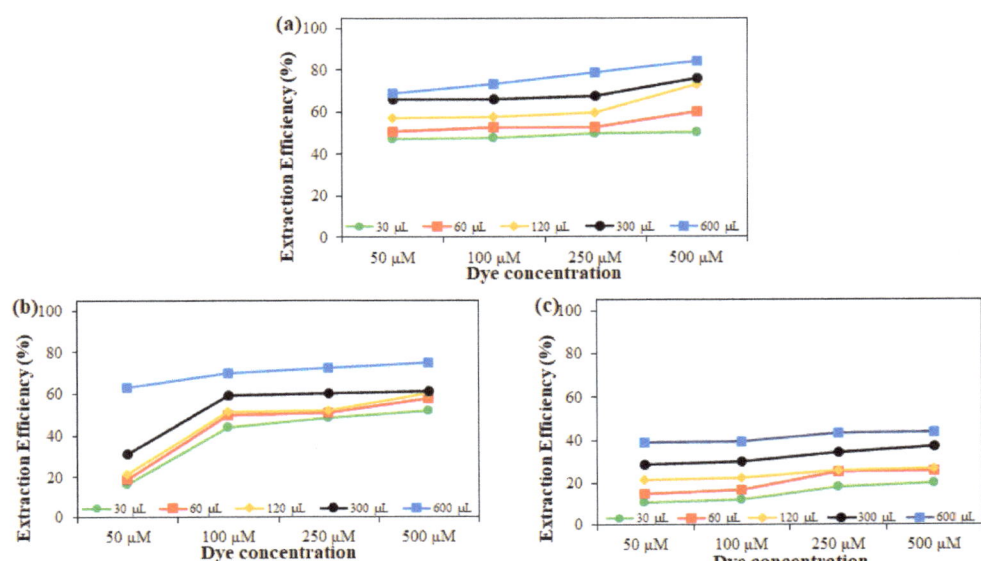

**Figure 5.** Dye extraction efficiencies after 5 min exposure of the dye solution with the C12:C6 mixture: (**a**) malachite green, (**b**) methyl violet, and (**c**) methylene blue at various dye concentrations (50 μmol/L, 100 μmol/L, 250 μmol/L, and 500 μmol/L) and extractant volume: 30 μL (green circle), 60 μL (red square), 120 μL (yellow diamond), 300 μL (black circle), and 600 μL (blue square). The error bars are the standard deviations from the three trials.

**Table 4.** Highest and lowest extraction efficiencies of the methylene blue (MB), methyl violet (MV), and malachite green (MG) dye solutions with the C12:C6 mixture at 5 min contact time.

|  | Extraction Efficiency (%) | |
| --- | --- | --- |
|  | Lowest | Highest |
| MB | 10.8 | 43.6 |
| MV | 16.7 | 75.0 |
| MG | 47.2 | 84.3 |

Even when contact time was greatly reduced, the extraction efficiencies, especially for MG and MV, were still high, showing the capacity of the fatty-acid mixture to extract dyes in aqueous solution. This could be attributed to the high affinity or solubility of the MG and MV dyes in the C12:C6 fatty-acid solvent even at the greatly reduced contact time of 5 min from 24 h, resulting in high mass transfer towards the hydrophobic phase.

Unlike the trend observed during the 24 h extraction wherein a higher volume range of the solvent could perform almost similarly for MV and MG, the gap in performance between the volume range increased when the time frame for the microextraction was reduced to 5 min. This shows that optimizing the C12:C6 volume was a favorable approach for a 5 min extraction of MG and MV, but the increase in solvent volume played only a small role at longer extraction periods. During a 24 h extraction, the difference in performance between 120 μL to 600 μL of solvent volume was ~5% for MG and ~3% for MV. However, in the case of reducing contact time to 5 min, a similar volume range varied in efficiency by as much as ~19% and ~42% for MG and MV, respectively.

The two dyes, along with a higher C12:C6 volume range, shared a common trend of minimal differences in efficiency during a 24 h extraction compared to a 5 min contact time, but the reverse was observed for MB. With the five tested MB concentrations, the extraction efficiency reflected a difference greater than ~37% when the C12:C6 solvent volume was varied between 120 µL to 600 µL for a 24 h extraction. However, the difference was reduced to ~18% when the extraction was shortened to 5 min. This indicates that increasing C12:C6 volume was a desirable optimization step during a 24 h extraction of MB, but the approach may not necessarily boost the recovery yield with as much magnitude at a shorter exposure.

Connecting these observations, the use of more resources by increasing extractant volume can be a practical approach during a 5 min extraction of relatively hydrophobic dyes such as MG and MV. However, it can be a counterintuitive approach for longer periods of extraction as increasing solvent volume was observed to impart minimal influence and would thus not require large resources to achieve high recovery yield. In contrast, for the removal of a more polar dye such as MB, adjustments to solvent volume contributed a higher boost in efficiency during the 24 h extraction than the 5 min period.

The described differences regarding the conditions in which C12:C6 can obtain an optimum recovery yield agree with earlier findings that the solubility of dyes is a significant criterion in determining optimal extraction conditions. The results illustrate that contact time, along with the other studied experimental parameters, could be optimized depending on the polarity of the dyes. Identifying how these factors interact may further improve the extraction process for resource-efficient and cost-effective future experiments that yield higher performance.

## 3. Materials and Methods

### 3.1. Materials

Hexanoic acid ($\geq$99%) and dodecanoic acid ($\geq$98%) were acquired from Sigma Aldrich (St. Louis, MO, USA). The dyes methylene blue, methyl violet, and malachite green were purchased from HiMedia (Mumbai, India), Sigma-Aldrich (St. Louis, MO, USA), and LabChem (Auckland, New Zealand), respectively. All reagents were used as received with no further purification.

### 3.2. FTIR and $^1$H NMR Measurements

$^1$H NMR studies were carried out with a High-Resolution JEOL 600 MHz spectrometer to confirm the constituent components of the fatty-acid mixture. With the addition of deuterated dimethyl sulfoxide (DMSO-d6), all binary mixtures and pure components were prepared on 5 mm NMR tubes properly sealed with a cap.

All Fourier transform infrared spectra for the samples and pure components were obtained using a Thermo Scientific Nicolet 6700 FTIR spectrometer, and the spectral data for each sample were attained at a 4 cm$^{-1}$ resolution. With 16 scans, all results were recorded over the wavenumber range of 500–4000 cm$^{-1}$.

### 3.3. Liquid–liquid Dye Microextraction Procedure

A binary mixture of dodecanoic acid (C12) and hexanoic acid (C6) in 1:3 mole ratio ($x_{C12}$ = 0.25) was used as dye microextractant solvent. The concentration of the dye stock solutions for methylene blue (MB), methyl violet (MV), and malachite green (MG) was varied at 50, 100, 250, and 500 µmol/L. Liquid–liquid microextraction (LLME) experiments were carried out by testing different solvent volumes at 30, 60, 120, 300, and 600 µL then loading the solvent dose into a 50 mL conical-bottom polypropylene tube containing 6 mL of dye solution.

To investigate the effect of contact time on dye removal, one batch was simply mixed and left in contact for 24 h at room temperature, while the other batch was centrifuged for 1 min at 7600$\times$ $g$ and left for another 4 min at room temperature. Following this, two distinct layers were observed, and the remaining concentration of dyes at the aqueous phase beneath the fatty-acid mixture was extracted with a syringe and analyzed using a Hitachi U-2900 UV-Vis Double Beam Spectrophotometer. Absorbances at the wavelengths

586 nm, 613 nm, and 663 nm were, respectively, used for the measurement of residual concentrations of MV, MG, and MB. The extraction efficiency was calculated as:

$$\% \text{ Extraction Efficiency } (E) = \frac{C_o - C_f}{C_o} \times 100\% \quad (1)$$

where $C_o$ and $C_f$ represent initial and final dye concentration in µM, respectively.

*3.4. Analytical Performance Parameters*

For the three sets of calibration curves used for methylene blue, methyl violet, and malachite green in the study, the acceptance criterion for the calibration curve's coefficient of determination was set to $\geq 0.98$, and the accuracy allowed was within 75–120%.

## 4. Conclusions

In this study, a low-volume fatty-acid mixture-based solvent was used for a simple and efficient microextraction method for the removal of dyes in aqueous solution. The solvent mixture was easily prepared by mixing dodecanoic acid and hexanoic acid, and the resulting mixture was added to the methylene blue, methyl violet, and malachite green dye solutions. The interaction between the variety of dye, C12:C6 volume, contact time, and dye concentration as well as how these factors relate to the solvent's extraction efficiency was explored and the fatty-acid mixture presented a green and economic procedure for the extraction of toxic dyes in wastewater treatment. This was corroborated by the high extraction efficiencies obtained for methylene blue (73.1%), methyl violet (98.6%), and malachite green (99.3%). The experimental results reveal that even at a microvolume solvent availability, a fatty-acid mixture performs efficiently even towards hydrophobic contaminants 200 times its volume.

The study also underscores the importance of the solvent being tailored according to desired function and polarity of the target micropollutant. As hydrophobicity contributes a large role in the extraction efficiency of the fatty-acid mixture, screening the match between the dye and the solvent would be beneficial for more targeted and hence more cost-efficient extraction. With a straightforward solvent preparation and with the advantage of a fast and high capacity to extract dyes in aqueous solution, the microextraction method using the fatty-acid-based mixture represents an emerging class of inexpensive green solvents for environmental remediation.

**Supplementary Materials:** The following are available online at https://www.mdpi.com/article/10.3390/separations8090135/s1, Table S1: Summary of the $^1$H NMR chemical shifts.

**Author Contributions:** Conceptualization, D.P.A. and F.C.F.J.; Data curation, D.P.A. and F.C.F.J.; Formal analysis, D.P.A. and F.C.F.J.; Funding acquisition, F.C.F.J.; Investigation, D.P.A. and F.C.F.J.; Methodology, D.P.A.; Project administration, D.P.A. and F.C.F.J.; Resources, F.C.F.J.; Software, F.C.F.J.; Supervision, F.C.F.J.; Validation, D.P.A. and F.C.F.J.; Visualization, D.P.A. and F.C.F.J.; Writing—original draft, D.P.A.; Writing—review & editing, D.P.A. and F.C.F.J. All authors have read and agreed to the published version of the manuscript.

**Funding:** This work was supported by the University Research Coordination Office (URCO) of De La Salle University—Manila under Project Number 24 F U 1TAY18-1TAY19.

**Institutional Review Board Statement:** Not applicable.

**Informed Consent Statement:** Not applicable.

**Data Availability Statement:** Not applicable.

**Conflicts of Interest:** The authors declare no conflict of interest.

## References

1. Janssens, I.; Tanghe, T.; Verstraete, W. Micropollutants: A Bottleneck in Sustainable Wastewater Treatment. *Water Sci. Technol.* **1997**, *35*, 13–26. [CrossRef]
2. Atas, M.S.; Dursun, S.; Akyildiz, H.; Citir, M.; Yavuz, C.T.; Yavuz, M.S. Selective Removal of Cationic Micro-Pollutants Using Disulfide-Linked Network Structures. *RSC Adv.* **2017**, *7*, 25969–25977. [CrossRef]
3. Margot, J.; Rossi, L.; Barry, D.P.A.; Holliger, C. A Review of the Fate of Micropollutants in Wastewater Treatment Plants. *Wiley Interdiscip. Rev. Water* **2015**, *2*, 457–487. [CrossRef]
4. Gopinathan, R.; Kanhere, J.; Banerjee, J. Effect of Malachite Green Toxicity on Non Target Soil Organisms. *Chemosphere* **2015**, *120*, 637–644. [CrossRef]
5. Schneider, K.; Hafner, C.; Jäger, I. Mutagenicity of Textile Dye Products. *J. Appl. Toxicol.* **2004**, *24*, 83–91. [CrossRef] [PubMed]
6. Dos Anjos, F.S.C.; Vieira, E.F.S.; Cestari, A.R. Interaction of Indigo Carmine Dye with Chitosan Evaluated by Adsorption and Thermochemical Data. *J. Colloid Interface Sci.* **2002**, *253*, 243–246. [CrossRef] [PubMed]
7. Chandran, C.B.; Singh, D.; Nigam, P. Remediation of Textile Effluent Using Agricultural Residues. *Appl. Biochem. Biotechnol.* **2002**, *102–103*, 207–212. [CrossRef]
8. Nguyen, T.A.; Juang, R.-S. Treatment of Waters and Wastewaters Containing Sulfur Dyes: A Review. *Chem. Eng. J.* **2013**, *219*, 109–117. [CrossRef]
9. Shi, B.; Li, G.; Wang, D.; Feng, C.; Tang, H. Removal of Direct Dyes by Coagulation: The Performance of Preformed Polymeric Aluminum Species. *J. Hazard. Mater.* **2007**, *143*, 567–574. [CrossRef] [PubMed]
10. Isosaari, P.; Srivastava, V.; Sillanpää, M. Ionic Liquid-Based Water Treatment Technologies for Organic Pollutants: Current Status and Future Prospects of Ionic Liquid Mediated Technologies. *Sci. Total Environ.* **2019**, *690*, 604–619. [CrossRef]
11. Verma, P.; Samanta, S.K. Microwave-Enhanced Advanced Oxidation Processes for the Degradation of Dyes in Water. *Environ. Chem. Lett.* **2018**, *16*, 969–1007. [CrossRef]
12. Kandelbauer, A.; Guebitz, G.M. Bioremediation for the Decolorization of Textile Dyes—A Review. In *Environmental Chemistry*; Springer: Berlin/Heidelberg, Germany, 2005; pp. 269–288.
13. Sharma, B.; Dangi, A.K.; Shukla, P. Contemporary Enzyme Based Technologies for Bioremediation: A Review. *J. Environ. Manag.* **2018**, *210*, 10–22. [CrossRef] [PubMed]
14. Kahwaji, S.; Johnson, M.B.; Kheirabadi, A.C.; Groulx, D.; White, M.A. Fatty Acids and Related Phase Change Materials for Reliable Thermal Energy Storage at Moderate Temperatures. *Sol. Energy Mater. Sol. Cells* **2017**, *167*, 109–120. [CrossRef]
15. Cellat, K.; Beyhan, B.; Güngör, C.; Konuklu, Y.; Karahan, O.; Dündar, C.; Paksoy, H. Thermal Enhancement of Concrete by Adding Bio-Based Fatty Acids as Phase Change Materials. *Energy Build.* **2015**, *106*, 156–163. [CrossRef]
16. Zhou, D.; Zhou, Y.; Liu, Y.; Luo, X.; Yuan, J. Preparation and Performance of Capric-Myristic Acid Binary Eutectic Mixtures for Latent Heat Thermal Energy Storages. *J. Nanomater.* **2019**, *2019*, 1–9. [CrossRef]
17. Florindo, C.; Romero, L.; Rintoul, I.; Branco, L.C.; Marrucho, I.M. From Phase Change Materials to Green Solvents: Hydrophobic Low Viscous Fatty Acid–Based Deep Eutectic Solvents. *ACS Sustain. Chem. Eng.* **2018**, *6*, 3888–3895. [CrossRef]
18. Arcon, D.P.; Franco, F.C. All-Fatty Acid Hydrophobic Deep Eutectic Solvents towards a Simple and Efficient Microextraction Method of Toxic Industrial Dyes. *J. Mol. Liq.* **2020**, *318*, 114220. [CrossRef]
19. Ma, W.; Row, K.H. PH-Induced Deep Eutectic Solvents Based Homogeneous Liquid-Liquid Microextraction for the Extraction of Two Antibiotics from Environmental Water. *Microchem. J.* **2021**, *160*, 105642. [CrossRef]
20. Shih, H.-K.; Shu, T.-Y.; Ponnusamy, V.K.; Jen, J.-F. A Novel Fatty-Acid-Based in-Tube Dispersive Liquid–Liquid Microextraction Technique for the Rapid Determination of Nonylphenol and 4-Tert-Octylphenol in Aqueous Samples Using High-Performance Liquid Chromatography–Ultraviolet Detection. *Anal. Chim. Acta* **2015**, *854*, 70–77. [CrossRef]
21. Gao, M.; Wang, J.; Song, X.; He, X.; Dahlgren, R.A.; Zhang, Z.; Ru, S.; Wang, X. An Effervescence-Assisted Switchable Fatty Acid-Based Microextraction with Solidification of Floating Organic Droplet for Determination of Fluoroquinolones and Tetracyclines in Seawater, Sediment, and Seafood. *Anal. Bioanal. Chem.* **2018**, *410*, 2671–2687. [CrossRef]
22. El-Deen, A.K.; Shimizu, K. A Green Air Assisted-Dispersive Liquid-Liquid Microextraction Based on Solidification of a Novel Low Viscous Ternary Deep Eutectic Solvent for the Enrichment of Endocrine Disrupting Compounds from Water. *J. Chromatogr. A* **2020**, *1629*, 461498. [CrossRef]
23. Ke, H. Phase Diagrams, Eutectic Mass Ratios and Thermal Energy Storage Properties of Multiple Fatty Acid Eutectics as Novel Solid-Liquid Phase Change Materials for Storage and Retrieval of Thermal Energy. *Appl. Therm. Eng.* **2017**, *113*, 1319–1331. [CrossRef]
24. He, H.; Yue, Q.; Gao, B.; Zhang, X.; Li, Q.; Wang, Y. The Effects of Compounding Conditions on the Properties of Fatty Acids Eutectic Mixtures as Phase Change Materials. *Energy Convers. Manag.* **2013**, *69*, 116–121. [CrossRef]
25. Van Osch, D.J.G.P.; Dietz, C.H.J.T.; van Spronsen, J.; Kroon, M.C.; Gallucci, F.; van Sint Annaland, M.; Tuinier, R. A Search for Natural Hydrophobic Deep Eutectic Solvents Based on Natural Components. *ACS Sustain. Chem. Eng.* **2019**, *7*, 2933–2942. [CrossRef]
26. Hao, L.; Su, T.; Hao, D.; Deng, C.; Ren, W.; Lü, H. Oxidative Desulfurization of Diesel Fuel with Caprolactam-Based Acidic Deep Eutectic Solvents: Tailoring the Reactivity of DESs by Adjusting the Composition. *Chin. J. Catal.* **2018**, *39*, 1552–1559. [CrossRef]
27. Liu, S.; Zhang, C.; Zhang, B.; Li, Z.; Hao, J. All-In-One Deep Eutectic Solvent Toward Cobalt-Based Electrocatalyst for Oxygen Evolution Reaction. *ACS Sustain. Chem. Eng.* **2019**, *7*, 8964–8971. [CrossRef]

28. Florindo, C.; Branco, L.C.; Marrucho, I.M. Development of Hydrophobic Deep Eutectic Solvents for Extraction of Pesticides from Aqueous Environments. *Fluid Phase Equilibria* **2017**, *448*, 135–142. [CrossRef]
29. Arora, K.; Singh, G.; Singh, G.; Kang, T.S. Aggregation Behavior of Sodium Dioctyl Sulfosuccinate in Deep Eutectic Solvents and Their Mixtures with Water: An Account of Solvent's Polarity, Cohesiveness, and Solvent Structure. *ACS Omega* **2018**, *3*, 13387–13398. [CrossRef]
30. Vilková, M.; Płotka-Wasylka, J.; Andruch, V. The Role of Water in Deep Eutectic Solvent-Base Extraction. *J. Mol. Liq.* **2020**, *304*, 112747. [CrossRef]
31. Shamsipur, M.; Zohrabi, P.; Hashemi, M. Application of a Supramolecular Solvent as the Carrier for Ferrofluid Based Liquid-Phase Microextraction for Spectrofluorimetric Determination of Levofloxacin in Biological Samples. *Anal. Methods* **2015**, *7*, 9609–9614. [CrossRef]
32. Dutta, B.K. *Principles of Mass Transfer and Separation Processes*; PHI Learning Private Limited: New Delhi, India, 2007.
33. Frisch, M.J.; Trucks, G.W.; Schlegel, H.B.; Scuseria, G.E.; Robb, M.A.; Cheeseman, J.R.; Scalmani, G.; Barone, V.; Petersson, G.A.; Nakatsuji, H.; et al. Gaussian16 B.01. Available online: www.gaussian.com (accessed on 18 August 2021).
34. Takeshita, J.; Hasegawa, Y.; Yanai, K.; Yamamoto, A.; Ishii, A.; Hasegawa, M.; Yamanaka, M. Organic Dye Adsorption by Amphiphilic Tris-Urea Supramolecular Hydrogel. *Chem. Asian J.* **2017**, *12*, 2029–2032. [CrossRef] [PubMed]
35. Tong, A.; Wu, Y.; Tan, S.; Li, L.; Akama, Y.; Tanaka, S. Aqueous Two-Phase System of Cationic and Anionic Surfactant Mixture and Its Application to the Extraction of Porphyrins and Metalloporphyrins. *Anal. Chim. Acta* **1998**, *369*, 11–16. [CrossRef]
36. Jiang, H.; Xu, D.; Zhang, L.; Ma, Y.; Gao, J.; Wang, Y. Vapor–Liquid Phase Equilibrium for Separation of Isopropanol from Its Aqueous Solution by Choline Chloride-Based Deep Eutectic Solvent Selected by COSMO-SAC Model. *J. Chem. Eng. Data* **2019**, *64*, 1338–1348. [CrossRef]
37. Jeliński, T.; Cysewski, P. Application of a Computational Model of Natural Deep Eutectic Solvents Utilizing the COSMO-RS Approach for Screening of Solvents with High Solubility of Rutin. *J. Mol. Model.* **2018**, *24*, 180. [CrossRef] [PubMed]

*Article*

# Removal of Pyridine, Quinoline, and Aniline from Oil by Extraction with Aqueous Solution of (Hydroxy)quinolinium and Benzothiazolium Ionic Liquids in Various Ways

Zhaojin Zhang [1], Yinan Li [1], Jing Gao [1], Alula Yohannes [2], Hang Song [1] and Shun Yao [1,*]

[1] School of Chemical Engineering, Sichuan University, Chengdu 650061, China; 2021223075227@stu.scu.edu.cn (Z.Z.); 2018141494035@stu.scu.edu.cn (Y.L.); oxfordys@yeah.net (J.G.); pharmposter2012@163.com (H.S.)

[2] College of Natural Science, Wolkite University, Wolkite P.O. Box 07, Ethiopia; alulayhnns@gmail.com

* Correspondence: cusack@scu.edu.cn; Tel.: +86-028-8540-5221

**Abstract:** Based on above background, quinolinium, 8-hydroxy-quinolinium, and benzothiazolium ionic liquids, containing the acidic anions of methanesulfonate ($[CH_3SO_3]^-$), phosphate ($[H_2PO_4]^-$), *p*-toluenesulfonate ($[p\text{-TSA}]^-$), and bisulfate ($[HSO_4]^-$) were synthesized. After comparison, the aqueous solution of benzothiazole bisulfate [HBth][$HSO_4$] was selected as the most ideal extractant for removing pyridine and aniline. Meanwhile, benzothiazole bisulfate [HBth][$HSO_4$] solution was found as the best one for removing quinoline from simulated oil. Then, the single stage extraction and two-step extraction were used in the extraction for the simulated oil containing pyridine, quinoline or aniline, and their mixture, respectively. Their denitrogenation performance on their N-removal effect was compared on the basis of structural features, and main extraction conditions were further investigated, including mass ratio of IL to water, mass ratio of IL to oil, and temperature. Furthermore, the extraction process was described by two kinetic equations. Recovery and reuse of IL were realized by back-extraction and liquid-liquid separation, and a related mechanism was speculated, according to all the experimental results. Finally, based on the developed method for preparing complex adsorbent tablets, corresponding immobilized IL was used to remove target objects, by solid phase extraction, in order to extend separation ways, which was more easily recovered after extraction.

**Keywords:** ionic liquids; denitrogenation; extraction; pyridine; quinoline; aniline; simulated oil

## 1. Introduction

With the in-depth study on sustainable chemistry, oil denitrogenation has aroused wide spread concern. Nitrogen exists in various forms in crude oil, which are mainly divided into non alkaline and alkaline nitrides, according to their alkalinity [1,2]. Basic nitrides mainly include aniline, pyridine, quinoline, and their derivatives; non-alkaline nitrides often include carbazole, pyrrole, indole, and their derivatives [3–5]. With the rapid development of the world economy, the demand for fuel oil is also increasing, but the nitrides contained in oil products will lead to some problems, especially in the following aspects:

(1) The nitrides in the oil has a great impact on the properties of the oil. Related compounds can play a catalytic role and accelerate the reaction of substances in the oil, which cause the deterioration of the oil and easily generate colloidal precipitation. All of these can result in the decrease in oil quality and storage stability meanwhile affect its performance in use. In addition, nitrides will also corrode the equipment and affect oil service life during storage or transportation [6].

(2) Nitrogen compounds in oil products are discharged into the atmosphere in the form of $NO_x$ after combustion, and the latter can form acid rain in combination with water molecules in the air, which is very harmful to crops, together with buildings, and

can cause water pollution [7,8]. More seriously, $NO_2$ can cause organ diseases, reduce the resistance of immune functions, and affect the health of human body. In addition, $NO_x$ and hydrocarbons exposed to strong sunlight and ultraviolet radiation in the atmospheric environment will lead to more severe environmental problems, such as photochemical smog and thinning of the ozone layer. It is advisable to remove nitrogen compounds in advance rather than let them discharge [9].

(3) In the oil secondary processing, including catalytic cracking and hydrofining, the competitive adsorption of basic nitrides and sulfides in oil products, on the active sites of the catalyst, leads to the degradation of the selectivity and working efficiency of precious metals and even to the poisoning and loss of the catalyst [10]. The result makes the deep hydrogenation process difficult and easy to form ammonium salt, which can block the pipelines in the reaction process.

The traditional ways to remove nitrogen compounds in oil products mainly include hydrogenation and non-hydrogenation denitrogenation technologies. The former refers to the conversion of nitrogen-containing compounds into ammonia and water by reaction with hydrogen under the action of catalyst (e.g., Co–Ni–Mo/$Al_2O_3$, Ni–W/$Al_2O_3$) [11]. At the initial stage, the catalytic denitrogenation efficiency of this technique was around 20%, which can reach higher than 70% after gradual improvement in recent years [12]. The oil quality produced by this technology has some problems, such as poor stability, short storage cycle, and so on. In the operation process, the conditions are harsh, and the amount of hydrogen consumed is large, so the relative risk is increased. In addition, energy consumption and cost are great. Due to these limitations of hydrodenitrogenation process, non-hydrodenitrogenation technologies are gradually developed as substitute means. It mainly includes extraction, acid refining, adsorption, microorganisms, and other methods [13,14]. Moreover, the hybrid of them is considered more effective; e.g., the denitrifier was prepared by acidic agent, complexant, and demulsifier. Table 1 describes the details of various non-hydrodenitrogenation methods for comparison.

Table 1. Comparison of non-hydrodenitrogenation methods.

| Ways | Principles | Advantages | Disadvantages | Denitrifiers | Ref |
|---|---|---|---|---|---|
| Extraction | Based on the principle of "like dissolves like", the purpose of separation and purification is achieved through selective dissolution | Nitrogen compounds can be removed under normal pressure, and the good resistivity and appearance of oil can be achieved | The reduction of hydrocarbon compounds limits the application of oil products, and the extraction efficiency needs to be improved | Furfural, phenol, dimethyl sulfoxide, and dimethyl formamide, etc. | [15,16] |
| Acid refining | Strong organic/inorganic acids and basic nitrogen-containing substances form insoluble salts in oil | Low processing cost and loss of oil, which can remove colloid and other impurities in oil | Acids can corrode the equipment and the acidic residue is difficult to treat, which is not friendly to the environment and operators | Inorganic acids, trifluoroacetic acid, oxalic acid and solid acid, etc. | [17,18] |
| Complexation separation | Lewis acid-base theory and complexation | The dosage of complexing agent is small and the removal efficiency is high | The complex can dissolve in the oil, and the subsequent separation is difficult | Transition metal halides such as ferric chloride | [19,20] |

Table 1. Cont.

| Ways | Principles | Advantages | Disadvantages | Denitrifiers | Ref |
|---|---|---|---|---|---|
| Photocatalysis denitrogenation | Most photocatalysts are semiconductor materials. Under light radiation, oil denitrogenation is realized by the redox of nitrides into harmless small molecules by photogenerated electrons and holes of valence bands | Safe, high stability, high catalytic activity and low energy consumption | Photogenerated charge carriers need to migrate to the photocatalyst surface and contact with the adsorbed nitrides before the redox reaction occurs, but their lifetime is only nanosecond | Visible light and photocatalysts such as nanocomposite metal oxide catalysts | [21,22] |
| Microwave denitrogenation | Non-thermal effect of electromagnetic field on reactive molecules | Short reaction time, high efficiency and simple process | Special equipment is needed, and the intensity, frequency, modulation mode of microwave and system composition have obvious impact on the denitrogenation results | UHF electromagnetic wave in the range of 0.1–100 cm and 300 MHz-300 GHz | [23] |
| Adsorption denitrogenation | Solid adsorbents have strong adsorption on polar nitrogen compounds | Solid-liquid separation is easier | Multiple steps and complex operation | Chalk, silica gel, resins, molecular sieve, etc. | [24,25] |
| Biological denitrogenation | The characteristic catalytic ability of microbial cultures or their enzymes is used to selectively remove nitrogen compounds from oils | For high selectivity, microorganisms have no effect on hydrocarbons in oil products, and energy consumption is low | Process is slow; and aldehydes, esters and other substances in oil products affect microbial growth | Various microorganisms such as nitroso bacteria, nitrobacteria, etc. | [26] |

As new green solvents, ionic liquids are usually composed of large organic cations and small inorganic anions; their different combinations can result in different types and properties of ILs. Compared with conventional solvents, ionic liquids have many advantages, such as low vapor pressure, non-volatility, good thermal stability, low melting point, strong solubility, etc. Due to their flexible designability and tailored properties, they are used in various chemical and engineering fields, especially in different extraction and separation processes, as well as basic researches [27–29]. Huh et al. used 1-ethyl 3-methylimidazole ethyl bisulfate and zinc chloride ([EMIm]EtSO$_4$–ZnCl$_2$) to remove the impurities in diesel oil. The experimental results show that the combination of ionic liquid and zinc has an obvious removal effect on nitrides [30]. For the denitrogenation of indole as a neutral nitrogen compound, the extraction was mostly governed by the interaction between the anion of IL and the H atom of N–H, rather than by the coordination of indole to the Zn center. Besides, Chen et al. studied the removal of pyridine from simulated oil by the ionic liquid of [Bmim]Cl/ZnCl$_2$ and the N-content was undetectable after 2-stage extraction. Under the condition of 25 °C and the ratio of IL to oil was 1:1, the removal efficiency of pyridine by the IL with ZnCl$_2$ anion was 97.8% after 30 min [31]. In another case, Liu et al. extracted pyridine from simulated oil with three kinds of ionic liquids in the experiment. The results show that the pyridine extraction efficiency followed the order of [C$_4$C$_1$im][NTf$_2$] (76.8 mol%) > [C$_4$C$_1$im][OTf] (72.5 mol%) > [C$_4$C$_1$im][PF$_6$] (67.5 mol%) > [C$_4$C$_1$im][BF$_4$] (58.8 mol%) [32]. When the cation was the same, it was found that the larger the anion was, the better the nitrogen removal effect of ILs was. Above research proves the feasibility of ionic liquids used in oil denitrogenation. However, there are only a few types of IL members used in individual investigation, and the nitrogen-containing object is relatively singular, so a lack of comparison exists. The separation kinetics are still unknown, which is not beneficial for mechanism exploration and process amplification. Besides that, the separation mode is only liquid-liquid biphasic extraction, which needs to be expanded in the current study.

Based on above background, quinolinium, 8-hydroxy-quinolinium, and benzothiazolium ionic liquids, containing the acidic anions of methanesulfonate ($[CH_3SO_3]^-$), phosphate ($[H_2PO_4]^-$), p-toluenesulfonate ($[p\text{-}TSA]^-$), and bisulfate ($[HSO_4]^-$) were synthesized. Then, the single stage extraction and 2-stage extraction were used in the extraction for the simulated oils containing pyridine, quinoline or aniline, and their mixture by using various IL solutions, respectively. Their denitrogenation performance was compared on the basis of structural features, and main extraction conditions were further investigated, including mass ratio of IL to water, mass ratio of IL to oil, and temperature. Furthermore, the extraction process was described by different kinetic equations. Recovery and reuse of IL were realized by back-extraction and liquid-liquid separation. Finally, based on the developed method for preparing adsorption tablets, corresponding immobilized IL was used to extract the target object in order to extend separation ways, which was more easily recovered after extraction.

## 2. Material and Methods

### 2.1. Materials and Reagents

All of the reagents and solvents used in the present study were of analytical-reagent grade or higher, which were used directly without special treatment unless otherwise specified. Benzothiazole, methanesulfonic acid, and aniline were purchased from TITAN Technology Co., Ltd. (Shanghai, China). Additionally, 8-hydroxyquinoline, quinoline, sulfuric acid, p-toluenesulfonic acid, phosphoric acid, ethanol, acetone, n-octane, ethyl cellulose, and pyridine are provided by Kelon Chemical Reagent Factory (Chengdu, China). All other unmentioned agents were purchased from Aladdin Chemical Reagent Company (Shanghai, China). Multi walled carbon nanotubes (length: 0.5–2 mm, diameter: 0–50 nm, >95%) were provided by XFNANO Materials Tech Co., Ltd. (Nanjing, China). Experimental water was ultrapure, which was obtained from the Milli-Q water purification system (Millipore, Bedford, MA, USA).

### 2.2. Synthesis of ILs

Twelve extractants with common cations and anions of ILs were prepared, and all of them have the melting points under 200 °C. For convenience, they are uniformly called ILs in previous studies [33–35]. Related products, including quinolinium methanesulfonate ($[Quli][CH_3SO_3]$), quinolinium phosphate ($[Quli][H_2PO_4]$), quinolinium p-toluenesulfonate ($[Quli][p\text{-}TSA]$), quinolinium bisulfate ($[Quli][HSO_4]$), 8-hydroxyquinolinium methanesulfonate ($[HHqu][CH_3SO_3]$), 8-hydroxyquinolinium phosphate ($[HHqu][H_2PO_4]$), 8-hydroxyquinolinium p-toluenesulfonate ($[HHqu][p\text{-}TSA]$), 8-hydroxyquinolinium bisulfate ($[HHqu][HSO_4]$), benzothiazolium methanesulfonate ($[HBth][CH_3SO_3]$), benzothiazolium phosphate ($[HBth][H_2PO_4]$), benzothiazolium p-toluenesulfonate ($[HBth][p\text{-}TSA]$), and benzothiazolium bisulfate ($[HBth][HSO_4]$) were prepared for the comparative research. Other ionic liquids were purchased from ZhongkeKete technology industry Co., Ltd. (Lanzhou, China).

For four quinolinium ILs, their synthetic route was based on the developed way [33], and their yield was 69~83%. Taking $[Quli][HSO_4]$ as an example, 6.458 g (0.05 mol) of quinoline was placed in a round bottom flask, and then 50 mL ethanol was added as solvent. The reaction was carried out under the condition of an ice bath, 4.92 g (slightly more than 0.05 mol) of concentrated sulfuric acid was slowly added into the solution with a constant pressure titration funnel; the whole system was stirred magnetically at room temperature for 2 h. After that, ethanol was removed, by rotary evaporation, to obtain a white solid product, which was washed with acetone several times. Finally, it was recrystallized in absolute ethanol and dried under vacuum for 24 h.

For four benzothiazolium ILs, their synthetic route was based on the developed way [34], and their yield was 73%~84%. Taking [HBth][CH$_3$SO$_3$] as an example, 25 mL of 0.05 mol of benzothiazole aqueous solution was placed in the flask, and then 0.05 mol of methane sulfonic acid and 25 mL of anhydrous ethanol mixed solution were added, dropwisely, with a constant pressure titration funnel under ice bath conditions. The reaction was carried out by magnetic stirring at room temperature for 3 h, and the solvent was removed, after the reaction, to obtain a white crude product. It was washed several times with ethyl acetate, then recrystallized in ethanol, and a white flake solid was obtained as the product after thorough dryness.

For four 8-hydroxyquinolinium ILs, their synthetic route was based on the developed way [35], and their yield was 66~79%. Taking [HHqu][p-TSA] as an example, 0.05 mol 8-hydroxyquinoline was mixed with 50 mL of absolute ethanol to form a homogeneous solution, and then, 25 mL anhydrous ethanol solution of 0.05 mol p-toluenesulfonic acid was slowly added to the solution with a constant pressure titration funnel under ice bath condition; the reaction was carried out by magnetic stirring at room temperature for 4 h. After that, ethanol was removed by rotary evaporation to obtain a light yellow crude product. Then, it was washed with acetone for several times and recrystallized in ethanol. Finally, a yellow powder solid product was obtained after being dried in vacuum for 24 h.

*2.3. Preparation of Simulated Oil for Separation*

According to the preparation of simulated oils and corresponding nitrogen content in current references [36], three kinds of simulated oils were obtained as follows:

(1) Preparation of quinoline simulated oil: 138.39 mg of analytical pure quinoline was weighed and completely dissolved in 100 mL of *n*-octane by ultrasound assistance (300 W) to prepare a solution. The concentration of quinoline was 1383.9 mg/L (the nitrogen content in the solution was 150 mg/L).

(2) Preparation of aniline simulated oil: 99.64 mg of analytical pure aniline was weighed and completely dissolved in 100 mL of *n*-octane by ultrasound assistance (300 W). The concentration of aniline was 996.4 mg/L (the nitrogen content in the solution was 150 mg/L).

(3) Preparation of pyridine simulated oil: 84.64 mg of analytical pure pyridine was weighed and completely dissolved in 100 mL of *n*-octane by ultrasound assistance (300 W). The concentration of pyridine was 846.4 mg/L (the nitrogen content in the solution was 150 mg/L).

*2.4. Quantitative Analysis*

Quinoline, aniline and pyridine all have characteristic ultraviolet spectra, and their maximum absorbance wavelength is very different from that of those selected ILs, so ultraviolet visible (UV-Vis) spectrophotometry was used to determine the change of their concentration in the simulated oil before and after extraction to determine corresponding N-removal efficiency. Here AOE380 UV spectrometer (Aoyi Instrument Co., Shanghai, China) was used for the following quantitation. For developing quantitative working curve, the oil samples with quinoline concentrations (x) of 6.92 mg/L, 13.84 mg/L, 27.70 mg/L, 41.52 mg/L, 55.36 mg/L, and 69.20 mg/L were prepared, respectively. Then, the absorbance (y) of quinoline in the oil was measured at the wavelength of 313 nm [37], and the standard curve was developed according to the relationship between absorbance (y) and the corresponding concentration (x). As the result, the standard curve equation was obtained as y = 0.0189x − 0.0007, $R^2$ = 0.9992.

Similarly, the standard oil samples were prepared with the aniline concentrations (y) of 9.96 mg/L, 19.93 mg/L, 29.89 mg/L, 39.86 mg/L, 49.82 mg/L, 59.78 mg/L, 69.75 mg/L, and 99.64 mg/L, respectively. Then, the absorbance (x) of aniline in the oil was measured at the wavelength of 280 nm [38], and its standard curve equation was developed as y = 0.0123x − 0.0258, $R^2$ = 0.9998. Moreover, the standard oil samples were prepared with the pyridine concentrations (y) of 11.30 mg/L, 22.60 mg/L, 36.00 mg/L, 45.00 mg/L, 56.50 mg/L, 71.00 mg/L, 85.00 mg/L, 98.00 mg/L, and 113.00 mg/L, respectively. Measure the absorbance (a) of pyridine in the solution at the wavelength of 270 nm [39], and its standard curve equation was developed, as y = 0.002371x − 0.0167, $R^2$ = 0.9997. These correlation coefficients proved the good linear relationship between concentration and UV absorbance. All experimental values were obtained from three parallel experiments.

### 2.5. Denitrogenation Experiments

A certain concentration of ionic liquid aqueous solution and simulated oil were added in a 50 mL conical flask and oscillated at a certain temperature for a period of time. After reaching the extraction end, the whole system stood for 30 min, and the upper simulated oil was sampled and analyzed by using the above spectroscopic way, together with working curves, to determine the concentration of quinoline, aniline and pyridine in the simulated oil after the removal of ionic liquid. Their initial concentration was $N_0$ (mg/L) and residual concentration was quantified as N (mg/L), and related denitrogenation efficiency could be calculated according to the following equation:

$$\text{N-removal efficiency (E)} = (N_0 - N)/N_0 \times 100\% \tag{1}$$

## 3. Results

### 3.1. Identification of Synthesized ILs

Here, 8-hydroxyquinolinium bisulfate ([HHqu][HSO$_4$]) and benzothiazolium bisulfate ([HBth][HSO$_4$]) are taken as two representatives for spectral analysis, which were the potential ideal extractants for target nitrogen compounds. Firstly, characterization of 8-hydroxyquinoline bisulfate ([HHqu[HSO$_4$]) and benzothiazole bisulfate ([HBth[HSO$_4$]) with infrared spectra was performed, which are shown in Figure 1. Obviously, the broad peak, with a wavenumber of nearly 3400 cm$^{-1}$, is formed by the stretching vibration of –OH, and the two peaks with wave number of 1600 cm$^{-1}$ and 1550 cm$^{-1}$ are mainly ascribed to the skeleton vibration of an aromatic ring. Besides that, the O=S=O asymmetric and symmetric stretching vibration peaks of HSO$_4^-$ anion appeared at about 1200 cm$^{-1}$ and 570 cm$^{-1}$. In detail, the signal assignment of related absorbance peaks and typical groups is summarized in Table 2.

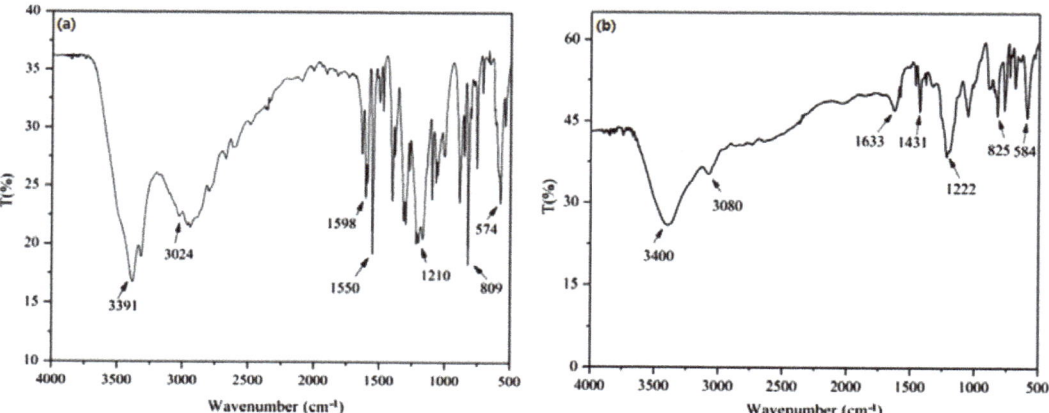

Figure 1. The IR spectra of [HHqu][HSO$_4$] (a) and [HBth][HSO$_4$] (b).

Table 2. FI-IR spectra analysis of [HHqu][HSO4] and [HBth][HSO4].

| | [HHqu][HSO4] | | | [HBth][HSO4] | |
|---|---|---|---|---|---|
| Group | Wavenumber | Vibration | Group | Wavenumber | Vibration |
| –OH | 3391 | $\nu_{O-H}$ | –OH | 3400 | $\nu_{O-H}$ |
| =CH | 3024 | $\nu_{=CH}$ | =CH | 3080 | $\nu_{=CH}$ |
| C=C | 1598, 1550 | $\nu_{C=C}$ | C=C | 1633, 1584 | $\nu_{C=C}$ |
| C–N | 1390 | $\nu_{C-N}$ | C–N | 1431 | $\nu_{C-N}$ |
| S=O | 1210, 574 | $\nu_{as\ S=O}$, $\nu_{s\ S=O}$ | S=O | 1222, 584 | $\nu_{as\ S=O}$, $\nu_{s\ S=O}$ |
| =CH | 809 | $\gamma_{=CH}$ | =CH | 825 | $\gamma_{=CH}$ |

Furthermore, the two ILs of [HHqu][HSO4] and [HBth][HSO4] were characterized by $^1$H NMR (500 MHz, MeOD), and their hydrogen atoms were numbered, respectively. The spectra are shown in Figure 2, and the NMR data are included as: 8.91 (H-1), 8.85 (H-2), 8.89 (H-3), 7.63 (H-4), 7.88 (H-5), 7.90 (H-6) for [HHqu][HSO4]; 9.74 (H-1), 8.80 (H-2), 7.59 (H-3), 7.61 (H-4), 7.68 (H-5) for [HBth][HSO4]. Every signal appears except active protons on O or N atom for the use of MeOD as solvent, and corresponding signals are clearly separated. The whole information of the two ILs have been compared with those in references [33–35], and it can be proved that the synthesized products are target ionic liquids, according to spectral analysis.

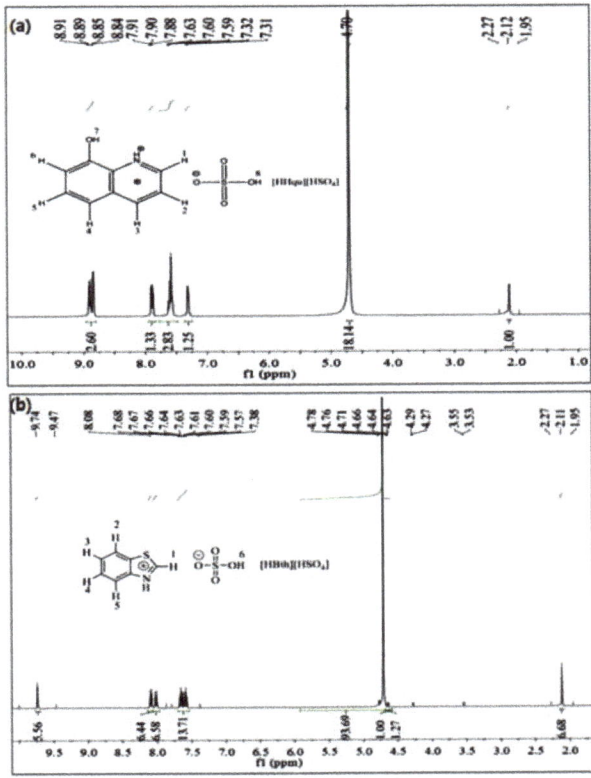

Figure 2. $^1$H NMR of [HHqu][HSO4] (a) and [HBth][HSO4] (b) (500 MHz, MeOD).

## 3.2. Comparison among ILs

Larger cations can offer more surface area for absorption through van der Waals interactions with targets [32], so quinolinium, 8-hydroxyquinolinium, and benzothiazolium cations were selected, instead of the most common imidazole, in this study. According to previous findings [40], acidic anions of ILs play very significant role in oil denitrogenation for related N-compounds, so the three kinds of cations were combined with them for comprehensive screening. In addition, considering the preparation cost, stability, and commonality, some reported ILs are not compared here. In this section, 0.25 g of twelve kinds of ionic liquids and 2 mL of water were added respectively into a 50 mL conical flask, and then they were mixed with 2.5 g of simulated oil containing pyridine, quinoline, or aniline. After constant temperature oscillation at 200 rpm and 30 °C for 30 min, and standing for 30 min, the sample of upper oil phase was taken to measure its UV absorbance; then, the concentration of related compound and denitrogenation efficiency of ILs were calculated for comparison. The experimental results are shown in Figure 3. Besides that, the volume of ionic liquid phase (water phase) and oil phase was observed during the whole process. As the result, no significant change occurred, suggesting that there was no obvious mutual solubility between the two phases.

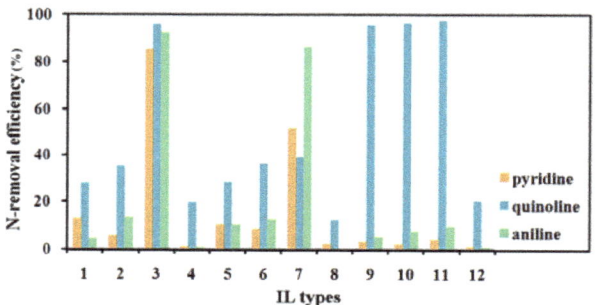

**Figure 3.** N-removal efficiency of twelve ILs (1: [HHqu][p-TSA], 2: [HHqu][CH$_3$SO$_3$], 3: [HHqu][HSO$_4$], 4: [HHqu][H$_2$PO$_4$], 5: [Quli][p-TSA], 6: [Quli][CH$_3$SO$_3$], 7: [Quli][HSO$_4$]; 8: [Quli][H$_2$PO$_4$]; 9: [HBth][p-TSA]; 10: [HBth][CH$_3$SO$_3$]; 11: [HBth][HSO$_4$]; 12: [HBth][H$_2$PO$_4$]).

### 3.2.1. Removal of Pyridine from Simulated Oil

It can be seen that the removal capacity of pyridine simulated oil by ionic liquids with different anions with the same cationic structure is in the order of HSO$_4^-$ > $p$-TSA$^-$ > CH$_3$SO$_3^-$ > H$_2$PO$_4^-$, indicating that the stronger the acidity, the better the removal effect of pyridine. When the anions of ionic liquids are the same, 8-hydroxyquinoline bisulfate shows the best N-removal performance (85.6%), followed by quinoline bisulfate (51.9%) and 8-hydroxyquinoline tosilate (12.7%). Their performance is significantly different, and the removal effect of more than half of tested ILs is not obvious. For the difference between [HHqu][HSO$_4$] and [Quli][HSO$_4$], the reason can be ascribed to the fact 8-hydroxyquinoline based cation contains an extra hydroxyl group, which can provide H-bonding effect; besides, the hydroxyl group is connected with the benzene ring and activates the latter, making its denitrogenation efficiency higher. Moreover, for comparison with reported data [41], the N-removal efficiency of four ILs including 1-butyl-3-methylimdazolium dicyanamide, 1-ethyl-3-methylimdazolium dicyanamide, ethylated tetrahydrothiophenium dicyanamide, and tetrahedral ethyldimethylsulfonium dicyanamide for pyridine-containing oil was 72.7%, 69.1%, 63.5%, and 59.8%, respectively; the extractive performance of [HHQu][HSO$_4$] is competitive. Therefore, it was selected as the ionic liquid for removing pyridine from simulated oil.

3.2.2. Removal of Quinoline from Simulated Oil

As a whole, the removal efficiency of tested ILs for quinoline is obviously higher than that for the other two objects except [Quli][HSO4], and there are four ILs above 90% ([HHqu][HSO$_4$]: 96.3%, [HBth][$p$-TSA]: 96.1%, [HBth][CH$_3$SO$_3$]: 96.9% and [HBth][HSO$_4$]: 97.8%). A little differently, the removal performance of quinoline from simulated oil by ionic liquids with different anions with the same cationic structure is in the order of HSO$_4^-$ > CH$_3$SO$_3^-$ > $p$-TSA$^-$ > H$_2$PO$_4^-$. It shows that the stronger the acidity, the better the removal effect of ILs for quinoline. Among 8-hydroxyquinoline, quinoline, and benzothiazole ionic liquids, the last type exhibits the best performance on the removal of quinoline and great superiority over the other two. Furthermore, it can be found the benzothiazolium series has high selectivity for quinoline, according to the comparison of the column height of three objects. Considering the combination of H$_2$PO$_4^-$ and the cation of 1-butyl-3-methylimidazolium ([Bmim]) has achieved more than 99% N-removal efficiency of quinoline in its $n$-heptane solution (the nitrogen content in the solution was also 150 mg/L) in the previous study [36], the performance of [Bmim][H$_2$PO$_4$] was investigated in an additional experiment under the same conditions as above, and its N-removal efficiency was finally determined as 97.0%. Therefore, benzothiazole bisulfate [HBth][HSO$_4$] was selected as the ionic liquid for removing quinoline from simulated oil.

3.2.3. Removal of Aniline from Simulated Oil

Generally, the basic order is pyridine > quinoline > aniline. For the removal of aniline, [HHqu][HSO$_4$] (92.7%) also shows the best performance among twelve ILs as for the removal of pyridine, followed by [Quli][HSO$_4$] (86.7%). It can be seen from, the comparison in Figure 3, that the removal efficiency of aniline from simulated oil by ionic liquids, with different anions but the same cation, is in the order of HSO$_4^-$ > CH$_3$SO$_3^-$ > $p$-TSA$^-$ > H$_2$PO$_4^-$ too. Similarly, it proves that the stronger the acidity of anions, the better the removal effect of aniline. Moreover, dihydrogen phosphate shows the worst performance not only in the four anions for aniline but also in the extraction of three objects combined with different cations (only 0.40~0.51% for aniline, compared with 0.74~1.80% for pyridine, 10.5~20.4% for quinoline). As for the cations, none of the four benzothiazolium ionic liquids had a removal efficiency of higher than 10%, which do not seem to be ideal choices for aniline.

## 4. Discussions

*4.1. Separation Conditions*

4.1.1. Mass Ratio of IL to Deionized Water

0.25 g of the best IL ([HHqu][HSO$_4$] for pyridine and aniline, and [HBth][HSO$_4$] for quinoline) was added into five 50 mL conical flasks, then 0~2.5 g of deionized water was mixed with the IL in the flasks, respectively. After that, 2.5 g of simulated oil, containing pyridine, quinoline, or aniline, was added in the system, which was shaken and reached equilibrium at 250 rpm and 30 °C. After standing for 30 min, the sample of upper oil phase was taken out to measure its UV absorbance, and denitrogenation efficiency was obtained according to the calculation. Figure 4 shows the results for the effect of IL-H$_2$O ratio ($w/w$, g/g) on the N-removal efficiency. Obviously, it can be found that the highest removal efficiency of all the three target components is reached when the mass-volume ratio of IL to water is 1:1; at this point, the removal efficiency of pyridine, quinoline, and aniline is 99.61%, 99.04%, and 97.21%, respectively. In addition, the overall trend in Figure 4a–c is similar. With the increase in water, the denitrogenation effect of different ionic liquids increases first and then decreases. When the dosage of water becomes greater, the concentration of ionic liquid will decrease, and the acidity of the system weakens, resulting in the reduction in N-removal capacity. On the contrary, the high viscosity of IL phase is not conducive to mass transfer in the extraction process if water is absent or too little. Compared with the mass ratio of [Bmim]H$_2$PO$_4$ and [PSmim]H$_2$PO$_4$/H$_2$O (1:0.5) to remove quinoline from the simulated oil (150 mg/L) [36], the mass ratio of 1:1

indicates the less consumption of IL and higher efficiency in this study. According to the experimental results, the mass ratio of IL to water was set at 1:1 for the subsequent removal of pyridine, quinoline, and aniline.

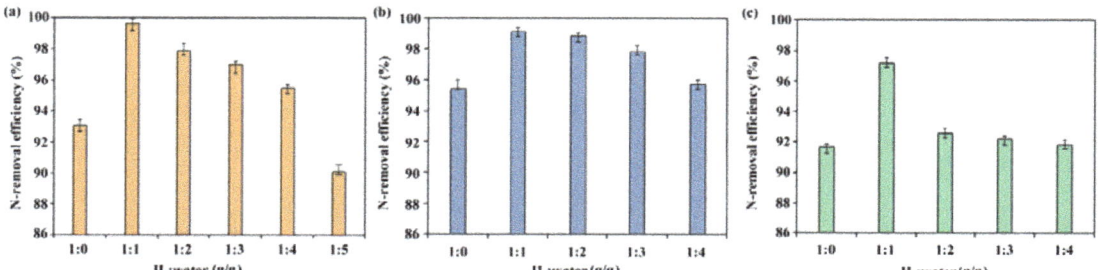

**Figure 4.** N-removal efficiency of the best IL with different mass ratio to water for (**a**) pyridine, (**b**) quinoline, and (**c**) aniline.

### 4.1.2. Mass Ratio of IL to Oil

In this section, 0.25 g of the selected IL and 0.25 g of water were first placed in the 50 mL conical flask, and then, 1.5 g (IL:oil = 1:6, $w/w$)~15 g (IL:oil = 1:60, $w/w$) of simulated oil containing pyridine, quinoline, or aniline was added in the system, respectively. After shaking at 250 rpm on 30 °C, the equilibrium was reached. After standing for 30 min, the sample of oil phase was taken to measure its UV absorbance, and the denitrogenation efficiency was determined. The experimental results are shown in Figure 5. The results indicate that the removal efficiency of pyridine in simulated oil gradually decreases with the decrease in mass-mass ratio of IL to oil. When the ratio is 1:10, the denitrogenation efficiency reaches 98.77%, and when the mass-mass ratio of [HHqu][HSO$_4$] to oil becomes 1:60, the efficiency of pyridine removal turns to 96.01%. In other words, when the processing amount of oil is expanded six times, the IL performance is only reduced by 2.76%; this shows the strong removal capacity of this method. Similarly, benzothiazole bisulfate also has a high extraction effect for quinoline in the simulated oil. When the mass-mass ratio of [HBth][HSO$_4$] to oil is changed from 1:10 to 1:80, the denitrogenation efficiency is only reduced from 99.49% to 96.65%, and when the agent oil ratio is 1:60, the removal efficiency of quinoline is 98.95%. With the gradual decrease in [HHqu][HSO$_4$]-oil mass ratio from 1:10 to 1:70, the removal effect of aniline also showed an obvious downward trend, and the denitrogenation efficiency decreased from 97.21% to 93.66%. At all experimental points, the removal efficiency of the three objective substances are higher than 90%, which provides a great space for the operator to choose conditions for treating different amounts of oil. In reported ways, 1:1 of IL to oil is very common [37,42], which results in much higher consumption of IL and lower economic efficiency. Moreover, as another kind of green medium, the deep eutectic solvents (DESs), composed of urea and a series of acids (i.e., citric acid/oxalic acid/malonic acid/$p$-toluenesulfonic acid), were applied to extract N-compounds with the mass ratio of 1:2 to oil recently [43], and here, the consumption of IL in the ratio of 1:10 is much lower than that of DESs, indicating the strong N-removal capacity of related ILs in this study.

### 4.1.3. Temperature

Next, 0.25 g of the selected IL, and 0.25 g of water, were mixed in a 50 mL conical flask, and 10 g of simulated oil containing pyridine, quinoline, or aniline was added into the system at 250 rpm. After constant temperature oscillation on 25 °C, 30 °C, 40 °C, 50 °C, or 60 °C for 20 min and standing for 30 min, the upper oil was sampled for measurement of its UV absorbance, and the denitrogenation efficiency was obtained after calculation. The experimental results are shown in Figure 6. It can be seen that, when the temperature is increased from 25 °C to 60 °C, the efficiency of pyridine removal will decrease from

98.90% to 87.20%, and the denitrogenation result reaches the best level at room temperature. The reason is that the removal of basic nitrogenous compounds by ionic liquids is an exothermic process, and higher temperature is not conducive to the progress of separation, so the denitrogenation efficiency decreases, obviously. Similarly, if the temperature rises from 25 °C to 60 °C, the efficiency of removing quinoline or aniline was reduced from 99.05% to 95.29% or from 97.61% to 93.33%, respectively. By comparison, the effect of temperature on pyridine removal is the most significant. Therefore, room temperature (25 °C) was selected for this experiment as the appropriate temperature for removing the three components. It is desirable for large scale applications since it guarantees extraction at, or below, ambient conditions, thus consuming less energy. For comparison, 50 °C was used in the reported method by using $CH_3CONH_2$–$0.3ZnCl_2$ as the extractant [44].

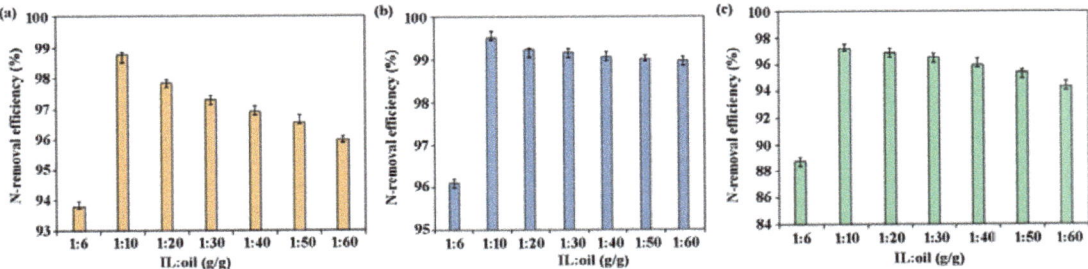

**Figure 5.** N-removal efficiency of the best IL with different mass ratio of IL to oil for (**a**) pyridine, (**b**) quinoline, (**c**) aniline.

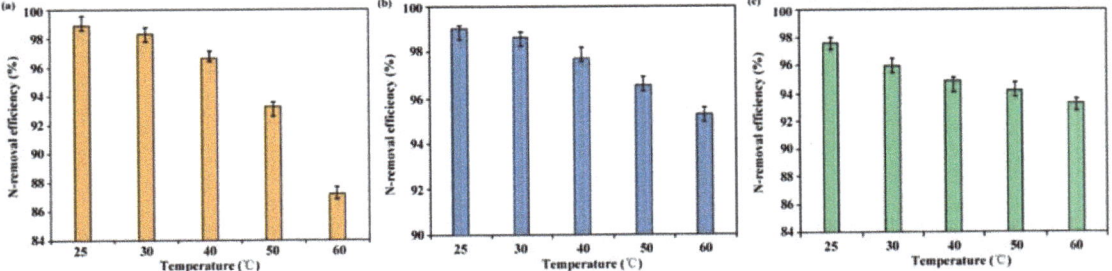

**Figure 6.** N-removal efficiency of the best IL under different temperatures for (**a**) pyridine, (**b**) quinoline, (**c**) aniline.

### 4.1.4. Oscillation Speed

Relatively, oscillation speed is not a very significant parameter, as it was above. Low speed can result in the unsatisfied efficiency of mass transfer during extraction, while excessive speed will be unnecessary and even cause severe emulsification together with difficult phase separation. According to current reports and pilot experiments, the effect of three levels of oscillation speed (low: 200, medium: 250, high: 300 rpm) on N-removal efficiency was explored under the same conditions (0.25 g IL, 0.25 g water, 10 g simulated oil, 25 °C and 20 min). The results in Figure 7 indicate the improvement is not obvious when the oscillation speed reaches 300 rpm. In order to reduce the power consumption and standing time, 250 rpm was chosen at last, which was much lower than 1000 rpm in previous processes [32]. At this level, ideal N-removal results for three kinds of oil samples can be obtained, and the two phases, composed of oil and IL-water, can be separated naturally.

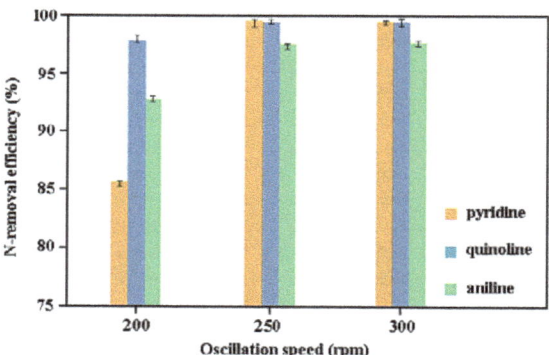

**Figure 7.** N-removal efficiency of the best IL with different oscillation speeds for pyridine, quinoline, and aniline.

### 4.2. Separation Kinetics

The study of extraction kinetics can make us understand the separation process more deeply. In some cases, extraction kinetics are a function of chemical reaction speed and diffusion speed; the chemical reaction can include bond destruction and formation, molecular polymerization or intermediation, and the reaction may occur in the phase or at the interface. These have an impact on the dynamic process. At the same time, the mass transfer of the diffusion process has a more complex effect on the kinetic process. For further analysis of related separation stages and mechanism, a kinetic study was carried out to explore the extraction process. Firstly, it was found the extraction speed of quinoline by [HBth][HSO$_4$] was higher than that of pyridine or aniline by [HHqu][HSO$_4$]. For the former, its concentration in oil remained unchanged after 10 min; for the latter two, their concentration in oil would not decrease after 20 min. Compared with the N-removal duration of [Bmim]H$_2$PO$_4$ and [PSmim]H$_2$PO$_4$ (30 min) [36], the time needed to reach extraction equilibrium is reduced by 1/3 to 2/3 in this study. Furthermore, the mass transfer process was described by pseudo-first and second order models, successively. The pseudo-first order model can be expressed according to [45] as:

$$\ln(C_s - C_t) = -k_1 t + \ln(C_s - C_0) \qquad (2)$$

where $k_1$ is the first-order kinetic constant; $C_S$ is the reduced concentration of the N-compound in oil phase at extraction equilibrium; $C_t$ is its reduced concentration in the oil phase at time $t$; $C_0$ is its initial concentration in the oil phase.

In addition, the pseudo-second order model can be expressed according to [46] as:

$$\frac{t}{C_t} = \frac{1}{k_2 C_s^2} + \frac{t}{C_s} \qquad (3)$$

where $k_2$ is the second-order kinetic constant; $C_S$ is the reduced concentration of the N-compound in oil phase at extraction equilibrium; $C_t$ is its reduced concentration in the oil phase at time $t$.

Generally, the first-order kinetic model shows that the extraction efficiency is directly proportional to the target concentration, and the initial concentration has no effect on the time to reach certain extraction efficiency. For the second-order kinetic model, the extraction efficiency is directly proportional to the square of the concentration, and the extraction speed is related to the initial concentration of the object. As shown in Table 3, the values of $R^2$ in a pseudo-first model are obviously higher than those in a pseudo-second model, indicating the kinetic data can be better fitted by the former. Compared with similar extraction process of other objects with ILs, the pseudo-first model is also more common than other models.

Table 3. The fitting results of pseudo-first order and pseudo-second order models for three objects.

| Object | Pseudo-First Model | | | Pseudo-Second Model | | |
|---|---|---|---|---|---|---|
| | Fitted Equation | $k_1$ (1/min) | $R^2$ | Fitted Equation | $k_2$ (L/mg·min) | $R^2$ |
| Pyridine | y = −0.20x − 2.58 | 0.20 | 0.996 | y = 58.28x + 407.59 | 8.33 | 0.970 |
| Quinoline | y = −0.35x − 4.02 | 0.35 | 0.994 | y = 58.99x + 254.55 | 14.17 | 0.974 |
| Aniline | y = −0.19x − 3.45 | 0.19 | 0.991 | y = 56.11x + 418.70 | 7.52 | 0.966 |

### 4.3. Denitrogenation of the Mixed Simulated Oils by ILs in Two-Step Extraction

Considering that an actual oil sample generally contains various nitrides at the same time, it is necessary to investigate the denitrogenation effect of ionic liquids on simulated oil under the coexistence of three target components in this section. Firstly, the three simulated oils were mixed according to the volume ratio of 1:1:1. Therefore, the nitrogen content of the mixed simulated oil was still kept as 150 mg/L, and pyridine, quinoline, and aniline coexisted in the oil. Secondly, a two-step extraction method was established here, that is, the aqueous solutions of two ionic liquids of [HBth][HSO$_4$] and [HHqu][HSO$_4$] were firstly prepared according to the ratio of 1:1 (IL/water, g/mL). Then, the extractant of [HHqu][HSO$_4$] solution was mixed with the oil phase, according to the mass ratio of ionic liquid to oil 1:10 (g/g); after extraction at 25 °C for 20 min, biphase separation was carried out with a separating funnel. Then, the [HBth][HSO$_4$] aqueous solution was mixed with the oil phase for secondary extraction. Similarly, the N-removal efficiency of the three objects was determined by measuring their final residual concentration in the oil phase. As the result, the removal efficiency of pyridine, quinoline, and aniline was determined as 99.03%, 98.86%, and 97.58%, respectively. If the [HBth][HSO$_4$] aqueous solution was used before [HHqu][HSO$_4$] solution, the removal efficiency of pyridine, quinoline, and aniline became 98.97%, 98.82%, and 97.50%, respectively. In summary, it can be found that there is no obvious difference between two kinds of extraction sequences, and two ILs can also perform when facing the samples of single simulated oil.

### 4.4. IL Recovery and Reuse Performance

In previous studies, ionic liquids could be recovered by back-extraction, autodetachment (heterogeneous system), adsorption, distillation, membranes, magnetic or electrical field after their applications. Besides, regeneration of the hydrophilic [Bmim]Ac and [Bmim]Ac/ZnAc$_2$ after oil denitrogenation was carried out by diluting them with water while those hydrophobic N-compounds were repelled from the system [47], and the dilution process was followed by simple distillation. Among these ways, back-extraction is regarded as an easy-to-use method free of complex equipment. Under the above separation conditions, the two ionic liquids of [HHqu][HSO$_4$] and [HBth][HSO$_4$] containing pyridine, quinoline, or aniline were further recovered by back-extraction in order to evaluate their reusability. Among the potential extractants, alcohols and ketones were first abandoned because of their high mutual solubility with ionic liquids; aromatic compounds (benzene or toluene) and haloalkanes were also not considered because they are more toxic. Compared with esters, ether has a lower boiling point, which is easy to recover and reduces the energy consumption in the post-treatment process. More importantly, it can not only form two phases with ionic liquids quickly, but it can also extract three kinds of target components well.

The specific recovery process was carried out as follows: after the denitrogenation experiment was completed, the oil sample and ionic liquid are separated with a separatory funnel, and then, the ionic liquid phase (lower phase) was collected and extracted with equivolume ether. Through the second bi-phase separation, the ionic liquid phase was obtained and distilled under reduced pressure to remove the residual ether. After thorough dryness, the recovered ILs were used for the next denitrogenation experiment after being diluted with water, and corresponding N-removal efficiency was determined in five cycles. The reusing investigation results of [HHqu][HSO$_4$] or [HBth][HSO$_4$] for removing pyridine, quinoline, or aniline are shown in Figure 8a–c, respectively. It can be seen that the effect of removing performance of the ILs for three target objects gradually decreases after being reused for five times. The experimental results showed that the denitrogenation efficiency of 8-hydroxyquinoline bisulfate for pyridine and aniline, together with benzothiazole bisulfate for quinoline, were 89.34%, 87.34%, and 84.34%, respectively. With the increase in reuse times, the amount of nitrogenous compounds extracted in the ionic liquid will increase, resulting in a reduction in the denitrogenation effect of the latter. If the N-removal efficiency reduces to be lower than 80%, operators can improve the purity of [HHqu][HSO$_4$] or [HBth][HSO$_4$] by enlarging the volume of ether or increasing the number of back-extraction times. Above results also prove that the interaction between acidic ionic liquids and weakly basic target substances is reversible. In the process of back extraction, with organic solvents such as ethers, the balance between ionic and molecular states of the three objects will move to the direction of molecular state because lowly polar solvents remove the target substances in the latter state continuously.

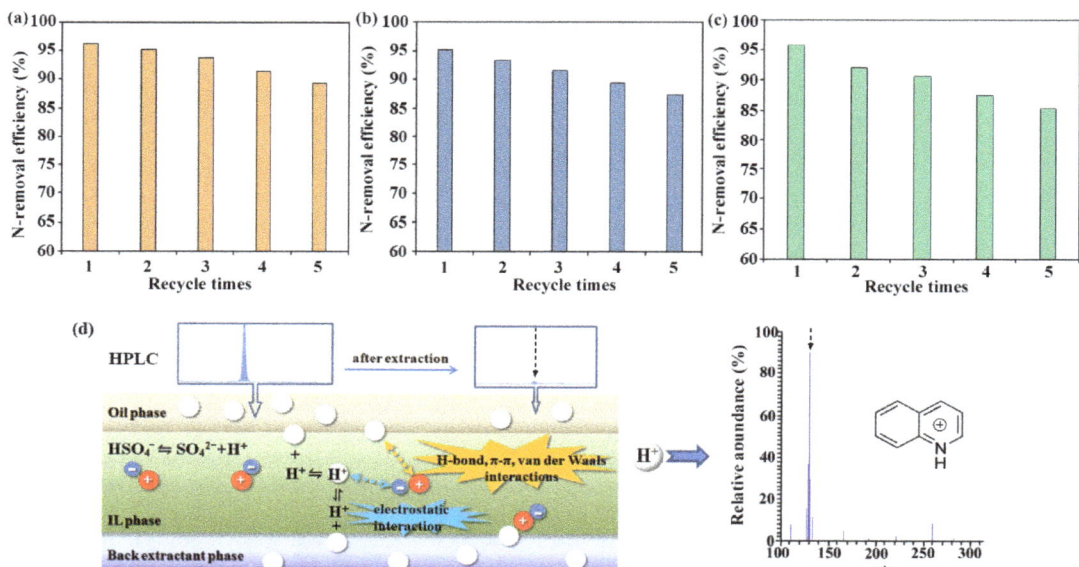

**Figure 8.** N-removal efficiency of the best IL in different recycle times for (**a**) pyridine, (**b**) quinoline, (**c**) aniline; (**d**) extraction mechanism in this study (white balls: targets, blue balls: IL anions, red balls: IL cations) under HPLC and MS analysis.

Finally, in order to investigate the possibility that the targets become protonated under the effect of acidic anion of the IL during extraction, the oil phase, containing quinoline before and after extraction, together with the mixture of quinoline and [HBth][HSO$_4$], were analyzed by a Waters Alliance 2695–2996 high-performance liquid chromatography (HPLC) system (Waters, Milford, MA, USA), coupled with UV detector under the developed conditions (Waters C$_{18}$ column, 3.9 × 150 mm, 5 µm; 60% methanol aqueous solution; 30 °C 10 µL; 225 nm), which were further identified by Waters Quattro Premier XE tri-quadrupole mass spectrometer (MS) in positive ion mode after HPLC analysis. It can be found from Figure 8d that the main chromatographic peak area belonging to target N-compound (retention time = 6.16 min) was significantly reduced, and the positive ion peak of protonated quinoline (C$_9$H$_8$N$^+$, $m/z$ = 130.16) has been checked in the mixture; no irreversible ion-exchange and structure transformation occur in such a mild N-removal process, and the protonation also exists in other similar applications of acidic ILs (e.g., substituted imidazole phosphates in [38]). Considering the above experimental results and similar extraction mechanisms in $H_2PO_4$-based ILs, the mechanism in this study can be depicted as Figure 8d.

*4.5. N-Removal by the Tablets of Immobilized IL*

In recent years, more and more immobilized ILs are being used in separations because of easy recovery and less loss/residue during operation [48,49]. In this section, the sorption tablets containing IL were prepared to remove N-compound through solid-phase extraction on the basis of our previous method [50] and above research results, which aimed to provide more programs and options for researchers. Taking the extraction of pyridine by [HBth][HSO$_4$] as an example, multi walled carbon nanotubes were used as the carriers for the immobilized IL, which was loaded in a ultrasound-assisted way [51]. In detail, 600 mg MWCNTs were thoroughly dispersed in 300 mL ethanol solution of [HBth][HSO$_4$] (1.5 mg/mL) by sonication (120 W) for 20 min, and then, the whole system was magnetically stirred under 30 °C for 12 h, and ethanol was totally removed under vacuum. After that, ethyl cellulose (EC) was mixed with the dried complex of IL-MWCNTs, evenly with the mass ratio of 0.05:0.10 (g/g, EC: IL-MWCNTs), and their powders (200 mesh) were pressed in a stainless steel mold with a diameter of 13 mm under the pressure of 15 MPa for 3 h. Finally, the immobilized IL complex tablets containing 42.9 mg IL each piece, with the thinness of 2 mm and weight of 0.15 g, could be obtained, which was placed in the oil sample for extraction. According to the mass ratio of IL to oil (1:10) in the Section 4.1.2, one tablet should treat 429 mg oil here; however, the volume of oil was somewhat small for such a tablet. Therefore, the concentration of simulated oil containing quinoline was diluted to be half of its original concentration (the nitrogen content in the solution was 75 mg/L), which was extracted by the tablet in the shaker, under room temperature, for 200 min. Besides, 130 rpm was selected because the tablet was stable in this shaking speed [49]. After measurement for the residual concentration of quinoline in the sample, its final N-removal efficiency was determined as 90.9%. Though it did not reach the levels of separation performance and speed as ideally as those in the liquid-liquid extraction by the same IL, the whole process, together with post-treatment, was more easily operated; moreover, another unused tablet could be conveniently added in the oil for further improvement of N-removal efficiency.

## 5. Conclusions

For the first time, the aqueous solutions of ILs were used in single extraction, two-step extraction and solid phase, simultaneously, for three representative N-compounds in the stimulated oils. A series of IL solutions were comprehensively screened for their extraction performance and compared with current reports. As for the results, benzothiazole bisulfate [HBth][HSO$_4$] was selected as the most ideal ionic liquid for removing pyridine and aniline, and benzothiazole bisulfate [HBth][HSO$_4$] was found as the best one for removing quinoline from simulated oil. Furthermore, the main operational parameters were investigated successively, including mass ratio of IL to water, mass ratio of IL to oil, and temperature. The separation kinetic data can be better fitted by pseudo-first model, and it was found that the extraction speed of quinoline by [HBth][HSO$_4$] solution was higher than that of pyridine or aniline by [HHqu][HSO$_4$] solution. After efficient recovery by back-extraction, the ILs can be reused in repeated experiments. At last, the immobilized IL in complex adsorbent tablets based on carbon nanotubes, can be successfully employed to remove target N-compound. As a whole, the research laid the foundation for further large-scale denitrogenation applications of ILs in various modes.

**Author Contributions:** This work presented here was carried out with collaboration among all authors. Methodology, Z.Z., Y.L. and J.G.; software, J.G., Y.L. and A.Y.; writing—original draft,

Z.Z. and J.G.; writing—review and editing, A.Y. and H.S.; validation, H.S.; funding acquisition and supervision, S.Y. All authors have read and agreed to the published version of the manuscript.

**Funding:** This work is supported by Sichuan Science and Technology Program (No. 2021YFG0276).

**Institutional Review Board Statement:** Not applicable.

**Data Availability Statement:** The data presented in this study are available on request from the corresponding author.

**Acknowledgments:** We are grateful to all employees of this institute for their encouragement and support of this research. Special thanks to School of Chemical Engineering, Sichuan University for the FT-IR technical assistance.

**Conflicts of Interest:** The authors declare no conflict of interest.

## References

1. Ahmed, I.; Jhung, S.H. Adsorptive desulfurization and denitrogenation using metal-organic frameworks. *J. Hazard. Mater.* **2016**, *301*, 259–276. [CrossRef]
2. Liu, Y.; Wang, W.; Hu, Q.; Zhu, Y.; Deng, J.; Tian, S. Characterization of basic nitrogen aromatic species obtained during fluid catalytic cracking by fourier transform ion cyclotron resonance mass spectrometry. *China Pet. Process. Petrochem. Technol.* **2012**, *14*, 18–24.
3. Klein, G.C.; Angström, A.; Rodgers, R.P.; Marshall, A.G. Use of saturates, aromatics, resins, and asphaltenes (sara) and non-basic nitrogen fractions analyzed by negative-ion electrospray ionization fourier transform ion cyclotron resonance mass spectrometry. *Energy Fuels* **2010**, *24*, 2545–2553.
4. Gu, X.H.; Mao, X.F.; Zhao, Y. Study on the basic nitrogen compounds from coal-derived oil. *J. Coal Sci. Eng.* **2013**, *19*, 83–89. [CrossRef]
5. Ahmed, I.; Jhung, S.H. Remarkable adsorptive removal of nitrogen-containing compounds from a model fuel by a graphene oxide/MIL-101 composite through a combined effect of improved porosity and hydrogen bonding. *J. Hazard. Mater.* **2016**, *314*, 318–325. [CrossRef]
6. Mu, L.; Wang, Y.; Cao, P.; Fan, Y.; Wang, L.; Liu, B.; Liu, M.; Yang, L.; Zhao, D. Removal of nitrogen and adsorption kinetics in shale oil by modified silica gel. *Ion Exch. Adsorpt.* **2019**, *35*, 260–269.
7. Fernando, S.; Hall, C.; Jha, S. $NO_X$ reduction from biodiesel fuels. *Energy Fuels* **2006**, *20*, 376–382. [CrossRef]
8. Lu, X.; Song, C.; Jia, S.; Tong, Z.; Tang, X.; Teng, Y. Low–temperature selective catalytic reduction of $NO_X$ with $NH_3$ over cerium and manganese oxides supported on $TiO_2$–graphene. *Chem. Eng. J.* **2015**, *260*, 776–784. [CrossRef]
9. Chand, V. Conservation of energy resources for sustainable development: A big issue and challenge for future. In *Environmental Concerns and Sustainable Development*; Shukla, V., Kumar, N., Eds.; Springer: Singapore, 2020.
10. Han, S.; Zhou, T.; Chai, Y.; Zhou, H.; Liu, C. Inhibition effects of quinoline, indole and carbazole on HDS of 4,6–DMDBT over Ni–Mo catalyst. *Acta Petrol. Sin.* **2010**, *26*, 177–183.
11. Furimsky, E.; Massoth, F.E. Hydrodenitrogenation of Petroleum. *Catal. Rev.* **2005**, *47*, 297–489. [CrossRef]
12. James, G.S. Desulfurization, denitrogenation, and demetalization. In *The Refinery of the Future*, 2nd ed.; Elsevier: London, UK, 2020.
13. Zhuang, S.; Guo, L.; Liang, J.; Zhang, Y. The development of non–hydrodenitrogenation technology of petroleum products. *Refin. Chem. Ind.* **2006**, *17*, 13–16.
14. Zhao, B.; Yuan, S.; Wang, Q. Research progress of shale oil composition and processing technology in China. *Chem. Eng. Des. Commun.* **2018**, *44*, 249–250.
15. Jin, K.S.; Jin, C.Y. Separation of nitrogen heterocyclic compounds from model coal tar fraction by solvent extraction. *Sep. Sci. Technol.* **2005**, *40*, 2095–2109.
16. Hwang, I.C.; Kim, K.L.; Park, S.J.; Han, K.J. Liquid liquid equilibria for binary system of ethanol +hexadecane at elevated temperature and the ternary systems of ethanol + heterocyclic nitrogen compounds + hexadecane at 298.15 K. *J. Chem. Eng. Data* **2007**, *52*, 1919–1924. [CrossRef]
17. Glaucia, H.C.P.; Yuan, R.; de Klerk, A. Nitrogen Removal from Oil: A Review. *Energy Fuels* **2017**, *31*, 14–36.
18. Jin, C.Y.; Yuan, L.; Li, N.; Pan, J. Removal of basic nitrogen compounds from diesel by solid superacids. *Ind. Catal.* **2010**, *18*, 51–54.
19. Zhang, J.; Xu, J.; Qian, J.; Liu, L. Denitrogenation of straight–run diesel with complexing extraction. *Pet. Sci. Technol.* **2013**, *31*, 777–782. [CrossRef]
20. Tang, X.; Hu, T.; Li, J.; Zhang, Y.; Chen, L. Progresses in the denitrogenation of diesel oil by complexation. *Petrochem. Technol.* **2014**, *43*, 843–847.
21. Zheng, L.; Yan, G.; Huang, Y.; Wang, X.; Long, J.; Li, L.; Xu, T. Visible–light photocatalytic denitrogenation of nitrogen–containing compound in petroleum by metastable $Bi_{20}TiO_{32}$. *Int. J. Hydrogen Energy* **2014**, *39*, 13401–13407. [CrossRef]
22. Zheng, L.P.; Huang, B. Photocatalytic denitrogenation of model oil over $Cu_2O/TiO_2$. *Chin. J. Struct. Chem.* **2013**, *32*, 1131–1138.

23. Qi, Q.; Li, P.; Zhang, Q.; Zhao, S. Progress of research in microwave technology application in petroleum processing. *Petrochem. Technol. Appl.* **2009**, *27*, 176–180.
24. Sun, J.; Xiu, P.; Cong, R.; Dong, H.; Wang, C.; Wang, L.; Liu, D. Reaction performance of nitrogen removal adsorption from the coking wax oil with activation resin. *Chem. Eng. Oil Gas* **2014**, *43*, 234–240.
25. Wang, Y.; Chi, Z. Research on nitrogen adsorption removal performance of mesoporous molecular sieve Al–MCM–41. *Petrochem. Technol. Appl.* **2014**, *32*, 113–117.
26. Morales, M.; Borgne, S.L. Protocols for the isolation and preliminary characterization of bacteria for biodesulfurization and biodenitrogenation of petroleum–derived fuels. In *Hydrocarbon and Lipid Microbiology Protocols*; McGenity, T., Timmis, K., Nogales, B., Eds.; Springer: Berlin/Heidelberg, Germany, 2014.
27. Seddon, K.R. Ionic liquids for clean technology. *J. Chem. Technol. Biotechnol.* **1997**, *68*, 351–356. [CrossRef]
28. Sheldon, R.A. Green solvents for sustainable organic synthesis: State of the art. *Green Chem.* **2005**, *7*, 267–278. [CrossRef]
29. Abro, R.; Abro, M.; Gao, S.; Bhutto, A.W.; Ali, Z.M.; Shah, A.; Chen, X.; Yu, G. Extractive denitrogenation of fuel oils using ionic liquids: A review. *RSC Adv.* **2016**, *6*, 93932. [CrossRef]
30. Huh, E.S.; Zazybin, A.; Palgunadi, J.; Ahn, S.; Hong, J.; Kim, H.S.; Cheong, M.; Ahn, B.S. Zn–containing ionic liquids for the extractive denitrogenation of a model oil: A mechanistic consideration. *Energy Fuels* **2009**, *23*, 3032–3038. [CrossRef]
31. Chen, X.; Yuan, S.; Abdeltawab, A.A.; Al-Deyab, S.S.; Zhang, J.; Yu, L.; Yu, G. Extractive desulfurization and denitrogenation of fuels using functional acidic ionic liquids. *Sep. Purif. Technol.* **2014**, *133*, 187–193. [CrossRef]
32. Lui, Y.Y.; Cattelan, L.; Player, L.C.; Masters, A.F.; Perosa, A.; Selva, M.; Maschmeyer, T. Extractive denitrogenation of fuel oils with ionic liquids: A systematic study. *Energy Fuels* **2017**, *31*, 2183–2189. [CrossRef]
33. Zhou, X.S.; Liu, J.B.; Luo, W.F.; Zhang, Y.W.; Song, H. Novel Brønsted–acidic ionic liquids based on benzothiazolium cations as catalysts for esterification reactions. *J. Serb. Chem. Soc.* **2011**, *76*, 1607–1615. [CrossRef]
34. Królikowska, M.; Królikowski, M.; Domańska, U. Effect of cation structure in quinolinium–based ionic liquids on the solubility in aromatic sulfur compounds or heptane: Thermodynamic study on phase diagrams. *Molecules* **2020**, *25*, 5687. [CrossRef]
35. Peng, L.; Yao, S.; Pu, L.; Yao, T.; Chen, C.; Song, H. New Brønsted–acidic ionic liquids based on 8–hydroxyquinoline cation as catalysts for the esterification reaction of n–hexylic acid. *Adv. Eng. Res.* **2015**, *9*, 257–260.
36. Zhou, Z.Q.; Li, W.S.; Liu, J. Removal of nitrogen compounds from fuel oils using imidazolium–based ionic liquids. *Petrol. Sci. Technol.* **2017**, *35*, 45–50. [CrossRef]
37. Asumana, C.; Yu, G.; Guan, Y.; Yang, S.; Zhou, S.; Chen, X. Extractive denitrogenation of fuel oils with dicyanamide–based ionic liquids. *Green Chem.* **2011**, *13*, 3300–3305. [CrossRef]
38. Wang, H.; Xie, C.; Yu, S.; Liu, F. Denitrification of simulated oil by extraction with $H_2PO_4$–based ionic liquids. *Chem. Eng. J.* **2014**, *237*, 286–290. [CrossRef]
39. Zhao, N.; Liu, D.; Li, Z.; Guo, Z.; Lou, B.; Yu, R.; Wang, F. Removal of nitrogen compounds from diesel fraction of coal tar with deep eutectic solvents. *Acta Petrol. Sin.* **2020**, *36*, 410–419.
40. Liu, J. Synthesis and denitrogenation performance of acetamide–based coordinated ionic liquid. *Fine Chem.* **2020**, *37*, 391–396.
41. Yang, H.L.; Chen, J.; Zhang, D.L.; Wang, W.; Cui, H.M.; Liu, Y. Kinetics of cerium(IV) and fluoride extraction from sulfuric solutions using bifunctional ionic liquid extractant (Bif–bi~nctional extractant Bit–ILE)[$A_{336}$][$P_{204}$]. *Trans. Nonferrous Met. Soc. China* **2014**, *24*, 1937–1945. [CrossRef]
42. Xie, L.L.; Favre-Reguillon, A.; Pellet-Rostaing, S.; Wang, X.X.; Fu, X.Z.; Estager, J.; Vrinat, M.; Lemaire, M. Selective extraction and identification of neutral nitrogen compounds contained in straight-run diesel feed using chloride based ionic liquid. *Ind. Eng. Chem. Res.* **2008**, *47*, 8801–8807. [CrossRef]
43. Tang, J.; Liang, S.W.; Dong, B.; Li, Y.; Yao, S. Extraction and quantitative analysis of tropane alkaloids in *Radix physochlainae* by emulsion liquid membrane with tropine–based ionic liquid. *J. Chromatogr. A* **2019**, *1583*, 9–18. [CrossRef]
44. Gao, S.; Fang, S.; Song, R.; Chen, X.; Yu, G. Extractive denitrogenation of shale oil using imidazolium ionic liquids. *Green Energy Environ.* **2020**, *5*, 173–182. [CrossRef]
45. Perez, A.M.G.; Rojas, M.F.; Díaz, L.A.C. Ionic liquid [BMIM][Cl] immobilized on cellulose fibers from pineapple leaves for desulphurization of fuels. *Adv. Mater. Lett.* **2019**, *10*, 334–340. [CrossRef]
46. Chen, C.; Li, X.; Yan, X.; Tian, M. Solid–phase extraction of aristolochic acid I from natural plant using dual ionic liquid–immobilized zif-67 as sorbent. *Separations* **2021**, *8*, 22. [CrossRef]
47. Chen, C.; Feng, X.T.; Yao, S. Ionic liquid–multi walled carbon nanotubes composite tablet for continuous adsorption of tetracyclines and heavy metals. *J. Clean. Prod.* **2021**, *286*, 124937. [CrossRef]
48. Hampel, S.; Kunze, D.; Haase, D.; Krmer, K.; Büchner, B. Carbon nanotubes filled with a chemotherapeutic agent: A nanocarrier mediates inhibition of tumor cell growth. *Nanomedicine* **2008**, *2*, 175–182. [CrossRef]
49. Zhu, D.Z.; Sun, D.M.; Wang, S.L.; Sun, X.Y.; Ni, Y.M. Absorption spectra analysis in the degradation process of quinoline in aqueous solution by VUV lights. *Spectrosc. Spectr. Anal.* **2009**, *29*, 1933–1936.
50. Li, H.P.; Wang, W.Y. Determination of the phenylamine concentration in waste water by ultraviolet differential spectrophotometry method. *Environ. Monit. China* **2011**, *27*, 28–29.
51. Christopher, S.F.; William, H.B.; Brent, L.I.; Eric, A.; Christopher, S.F. *Organic Chemistry*, 8th ed.; Brooks Cole: Boston, MA, USA, 2018.

*Article*

# Gum Arabic-Magnetite Nanocomposite as an Eco-Friendly Adsorbent for Removal of Lead(II) Ions from Aqueous Solutions: Equilibrium, Kinetic and Thermodynamic Studies

Ismat H. Ali [1,*], Mutasem Z. Bani-Fwaz [1], Adel A. El-Zahhar [1,2], Riadh Marzouki [1,3], Mosbah Jemmali [4,5] and Sara M. Ebraheem [6]

1. Department of Chemistry, College of Science, King Khalid University, Abha 61413, Saudi Arabia; mbanifawaz@kku.edu.sa (M.Z.B.-F.); elzahhar@kku.edu.sa (A.A.E.-Z.); rmarzouki@kku.edu.sa (R.M.)
2. Department of Nuclear Chemistry, Egyptian Atomic Energy Authority, Cairo 13759, Egypt
3. Laboratory of Materials, Crystal Chemistry and Applied Thermodynamics, LR15ES01, Faculty of Sciences of Tunis, University of Tunis El Manar, Tunis 1608, Tunisia
4. Faculty of Science, University of Sfax, LSME, BP1171, Sfax 3018, Tunisia; jmosbah@yahoo.fr
5. Department of Chemistry, College of Science and Arts, Ar-Rass, Qassim University, P.O. Box 53, Buraydah 51921, Saudi Arabia
6. Department of Chemistry, College of Science and Arts, King Khalid University, Saratabida 61914, Saudi Arabia; saraabdulgader7@gmail.com
* Correspondence: ismathassanali@gmail.com

**Abstract:** In this study, a gum Arabic-magnetite nanocomposite (GA/MNPs) was synthesized using the solution method. The prepared nanocomposite was characterized by Fourier transform infrared spectroscopy (FTIR), scanning electron microscopy (SEM), transmission electron microscopy (TEM), X-ray diffraction (XRD), vibrating sample magnetometer (VSM), and thermogravimetric analysis (TGA). The prepared composite was evaluated for the adsorption of lead(II) ions from aqueous solutions. The controlling factors such as pH, contact time, adsorbent dose, initial ion concentration, and temperature were investigated. The optimum adsorption conditions were found to be 0.3 g/50 mL, pH = 6.00, and contact time of 30 min. The experimental data well fitted the pseudo-second-order kinetic model and the Langmuir isotherm model. The maximum adsorption capacity was determined as 50.5 mg/g. Thermodynamic parameters were calculated postulating an endothermic and spontaneous process and a physio-sorption pathway.

**Keywords:** adsorption; gum Arabic; magnetite; nano-composite; lead(II)

## 1. Introduction

There has been growing anxiety regarding public health and ecological contamination problems related to the presence of heavy metals. The causes of accumulation of heavy metals in soil and water have increased intensely to involve agriculture, the industrial sector, mining, drug manufacture, and many other activities [1]. Amongst the heavy metals that can potentially lead to severe health issues are Pb, Cd, As, Cr, and Hg, and hence have received special attention [2]. One of the major sources of heavy metals in drinking and wastewater are household chemicals and industrial release. Heavy metals in all types of water are typically known to exist as inorganic complexes.

Several procedures for heavy metal removal from water have been established. Among these are adsorption, ion exchange, electrocoagulation, ion-exchange, and precipitation [3–5]. Adsorption is considered one of the most important techniques because it is a low-cost, effective, and simple technique [6]. One of the most toxic heavy metals, lead, has drawn the most concern and attention, since it is extremely poisonous and can cause significant problems to human health [2,7–9]. Lead is usually present in two oxidation states of +2 and +4. The most widespread and concerning species of lead is $Pb^{2+}$ [10].

A wide spectrum of adsorbents have been reported in the literature for the removal of heavy metals from wastewater. Among these, composites are promising systems for the removal of heavy metals from water. Composites usually have large surface areas that increase their efficiency [11]. GA is a natural substance that has many applications. It is extensively used in the food, cosmetic, and pharmaceutical industries. It is also used as an additive and emulsifier. In some societies, GA is commonly used as a treatment for some chronic diseases such as diabetes mellitus [12]. GA is a combination of inorganic salts and polysaccharides. The inorganic salts are usually composed of K, Ca, and Mg. The polysaccharide portion is composed of repeated units β-d-galactopyranosyl and glucuronic acid [13].

The use of gum Arabic-magnetite has been reported for the removal of Cu(II) ions from aqueous solutions. The maximum adsorption capacity was determined as 38.5 mg/g and the adsorption process was found to be endothermic [14]. It is also reported that GA/MNPs have been used to adsorb methylene blue dye from synthetic wastewater. Results showed that the maximum adsorption capacity was 8.8 mg/g and the adsorption process followed the Langmuir isotherm model [15]. Removal of Pb(II) ions from water was reported using several types of nanocomposites such as a polymer-based graphene oxide nanocomposite [16], nanocomposite of ZnO with montmorillonite [17], and cellulose acetate/titanium oxide nanocomposite [18].

In this study, a gum Arabic-magnetite composite (GA/MNPs) was synthesized, characterized, and assessed by several spectroscopic and analytical methods as an adsorbent for Pb(II) ions from synthetic wastewater. The GA/MNP composite is a partially bio-based material and has unique properties. Furthermore, GA is a low cost and available material. It is reported that the mixing of the magnetite nanoparticles with GA improves particle stability in aqueous suspensions and resulted in the formation of smaller agglomerates compared to the untreated samples [19]. These properties make it appropriate for many applications such as the removal of pollutants from water. The composite was prepared by the solution method and characterized using several techniques such as Fourier transform infrared spectroscopy (FTIR), X-ray diffraction (XRD), scanning electron microscopy (SEM), transmission electron microscopy (TEM), and thermogravimetric analysis (TGA).

## 2. Materials and Methods

### 2.1. Materials

$FeCl_2 \cdot 4H_2O$ (>99%) and $FeCl_3 \cdot 6H_2O$ (>99%) used for MNP preparation were obtained from Sigma-Aldrich (Saint Louis, MO, USA), $NH_3 \cdot H_2O$ ($NH_3$ content, 28–30%) and GA was purchased from Sigma-Aldrich.

### 2.2. Preparation of GA/MNPs

First, MNPs were prepared via co-precipitation as previously reported [20]. Briefly, 3.0 g GA was suspended in 100 mL of distilled water, then treated with 100 mL ferric-ferrous solution containing 0.01 M ferrous chloride tetrahydrate and 0.02 M of anhydrous ferric chloride (stoichiometric ratio of 1:2). The reaction mixture was titrated very slowly (150 rpm) with ammonium hydroxide solution (9 M) until its pH reached 12. The dark composite (GA/$Fe_3O_4$) was washed using distilled water until the pH of the washing solution was neutral. The produced composite was collected by a magnet and air dried at 70 °C for 12 h.

### 2.3. Characterization of the Prepared Composite

Fourier transform infrared spectra of GA, GA/MNPs, and GA/MNPs-Pb(II) were analyzed using a NICOLET 6700 Thermo Scientific spectrometer (Thermo Fisher Scientific, Waltham, MA, USA). The materials' morphologies were analyzed via scanning electron microscopy (SEM-FEI; Quanta 200, Thermo Fisher Scientific, Waltham, MA, USA) and transmittance electron microscopy (TEM) using JEM-2100 (JEOL, Boston, MA, USA) at 200 kV. The magnetic behavior of the sample was studied using a SQUID-Vibrating sample magnetometer (SVSM, Quantum design. Akron, OH, USA). The material's thermal stability was analyzed using the TGA-50H thermal analyzer (Shimadzu, Kyoto, Japan) with a temperature range of 25–600 °C with a heating rate of 10 °C/min under a nitrogen atmosphere.

### 2.4. Adsorption Experiments

Due to the easiness of the process and the complete control of all factors, adsorption experiments were performed using the batch technique. All variables affecting the adsorption efficiency such as pH of the media (1.00–11.00), adsorbent mass (0.3 g/50 mL), contact time (5–60 min), initial Pb(II) concentration (50–300 mg/L), and temperature (298–328 K) were studied by changing one variable when all other factors (pH, adsorbent mass, contact time, and initial Pb(II) concentration) were kept constant. Each experiment was repeated three times to ensure the reproducibility of the results. Data showing standard deviation greater than 5% were rejected.

The removal percentage (R%) of Pb(II) ions from aqueous solutions can be calculated using Equation (1):

$$R\% = \frac{C_o - C_e}{C_o} \times 100 \qquad (1)$$

where $C_o$ and $C_e$ are the initial and equilibrium concentration of Pb(II) ions.

### 2.5. Kinetic Studies

The kinetic behavior of the adsorption process was investigated using various kinetic models. Equation (2) gives the Lagergren pseudo-first-order model [7].

$$\ln(q_e - q_t) = \ln q_e - k_1 t \qquad (2)$$

The pseudo-second-order equation is presented in Equation (3):

$$\frac{t}{q_e} = \frac{1}{k_2 q_e^2} + \frac{t}{q_e} \qquad (3)$$

where $q_e$ and $q_t$ (mg/g) are the adsorbed amounts of Pb(II) ions by 0.30 g of the composite at equilibrium and time t, respectively. $k_1$ (min$^{-1}$) is the pseudo-first-order constant for the adsorption, $k_2$ (g mg$^{-1}$ min$^{-1}$) is the pseudo-second-order constant.

The intra-particle diffusion kinetic model is displayed by Equation (4)

$$q_t = k_{id} t^{1/2} + I \qquad (4)$$

where I is a boundary layer thickness constant (mg/g) and $k_{id}$ is the diffusion rate constant (mg/g. min).

Elovich model was also used to explore the kinetics of the adsorption process. The linear equation of this model is given in Equation (5):

$$q_t = \left(\frac{1}{\beta}\right) \ln(\alpha\beta) + \left(\frac{1}{\beta}\right) \ln t \qquad (5)$$

where $\alpha$ and $\beta$ are Elovich constants; $q_t$ (mg/g) is the adsorbed quantity of Pb(II) ions onto GA/MNPs; and t (min) is the time.

### 2.6. Adsorption Isotherms Models

Among the various isotherm models, four were adopted in this work: Langmuir, Freundlich, Temkin, and Dubinin-Radushkevich.

Equation (6) illustrates the Langmuir isotherm model.

$$\frac{C_e}{q_e} = \frac{C_e}{q_{max}} + \frac{1}{b\, q_{max}} \qquad (6)$$

where $q_e$ is the quantity of Pb(II) ions removed (mg/g); $C_e$ is the remaining Pb(II) ion concentration (mg/L); $q_{max}$ is the maximum adsorption capacity of the composite (mg/g); and b is the Langmuir constant (L/mg).

$$R_L = \frac{1}{1 + b\, C_o} \quad (7)$$

where $C_o$ is the maximum initial concentration of Pb(II) ions.

The linear Freundlich model is presented by Equation (8)

$$\ln q_e = \ln k_f + n \ln C_e \quad (8)$$

where $k_f$ (mg/g)/(mg/L) and n (dimensionless) are the Freundlich constants.

The Temkin model was also investigated and is given by Equation (9):

$$q_e = B \ln A + B \ln C_e \quad (9)$$

where T is the temperature in Kelvin; $B = (RT)/b_t$ and R is the gas constant; and $b_t$ is a constant associated with the adsorption heat (J/mol). A is the equilibrium binding constant corresponding to the maximal energy of binding.

The Dubinin–Radushkevich (DR) model was exploited in this study. The DR equation is displayed in Equation (10):

$$\ln q_e = \ln q_m - \beta\, \varepsilon^2 \quad (10)$$

where $\varepsilon$ is represented by Equation (11)

$$\varepsilon = RT \ln\left(1 + \frac{1}{C_e}\right) \quad (11)$$

where $\beta$ is a constant related to the adsorption free energy, and $q_m$ (mg/g) is the theoretical saturation capacity obtained from the DR model.

### 2.7. Reusability of the Composite

Desorption and reactivation of the adsorbent is a crucial concern, especially from the cost-effective point of view. In this study, GA/MNPs were reactivated by rinsing the used composite with deionized water and then treated with 0.01 M sulfuric acid under continuous shaking for 2 h. Then, the composite was filtered and washed with deionized water and reused for Pb(II) ion removal. Experiments were performed at the optimum conditions (0.30 g of GA/MNPs per 50 mL of 50 mg/L of Pb(II) solutions, pH = 6.00, contact time = 30 min, and T = 298 K).

## 3. Results

### 3.1. Characterization

#### 3.1.1. FTIR Spectroscopy

The FTIR spectra of GA, GA/MNPs, and GA/MNPs-Pb(II) are given in Figure 1. The spectrum of GA showed the peaks at 3412, 2931, and 1608 cm$^{-1}$ assigned for the –OH stretching, –CH$_2$ stretching vibration [21], and C-O asymmetric stretching vibrations, respectively. The peak at 1420 cm$^{-1}$ is assigned to the wagging vibrations for CH and –CH [22]. The C–O–C linkage appeared at 1073 cm$^{-1}$ [23], where the peak at 603 cm$^{-1}$ could be assigned for the C–H out-of-plane bending vibration [21]. The spectrum of GA/MNPs showed increased intensity for the peak at 1607 cm$^{-1}$ with a very slight shift, indicating the formation of hydrogen bonding between the GA and the MNPs [24]. The GA/MNP-Pb spectrum showed a slight change in the intensities and positions of the major bands as the C=O band at 1600 cm$^{-1}$ and the C–O bond appeared at 1068 and 1420 cm$^{-1}$, reflecting the effect of Pb adsorption.

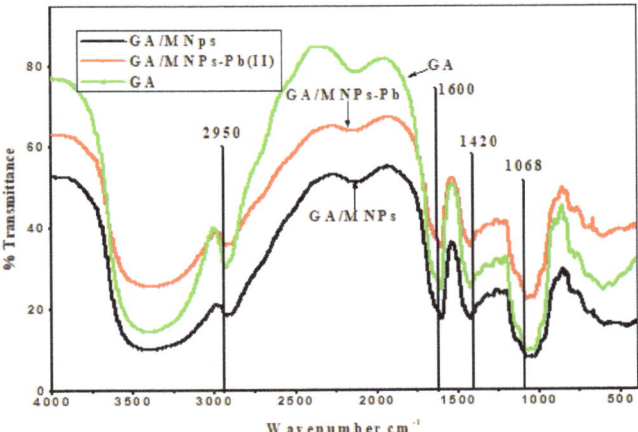

**Figure 1.** FTIR spectra of GA, GA/MNPs, and GA/MNPs−Pb(II).

3.1.2. SEM

The morphologies of the materials are presented in the SEM micrographs in Figure 2. The micrographs show the homogenous dispersion of MNPs within the composite with a small particle size compared to the GA particles. The results also showed evident variation in the particle surface morphology with clear surface pore and homogeneous MNP distribution within the GA/MNPs, which could provide synergistic adsorption for metal ions.

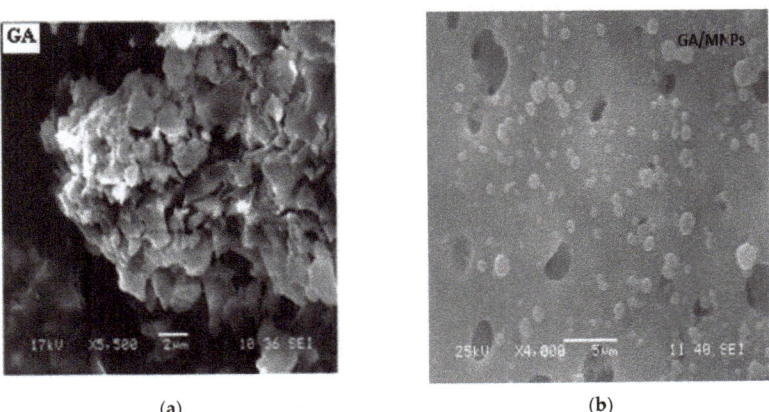

**Figure 2.** Morphology of (**a**) GA alone and (**b**) GA/MNPs.

3.1.3. TEM

The TEM images (Figure 3) confirm that the prepared GA/MNPs is a nanocomposite material and has a core shell structure as the MNPs appeared coated by GA. The preparation procedure may not achieve any particle accumulation. It was reported that magnetic nanoparticles with small particle sizes and reduced accumulation may have essential magnetic properties for many applications [25,26]. The TEM images also revealed well-dispersed magnetite nanoparticle MNPs within the GA with uniform size and shape. The content of magnetite particles and the formation of the core shell structure may indicate high magnetization properties of the GA/MNPs composite.

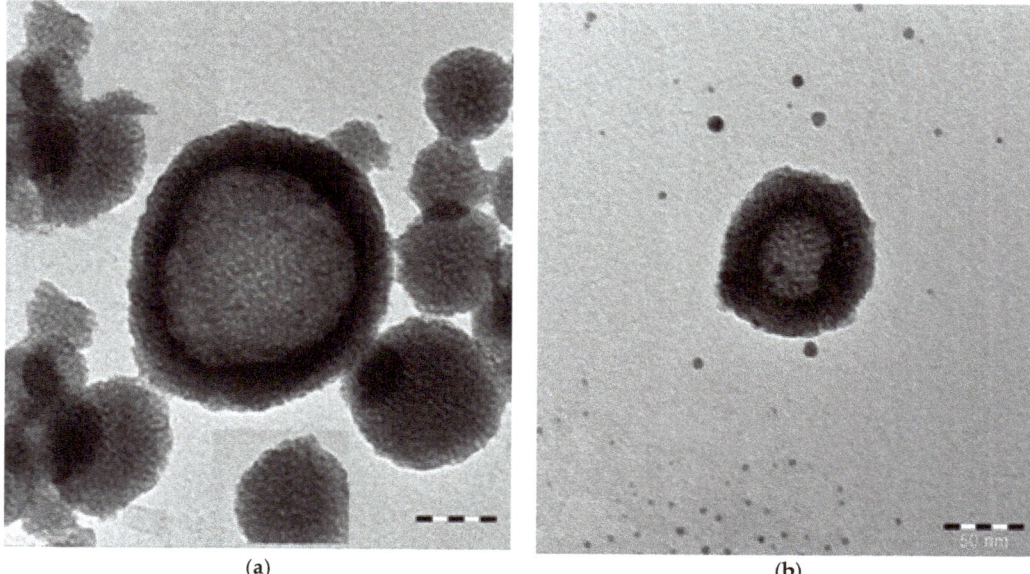

**Figure 3.** TEM images for the (**a**) GA/MNP and (**b**) MNP distribution.

### 3.1.4. XRD

X-ray diffraction pattern provides structural information including the amorphous and crystalline structure of the material. The XRD pattern of the pure GA showed only one diffused peak near 2θ = 20.3° due to its poor crystallinity [22,23]. The XRD image of GA/MNPs is shown in Figure 4. GA/MNPs showed five sharp peaks at 2θ = 30.1°, 35.8°, 43.0°, 57.5°, and 63.2° due to the good crystallinity of GA/MNPs, proving the incorporation of the MNPs into the GA backbone. Moreover, the well crystallized peak at 35.8° proves the existence of $Fe_3O_4$ [25,26].

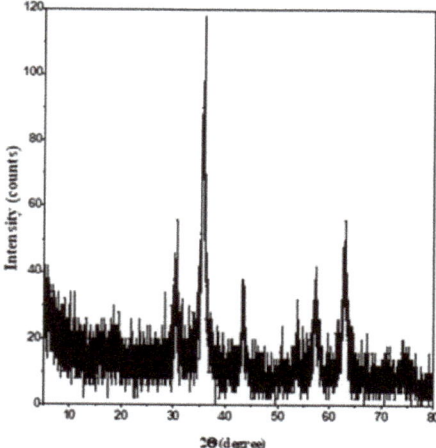

**Figure 4.** XRD pattern for GA/MNPs.

### 3.1.5. TGA

The TGA weight loss percentage is given in Figure 5 for GA and GA/MNPs. The results showed that the inclusion of MNPs within the composite GA/MNPs improved the thermal stability of the produced composite. The weight loss percentage reached about 80% for GA. The first weight loss appeared between 100 to 300 °C for GA, which involved the removal of absorbed water molecules by gradual dehydration. The second weight loss appeared between 300 °C and 450 °C, which could be assigned for the decomposition of hydrocarbons. The TGA of the GA/MNPs showed lower weight loss with delayed thermal degradation. This finding could be due to the formation of an expanded layer on the GA surface and the MNPs, which highly affect the thermal stability of the GA/MNP composite. The MNPs could act as a barrier layer, protect GA from thermal degradation, and decrease the heat transfer within the composite [27].

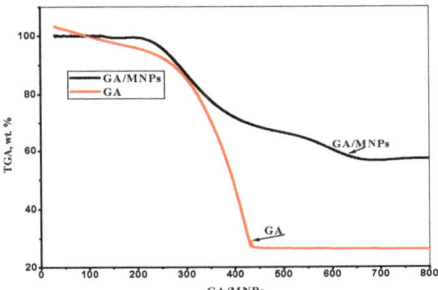

**Figure 5.** TGA analyses for GA and GA/MNPs.

### 3.1.6. Magnetic Properties

The magnetic properties of the GA/MNP composite were investigated using the vibrating sample magnetometer (VSM) technique. The observed magnetic properties of the prepared particles reflected their dispersion/aggregation and morphological properties. Figure 6 indicates that the magnetic properties reflect the inclusion of MNP within the composite adsorbent as polydisperse particles or monodisperse. The observed magnetic properties confirm the formation of particles with consistent MNP distribution within the composite. The adsorbent particles showed superparamagnetic behavior at ambient temperature.

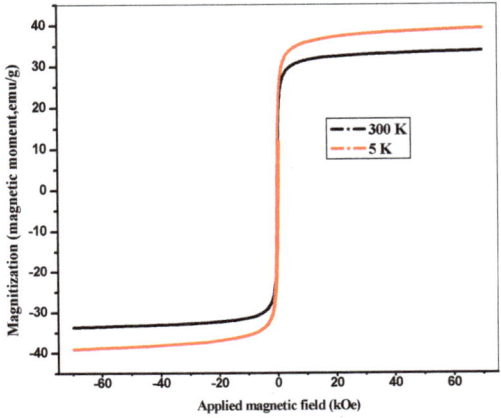

**Figure 6.** VSM analysis of the GA/MNPs.

## 3.2. Adsorption Studies

### 3.2.1. GA Capacity to Adsorb Pb(II) Ions

Several experiments were conducted to examine the ability of GA alone to remove the Pb(II) ions from aqueous solutions. Results showed that the efficiency of GA in all experiments did not exceed 5%.

### 3.2.2. Influence of pH

All adsorption systems are likely to be affected by the pH value of the media, which can control the binding of the adsorbate onto the adsorbent surface. In this work, the GA/MNP composite was tested for the removal of Pb(II) ions from aqueous solutions in the pH range of 1.00 to 11.00. The adsorption profile is presented in Figure 7. As expected, in all studied ranges, the effect of the pH value was profound. The maximum adsorption capacity increased with pH increase, reaching a maximum and then decreasing with pH increase. The low adsorption efficiency at low pH values is most likely due to the competition between $H^+$ ions and $Pb^{2+}$ ions on adsorption sites. As the pH value increases, more carboxylate groups (–COOH) from the composite surface could be dissociated into a negatively charged group (–COO−), leading to high adsorption capacity. The adsorption efficiencies decreased sharply above pH 6.00 because lead hydroxides precipitate above this pH value [28,29].

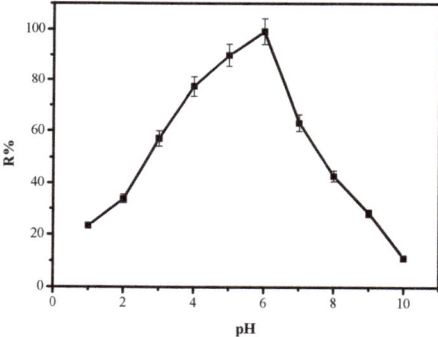

**Figure 7.** Effect of pH on the efficiency of the GA/MNP composite for the removal Pb(II) from an aqueous solution at 298 K.

### 3.2.3. Effect of Contact Time

In all adsorption experiments, contact time is a significant factor as it strongly influences the adsorbent efficiency. Figure 8 shows the results from the experiments performed under different contact time while keeping all of the other factors constant (pH = 6.00, adsorbent mass = 0.30 g, Pb(II) ion concentration = 50 mg/L, and temperature = 298K). It was found that the composite reached the maximum removal efficiency (98.8%) in 30 min. A contact time of 30 min was chosen for all other experiments.

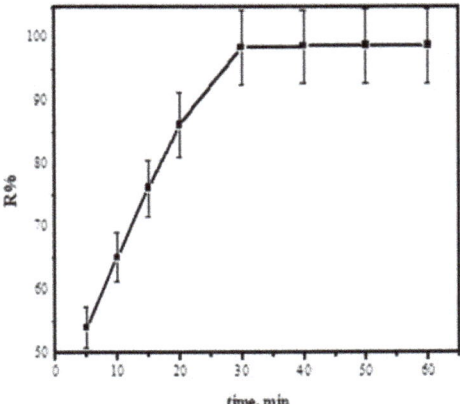

**Figure 8.** Effect of the contact time of removal efficiency of Pb(II) ions by the GA/MNP composite at 298 K.

3.2.4. Kinetic Behavior

The kinetics and mechanism of the Pb(II) ion uptake were assessed by testing the experimental data using various kinetic models.

Pseudo-First and Second-Order Models

The linear plots for both models were exploited to assess the applicability of both models. The kinetic parameters are displayed in Table 1. Based on the obtained correlation coefficient ($R^2$), the experimental data showed well-fitting with the pseudo-second-order model (Figure 9b). The theoretical adsorption capacity obtained from the pseudo-second-order model (55.2 mg/g) is in good agreement with that obtained from the Langmuir model (50.5 mg/g).

**Table 1.** Kinetic data of the adsorption of Pb(II) ions onto GA/MNPs.

| Kinetic Model | Parameters | |
|---|---|---|
| first order | $q_e$ (mg/g) | 25.1 |
|  | $k_1$ (min$^{-1}$) | 0.0792 |
|  | $R^2$ | 0.8595 |
| second order | $q_e$ (mg/g) | 55.2 |
|  | $k_2$ (g/mg·min) | 0.0181 |
|  | $R^2$ | 0.9937 |
| Intra-particle diffusion | $k_{id}$ (mg/g·min) | 4.45 |
|  | I | 23.5 |
|  | $R^2$ | 0.9043 |
| Elovich | A | $30 \times 10^{58}$ |
|  | B | 0.121 |
|  | $R^2$ | 0.9378 |

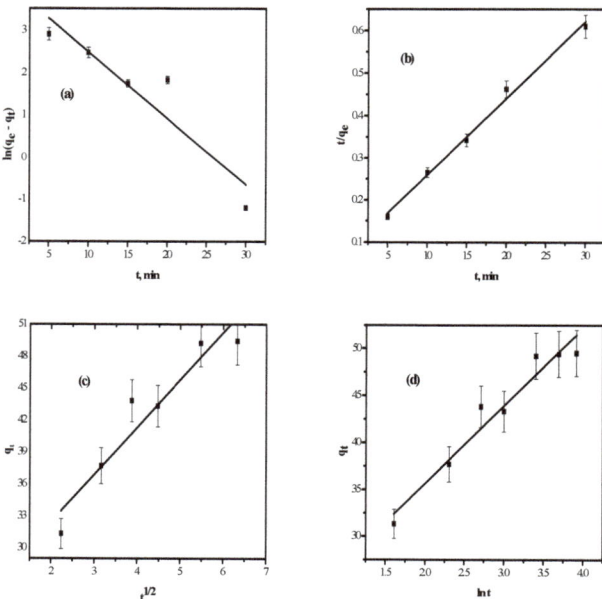

**Figure 9.** Kinetic models for adsorption system of removal of Pb(II) ion using GA/MNPs: (**a**) pseudo−first−order; (**b**) pseudo-second-order; (**c**) intra-particle diffusion; (**d**) Elovich.

Intra-Particle Diffusion Kinetic Model

Equation (4) was applied to study the kinetic behavior of the adsorption process using the intra-particle diffusion model. The $k_{id}$ value was calculated from the slope of Equation (4) and Figure 9c and is displayed in Table 1. The plot of qt against $t^{1/2}$ was curved, showing that numerous processes control the adsorption system. The non-linear portion of the plot is attributable to the effect of boundary layer diffusion and means that the rate constant ($k_{id}$) is controlling the intra-particle diffusion [7].

Elovich Model

The modified form of the Elovich model [30] was also used to investigate the kinetic behavior of adsorption of Pb(II) ions onto GA/MNPs. The Elovich equation [30,31] was frequently exploited to examine the kinetic behavior of the chemical adsorption of gases on solid surfaces. Currently, it was stated [30] that this model can also be used to study the adsorption from aqueous solutions. The importance of α and β constants has not been undoubtedly determined [32]. The plot of $q_t$ against ln t is shown in Figure 9d, and the values of α and β constants as well as the correlation coefficient ($R^2$) are given in Table 1. Results in Table 1 and Figure 9d confirmed that Elovich model did not fit linearly with the experimental data ($R^2$ = 0.9043), indicating that this adsorption system cannot be defined by the Elovich model.

### 3.2.5. Effect of Initial Pb(II) Concentration

Figure 10 illustrates the effect of Pb(II) ion initial concentration on the removal percentage. The efficacy decreased from 98.8% to 62.7% as the initial Pb(II) ion concentration increased from 50 to 300 mg/L. The decrease in adsorption efficiency can be ascribed to a lack of enough adsorbent surface area to gather the obtainable Pb(II) ions from the solution [26]. However, the uptake capacity improved from 12.4 to 47.2 mg/g as the preliminary Pb(II) ion concentration increased from 50 to 300 mg/L.

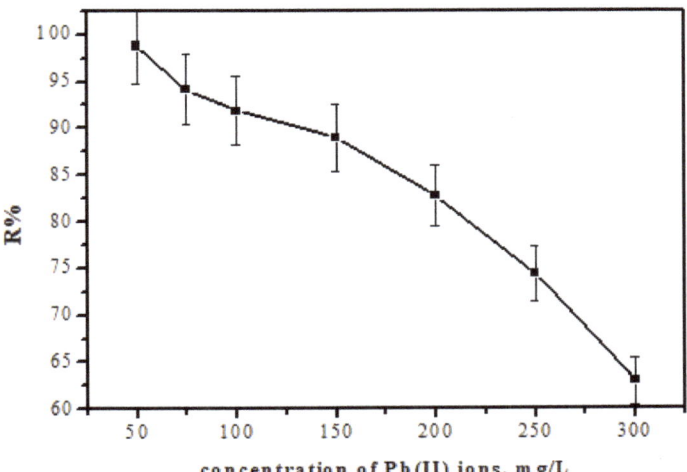

**Figure 10.** Effect of Pb(II) concentration on the efficiency of the GA/MNP composite for the removal of Pb(II) ions from synthetic wastewater at 298 K.

*3.3. Isotherm Models*

The way by which the interface of Pb(II) ions and the adsorption sites of the composite occurs can be understood by adsorption isotherm models [33]. The adsorption isotherms are characterized by certain factors, whose values express the surface properties and affinity of the adsorbent for Pb(II) ion adsorption [34]. In the current study, these models were verified and several isotherms are displayed in Figure 8.

3.3.1. Langmuir Model

Figure 11a illustrates the plot of the Langmuir model. It is clear that the plot exhibited a straight line with the slope expressing the reciprocal of the maximum adsorption capacity and the intercept expressing $1/bq_{max}$. A high b value indicates a more binding affinity between the adsorbate and adsorbent. The parameters obtained from this model are listed in Table 2. It can be deduced from the results shown in Table 2 and Figure 11a that the adsorption system obeyed the Langmuir Model ($R^2 = 0.996$). The obtained $q_{max}$ value was 50.5 mg/g, which was very close to the values found at the optimal pH value and from the second-order kinetic model. The obtained b value (0.871) relative to the reported values [7,26] indicated a high attraction between the GA/MNPs and Pb(II) ions.

Additionally, substantial indication linked to adsorption nature can be obtained from the values of separation factor ($R_L$), which is a significant feature of the Langmuir model. $R_L$ value is usually used to realize the nature of adsorbent/adsorbate affinity. $R_L$ values are determined by Equation (7).

The $R_L$ values designated the adsorption type as follows: (i) unfavorable ($R_L > 1$); (ii) linear ($R_L = 1$); and (iii) favorable ($0 < R_L < 1$) or irreversible ($R_L = 0$). In this study, as shown in Table 2, all $R_L$ values were between 0 and showed that the adsorption system was a favorable process.

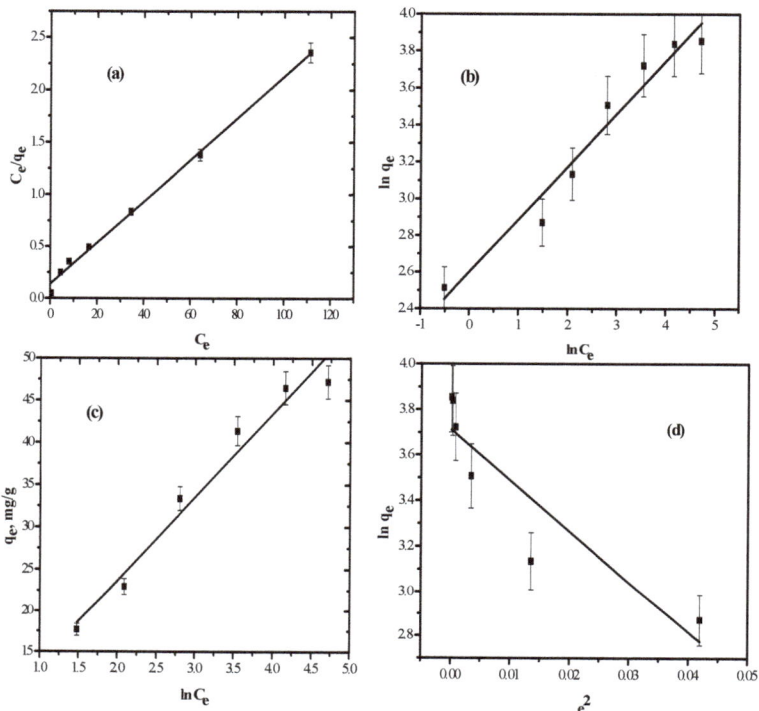

**Figure 11.** Adsorption isotherm: (**a**) Langmuir; (**b**) Freundlich; (**c**) Temkin; (**d**) DR models.

**Table 2.** Parameters of the adsorption isotherm models.

| Adsorption Model | Isotherm Parameters | Values |
|---|---|---|
| Langmuir | $q_{max}$ (mg/g)<br>$R_L$ (L/g)<br>$R^2$ | 50.5<br>0.029<br>0.887 |
| Freundlich | n<br>$K_f$ (mg/g)/(mg/L)<br>$R^2$ | 0.67<br>4.6<br>0.9981 |
| Temkin | A (L/g)<br>$b_t$ (kJ/mol)<br>$R^2$ | 2.1<br>370.3<br>0.899 |
| DR | $\beta$<br>$q_m$ (mg/g)<br>$E_f$ (kJ/mol)<br>$R^2$ | 22.4<br>40.4<br>0.15<br>0.7864 |

3.3.2. Freundlich Model

The values of the parameters extracted from Equation (8) and Figure 11b are displayed in Table 2. It was found that the n value was greater than 1, showing a favorable adsorption process [35]. This is in good agreement with the conclusions obtained from the $R_L$ values calculated from the Langmuir model.

### 3.3.3. Temkin Isotherm

In the Temkin isotherm model, the heat of adsorption of all adsorbed particles in a single layer is expected to decrease sharply with the adsorbent surface coverage due to a reduction in the adsorbent/adsorbate interactions.

The values of the constants A and B are determined from the slope and intercept of Equation (9) and Figure 11c, as shown in Table 2. The $b_t$ value was found to be 251 J/mol, representing a physical adsorption system [36].

### 3.3.4. Dubinin–Radushkevich Model

The adsorption energy of the system can be deduced from the Dubinin–Radushkevich (DR) model. This model is usually exploited to obtain information about the adsorption mechanism [32]. The DR model is usually proposed only for both homogeneous and heterogonous adsorption systems.

Values of $q_m$, $\beta$, and $R^2$ were determined using Figure 11d and displayed in Table 2. The free energy of adsorption $E_f$ is the free energy change when one mole of adsorbate is moved to the solid adsorbent surface and is determined using Equation (12):

$$E_f = \frac{1}{\sqrt{2\beta}} \qquad (12)$$

The type of the adsorption process can be obtained from the value of $E_f$. The adsorption system is considered chemical when the $E_f$ value lies between 8.0 and 16.0 kJ/mol and physical when $E_f$ is smaller than 8.0 kJ/mol. The value of $E_f$ displayed in Table 2 proves that the adsorption system is physical in nature. This conforms with the conclusions obtained from the Elovich kinetic model [7].

### 3.4. Effect of Temperature

The influence of temperature on the efficiency of the adsorption system was examined by conducting various experiments in the temperature range from 298 to 328 K while keeping the other optimal experimental circumstances constant (pH = 6.00, adsorbent mass = 0.3 g, contact time = 30 min, and Pb(II) ion concentration = 150 mg/L). Results shown in Figure 12 show that the adsorption efficacy is directly proportional to temperature. This can be attributed to the activation of more adsorption sites on the adsorbent surface and the increase in the speed of Pb(II) ions at elevated temperatures.

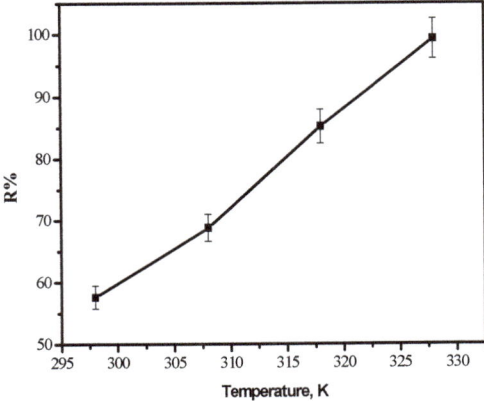

**Figure 12.** Effect of temperature on the efficacy of the removal of Pb(II) ions from synthetic wastewater by the GA/MNP composite.

## 3.5. Thermodynamic Parameters

Some thermodynamic activation parameters, viz., entropy change of activation ΔS, enthalpy change of activation ΔH, and free energy change of activation ΔG, were determined using Equations (13)–(15):

$$K_D = \frac{C_o}{C_e} \quad (13)$$

$$\Delta G° = -RT \ln K_c \quad (14)$$

$$\ln K_D = \frac{\Delta S°}{R} - \frac{\Delta H°}{RT} \quad (15)$$

where $K_D$ is the distribution coefficient, R is the universal gas constant (8.314 JK$^{-1}$ mol$^{-1}$) and T(K) is the temperature.

The plot of ln $K_D$ against 1/T as shown in Figure 13 was exploited to determine the values of both ΔH° and ΔS°. The results are displayed in Table 3. The obtained positive value of ΔH° proves the endothermic nature of the adsorption system, while the positive value of ΔS° suggests the spontaneous nature of the Pb(II) ion removal process. Furthermore, the high value of ΔS° indicates an irregular and permeable adsorbent. Moreover, in all experiments, ΔG° showed negative values that increased by rising temperature, suggesting a favorable adsorption process at elevated temperatures that conforms with the observation that the adsorption efficiency increases with temperature. It is reported that the adsorption process is considered physical when the adsorption free energy change (ΔG°) ranges between −20 and 0 kJ/mol and chemical when ΔG° ranges between −400 and −80 kJ/mol. Thus, the adsorption nature for the current process is physical [7,37–39].

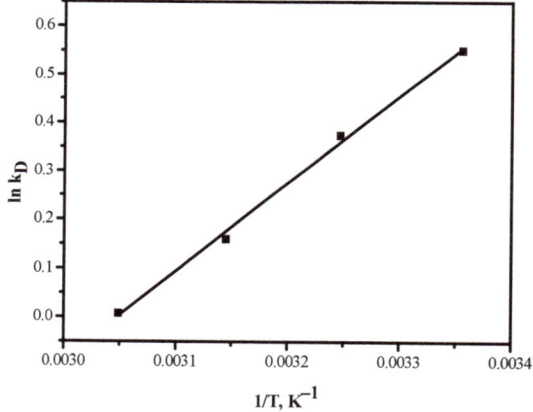

**Figure 13.** Plot of ln $k_D$ against 1/T.

**Table 3.** Thermodynamic parameters of the adsorption of Pb(II) ions on the GA/MNP surface.

| T, K | ΔG, kJ/mol | ΔS, J/mol K | ΔH, kJ/mol |
|---|---|---|---|
| 298 | −6.1 | | |
| 308 | −6.5 | 26.6 | 15.0 |
| 318 | −6.8 | | |
| 328 | −7.3 | | |

## 3.6. Selectivity

The selective adsorption investigations were performed using an aqueous solution containing a mixture of multi-interfering ions including Cr(III), Cu(II), and Ni(II). The competing metal ions were selected based on either similar mode coordination or charge, besides the co-occurrence of such solutions within the real effluent from various industrial sectors. Results displayed in Table 4 show that the efficiency of removal of Pb(II) by GA/MNPs slightly decreased with the increase in concentration of these interfering ions, indicating a moderate selectively trend of GA/MNPs toward Pb(II) ions.

**Table 4.** The removal efficiency of Pb(II) by GA/MNPs in the presence of some interfering ions.

| Existing Metal Ion | Concentration of Each Metal | R% of Pb(II) Ions |
|---|---|---|
| Cr(III), Ni(II), Cu(II) | 5.0 mg/L | 92 |
| Cr(III), Ni(II), Cu(II) | 10.0 mg/L | 88 |
| Cr(III), Ni(II), Cu(II) | 15 mg/L | 84 |

### 3.7. Reusability of the Composite

The adsorption and reactivation processes were performed four times using the same composite. Figure 14 shows the removal efficiency in each case. The removal efficiency decreased gradually from 95.6 when the GA/MNPs were used for the first time until it reached, in the last cycle, 81% of its initial efficiency. These results confirm the high reusability performances of the composite at least three times.

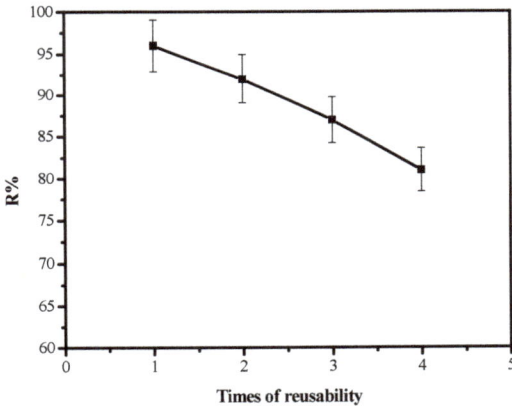

**Figure 14.** Reusability of the GA/MNPs for the removal of Pb(II) ions.

### 3.8. Comparison of GAMNPs with Other Materials Used for Pb(II) Removal

The maximum uptake capacity and some experimental optimum conditions for the adsorption process of Pb(II) ions from aqueous solutions are shown in Table 5. The $q_{max}$ value of the current study is good, relative to the other adsorbents. The differences between the reported values of $q_{max}$ can be attributed to the variances in adsorbent properties such as functional groups, porosity, and surface area. The values displayed in Table 4 indicate that GA/MNPs are an efficient adsorbent because it took only 30 min to remove 50 mg/L of Pb(II) ions from the aqueous solutions. Additionally, GA/MNPs achieved the shortest time to remove the Pb(II) ions. It is also obvious that the required GA/MNP mass to remove the same concentration of Pb(II) ions is less or equal to most of the reported materials except $Cu_{0.5}Mg_{0.5}Fe_2O_4$, nano-silica hollow spheres, *S. Oleasea* bark, and zeolite, as shown in Table 5. Table 5 also shows that the optimum pH value used in this study was in good agreement with most of the reported values. It is also clear that the maximum adsorption in most cases occurred at pH a range of 5.50–6.00, which is in good agreement with this study.

Table 5. Comparison between the maximum uptake capacity and the optimum conditions for various adsorbents.

| Material | $q_{max}$ (mg/g) | Experimental Optimum Conditions | | | | | Ref. |
|---|---|---|---|---|---|---|---|
| | | pH | T (K) | Contact Time (min) | $C_o$ (mg/L) | Adsorbent Mass (g/L) | |
| *Juniperus procera* AC | 30.3 | 4.6 | 298 | 100 | 50 | 8 | [7] |
| Apricot stone | 21.4 | 6.0 | 293 | 300 | 50 | 2 | [40] |
| *Oryza sativa* husk | 6.1 | 8.0 | 333 | 70 | 10 | 12 | [41] |
| Nanocellulose fibers | 9.4 | 5.0 | 298 | 90 | 25 | 8 | [42] |
| Polypyrrole-based AC | 50.0 | 5.5 | 298 | 120 | 100 | 5 | [43] |
| $Cu_{0.5}Mg_{0.5}Fe_2O_4$ | 57.7 | 6.0 | 298 | 120 | 10 | 0.1 | [44] |
| Waste tire rubber ash | 22.4 | 6.0 | 303 | 90 | 400 | 2 | [45] |
| Nanosilica hollow spheres | 200.0 | 5.0 | 333 | 40 | 300 | 0.05 | [46] |
| *S. Oleasea* bark | 69.40 | 6.0 | 323 | 60 | 100 | 5.0 | [47] |
| Zeolite | 65.8 | 5.7 | 303 | 50 | 40 | 5.0 | [48] |
| GA/MNPs | 50.5 | 6.0 | 298 | 30 | 50 | 6 | This study |

## 4. Conclusions

In this study, the adsorption characteristics of GA/MNP nanocomposite were evaluated. The investigation revealed that the prepared nanocomposite may be considered as an efficient adsorbent for lead(II) ion removal from an aqueous solution. The removal efficiency was found to be 99.3% at the optimum conditions. The removal % increased with the increase in the adsorbent dosage and the decrease in the adsorbate concentration. The maximum adsorption capacity was noticed to increase with increasing temperature, suggesting that the nature of the adsorption process is endothermic. This was further consolidated by the thermodynamic parameters calculated from experiments carried out at various temperatures. The adsorption process was found to follow the Langmuir isotherm model and pseudo-second-order kinetics. Moreover, the efficiency of the removal of Pb(II) ions was noticed to decrease slightly in the presence of interfering ions such as Cr(III), Ni(II), and Cu(II) ions. The composite was regenerated and used successfully at least three times to remove the Pb(II) ions from the synthetic wastewater.

**Author Contributions:** Conceptualization, I.H.A. and A.A.E.-Z.; Methodology, I.H.A., A.A.E.-Z., and S.M.E.; Formal analysis, A.A.E.-Z. and R.M.; Investigation, S.M.E., R.M., and M.J.; Resources, M.Z.B.-F.; Data curation, I.H.A.; Writing—original draft preparation, I.H.A.; Writing—review and editing, M.Z.B.-F.; Visualization, A.A.E.-Z. and M.J.; Supervision, I.H.A.; Project administration, M.Z.B.-F.; Funding acquisition, M.Z.B.-F. All authors have read and agreed to the published version of the manuscript.

**Funding:** The authors extend their appreciation to the Deanship of Scientific Research at King Khalid University for funding this work through a research group program under grant number R.G.P.1/230/41.

**Data Availability Statement:** Data are contained within the article.

**Conflicts of Interest:** The authors declare no conflict of interest.

## References

1. Malkoc, E.; Nuhoglu, Y.; Abali, Y. Cr (VI) Adsorption by Waste Acorn of *Quercus ithaburensis* in Fixed Beds: Prediction of Breakthrough Curves. *Chem. Eng. J.* **2006**, *119*, 61–68. [CrossRef]
2. Gebretsadik, H.; Gebrekidan, A.; Demlie, L.; Suvarapu, N. (Reviewing editor)Removal of heavy metals from aqueous solutions using *Eucalyptus Camaldulensis*: An alternate low cost adsorbent. *Cogent Chem.* **2020**, *6*, 1720892. [CrossRef]
3. Eltayeb, N.; Khan, A. Design and Preparation of a New and Novel Nanocomposite with CNTs and Its Sensor Applications. *J. Mater. Res. Technol.* **2019**, *8*, 2238–2246. [CrossRef]
4. Khan, A.; Asiri, A.; Khan, A.; Abdul Rub, M.; Azum, N.; Rahman, M.M.; Khan, S.B.; Alamry, K.A.; Ab Ghani, S. Sol-gel synthesis and characterization of conducting polythiophene/tin phosphate nano tetrapod composite cation-exchanger and its application as Hg (II) selective membrane electrode. *J. Sol-Gel Sci. Technol.* **2013**, *65*, 160–169. [CrossRef]
5. Demirbaş, Ö.; Çalımlı, M.H.; Demirkan, B.; Alma, M.H.; Nas, M.S.; Khan, A.; Asiri, A.M.; Sen, F. The Kinetic Parameters of Adsorption of Enzymes Using Carbon-Based Materials Obtained from Different Food Wastes. *BioNanoScience* **2019**, *9*, 749–757. [CrossRef]
6. Al Mesfer, M.K.; Danish, M.; Ali, I.H.; Khan, M.I. Adsorption behavior of molecular sieve 3 Å and silica gel for $CO_2$ separation: Equilibrium, breakthrough and mass transfer zone. *Heat Mass Transf.* **2020**, *56*, 3243–3259. [CrossRef]
7. Ali, I.H.; Al Mesfer, M.K.; Khan, M.I.; Danish, M.; Alghamdi, M.M. Exploring Adsorption Process of Lead (II) and Chromium (VI) Ions from Aqueous Solutions on Acid Activated Carbon Prepared from *Juniperus procera* Leaves. *Processes* **2019**, *7*, 217. [CrossRef]
8. Jun, B.; Her, N.; Park, C.M.; Yoon, Y. Effective removal of Pb(ii) from synthetic wastewater using Ti3C2Tx MXene. *Environ. Sci. Water Res. Technol.* **2020**, *6*, 173–180. [CrossRef]
9. Tao, Y.; Zhang, C.; Lü, T.; Zhao, H. Removal of Pb (II) Ions from Wastewater by Using Polyethyleneimine-Functionalized $Fe_3O_4$ Magnetic Nanoparticles. *Appl. Sci.* **2020**, *10*, 948. [CrossRef]
10. Greenwood, N.N.; Earnshaw, A. Chemistry of the elements. In *Applied Organometallic Chemistry*, 2nd ed.; Wiley & Sons: Oxford, UK, 1998; Volume 9, p. 3365.
11. Akartasse, N.; Mejdoubi, E.; Razzouki, B.; Azzaoui, K.; Jodeh, S.; Hamed, O.; Ramdani, M.; Lamhamdi, A.; Berrabah, M.; Lahmass, I.; et al. Natural product based composite for extraction of arsenic (III) from waste water. *Chem. Cent. J.* **2017**, *11*, 33–45. [CrossRef]
12. Montenegro, M.A.; Boiero, M.L.; Valle, L.; Borsarelli, C.D. *Gum Arabic: More Than an Edible Emulsifier*, 1st ed.; IntechOpen: London, UK, 2012; pp. 1–26.
13. Abdul Khalil, H.P.S.; Chong, E.W.N.; Owolabi, F.A.T.; Asniza, M.; Tye, Y.Y.; Rizal, S.; Nural Fazita, M.R.; Haafiz, M.K.M.; Nurmiati, Z.; Paridah, M.T. Enhancement of basic properties of polysaccharide-based composites with organic and inorganic fillers: A review. *J. Appl. Polym. Sci.* **2019**, *136*, 47251–47270. [CrossRef]
14. Banerjee, S.S.; Chen, D.-H. Fast removal of copper ions by gum arabic modified magnetic nano-adsorbent. *J. Hazard. Mater.* **2007**, *147*, 792–799. [CrossRef] [PubMed]
15. Alzahrani, E. Gum Arabic-coated magnetic nanoparticles for methylene blue removal. *Int. J. Innov. Res. Sci. Eng. Technol.* **2014**, *3*, 15118–15129. [CrossRef]
16. Musico, Y.L.F.; Santos, C.M.; Dalida, M.L.P.; Rodrigues, D.F. Improved removal of lead (ii) from water using a polymer-based graphene oxide nanocomposite. *J. Mater. Chem. A* **2013**, *1*, 3789–3796. [CrossRef]
17. Sani, H.A.; Ahmad, M.B.; Hussein, M.Z.; Ibrahim, N.A.; Musa, A.; Saleh, T.A. Nanocomposite of ZnO with montmorillonite for removal of lead and copper ions from aqueous solutions. *Process Saf. Environ. Prot.* **2017**, *109*, 97–105. [CrossRef]
18. Alebel Gebru, K.; Das, C. Removal of Pb (II) and Cu (II) ions from wastewater using composite electrospun cellulose acetate/titanium oxide ($TiO_2$) adsorbent. *J. Water Process. Eng.* **2017**, *16*, 1–13. [CrossRef]
19. Williams, D.N.; Gold, K.A.; Holoman, T.R.P.; Ehrman, S.H.; Wilson, O.C., Jr. Surface modification of magnetic nanoparticles using gum Arabic. *J. Nanopart. Res.* **2006**, *8*, 749–753. [CrossRef]
20. Keshk, M.A.S.S.; El-Zahhar, A.A.; Youssef, M.S.A.; Bondock, S. Novel synthesis of flame-retardant magnetic nanoparticles/hydroxy acid cellulose-6-phosphate composite. *Mater. Res. Express* **2019**, *6*, 85310. [CrossRef]
21. Sharma, G.; Kumar, A.; Devi, K.; Sharma, S.; Naushad, M.; Ghfar, A.A.; Ahamad, T.; Stadler, F.J. Guar gum-crosslinked-Soya lecithin nanohydrogel sheets as effective adsorbent for the removal of thiophanate methyl fungicide. *Int. J. Biol. Macromol.* **2018**, *114*, 295–305. [CrossRef]
22. Farooq, M.; Sagbas, S.; Sahiner, M.; Siddiq, M.; Turk, M.; Aktas, N.; Sahiner, N. Synthesis, characterization and modification of Gum Arabic microgels for hemocompatibility and antimicrobial studies. *Carbohydr. Polym.* **2017**, *156*, 380–389. [CrossRef]
23. Keshk, S.M.A.S.; El-Zahhar, A.A.; Alsulami, Q.A.; Jaremko, M.; Bondock, S.; Heinze, T. Synthesis, characterization and ampyrone drug release behavior of magnetite nanoparticle/2,3-dialdehyde cellulose-6-phosphate composite. *Cellulose* **2020**, *27*, 1603. [CrossRef]
24. Zhuang, J.; Li, M.; Pu, Y.; Ragauskas, A.J.; Yoo, C.G. Observation of Potential Contaminants in Processed Biomass Using Fourier Transform Infrared Spectroscopy. *Appl. Sci.* **2020**, *10*, 4345. [CrossRef]
25. Aguilera, G.; Berry, C.; West, R.; Gonzalez-Monterrubio, E.; Angulo-Molina, A.; Arias-Carrion, O.; Angel Mendez-Rojas, M. Carboxymethyl cellulose coated magnetic nanoparticles transport across a human lung microvascular endothelial cell model of the blood–brain barrier. *Nanoscale Adv.* **2019**, *2*, 671–685. [CrossRef]
26. Bhakat, D.; Barik, P.; Bhattacharjee, A. Electrical conductivity behavior of Gum Arabic biopolymer-$Fe_3O_4$ nanocomposite. *J. Phys. Chem. Solids* **2018**, *112*, 73–97. [CrossRef]

27. Hedayati, K.; Goodarzi, M.; Ghanbari, D. Hydrothermal Synthesis of $Fe_3O_4$ Nanoparticles and Flame Resistance Magnetic Poly styrene Nanocomposite. *J. Nanostruct.* **2017**, *7*, 32–39.
28. Amer, M.W.; Ahmad, R.A.; Awwad, A.M. Biosorption of Cu (II), Ni (II), Zn (II) and Pb (II) ions from aqueous solution by Sophora japonica pods powder. *Int. J. Ind. Chem.* **2015**, *6*, 67–75. [CrossRef]
29. Ouyang, D.; Zhuo, Y.; Hu, L.; Zeng, Q.; Hu, Y.; He, Z. Research on the Adsorption Behavior of Heavy Metal Ions by Porous Material Prepared with Silicate Tailings. *Minerals* **2019**, *9*, 291. [CrossRef]
30. Elkhaleefa, A.; Ali, I.H.; Brima, E.I.; Elhag, A.B.; Karama, B. Efficient Removal of Ni (II) from Aqueous Solution by Date Seeds Powder Biosorbent: Adsorption Kinetics, Isotherm and Thermodynamics. *Processes* **2020**, *8*, 1001. [CrossRef]
31. Alfaro-Cuevas-Villanueva, R.; Hidalgo-Vázquez, A.R.; Penagos, C.J.C.; Cortés-Martínez, R. Thermodynamic, Kinetic, and Equilibrium Parameters for the Removal of Lead and Cadmium from Aqueous Solutions with Calcium Alginate Beads. *Sci. World J.* **2014**. [CrossRef]
32. Kithome, M.; Paul, J.W.; Lavkulich, L.M.; Bomke, A.A. Kinetics of Ammonium Adsorption and Desorption by the Natural Zeolite Clinoptilolite. *Soil Sci. Soc. Am. J.* **1998**, *62*, 622–629. [CrossRef]
33. Papageorgiou, S.K.; Katsaros, F.K.; Kouvelos, E.P.; Nolan, J.W.; Deit, H.L.; Kanellopoulos, N.K. Heavy metal sorption by calcium alginate beads from *Laminaria digitata*. *J. Hazard. Mater.* **2006**, *137*, 1765–1772. [CrossRef]
34. Ali, I.H.; Alrafai, H.A. Kinetic, isotherm and thermodynamic studies on biosorption of chromium (VI) by using activated carbon from leaves of *Ficus nitida*. *Chem. Cent. J.* **2016**, *10*, 36. [CrossRef] [PubMed]
35. Mehrmand, N.; Moraveji, M.K.; Parvareh, A. Adsorption of Pb(II), Cu(II) andNi (II) ions from aqueous solutions by functionalised henna powder (*Lawsonia Inermis*); isotherm, kinetic and thermodynamic studies. *Int. J. Environ. Anal. Chem.* **2020**. [CrossRef]
36. El-Naggar, I.M.; Ahmed, S.A.; Shehata, N.E.S.; Fathy, S.M.; Shehata, A. A novel approach for the removal of lead (II) ion from wastewater using Kaolinite/Smectite natural composite adsorbent. *Appl. Water Sci.* **2019**, *9*, 7–19. [CrossRef]
37. Khan, M.I.; Almesfer, M.K.; Danish, M.; Ali, I.H.; Shoukry, H.; Patel, R.; Gardy, J.; Nizami, A.S.; Rehan, M. Potential of Saudi natural clay as an effective adsorbent in heavy metals removal from wastewater. *Desalin. Water Treat.* **2019**, *158*, 140–151. [CrossRef]
38. Ali, I.H.; Sulfab, Y. Concurrent two one-electron transfer in the oxidation of chromium (III) complexes with trans-1,2-diaminocyclohexane-$N,N,N',N'$-tetraacetate and diethylenetriaminepentaacetate ligands by periodate ion. *Int. J. Chem. Kinet.* **2012**, *44*, 729–735. [CrossRef]
39. Ali, I.H.; Sulfab, Y. One-step, two-electron oxidation of cis-diaquabis(1,10-phenanthroline)chromium (III) to cis-dioxobis (1,10-phenanthroline) chromium (V) by periodate in aqueous acidic solutions. *Int. J. Chem. Kinet.* **2011**, *43*, 563–568. [CrossRef]
40. Mouni, L.; Merabet, D.; Bouzaza, A.; Belkhiri, L. Adsorption of Pb (II) from aqueous solutions using activated carbon developed from Apricot stone. *Desalination* **2011**, *276*, 148–153. [CrossRef]
41. Kaur, M.; Kumari, S.; Sharma, P. Removal of Pb (II) from aqueous solution using nanoadsorbent of *Oryza sativa* husk: Isotherm, kinetic and thermodynamic studies. *Biotechnol. Rep.* **2020**, *25*, e00410. [CrossRef] [PubMed]
42. Kardam, A.; Raj, K.R.; Srivastava, S.; Srivastava, M.M. Nanocellulose fibers for biosorption of cadmium, nickel, and lead ions from aqueous solution. *Clean Technol. Environ. Policy* **2014**, *16*, 385–393. [CrossRef]
43. Alghamdi, A.A.; Al-Odayni, A.-B.; Saeed, W.S.; Al-Kahtani, A.; Alharthi, F.A.; Aouak, T. Efficient Adsorption of Lead (II) from Aqueous Phase Solutions Using Polypyrrole-Based Activated Carbon. *Materials* **2019**, *12*, 2020. [CrossRef]
44. Tran, C.V.; Quang, D.V.; Thi, H.P.N.; Truong, T.N.; La, D.D. Effective Removal of Pb (II) from Aqueous Media by a New Design of Cu−Mg Binary Ferrite. *ACS Omega* **2020**, *5*, 7298–7306. [CrossRef] [PubMed]
45. Mousavi, H.Z.; Hosseynifar, A.; Jahed, V.; Dehghani, S.A.M. Removal of lead from aqueous solution using waste tire rubber ash as an adsorbent. *Braz. J. Chem. Eng.* **2010**, *27*, 79–87. [CrossRef]
46. Manyangadze, M.; Chikuruwo, N.M.H.; Narsaiah, T.B.; Chakra, C.H.; Charis, G.; Danha, G.; Mamvura, T.A. Adsorption of lead ions from wastewater using nano silica spheres synthesized on calcium carbonate templates. *Heliyon* **2020**, *6*, e05390. [CrossRef] [PubMed]
47. Khatoon, A.; Kashif Uddin, M.; Rao, R.A.K. Adsorptive remediation of Pb (II) from aqueous media using *Schleichera oleosa* bark. *Environ. Technol. Innov.* **2018**, *11*, 1–14. [CrossRef]
48. He, K.; Chen, Y.; Tang, Z.; Hu, Y. Removal of heavy metal ions from aqueous solution by zeolite synthesized from flyash. *Environ. Sci. Pollut. Res.* **2016**, *23*, 2778–2788. [CrossRef] [PubMed]

Article

# Study on the Preparation of Magnetic Mn–Co–Fe Spinel and Its Mercury Removal Performance

Jiawei Huang [1], Zhaoping Zhong [1,*], Yueyang Xu [1,2,3] and Yuanqiang Xu [1]

[1] Key Laboratory of Energy Thermal Conversion and Control of the Ministry of Education, Southeast University, Nanjing 210096, China; 220190482@seu.edu.cn (J.H.); xyy_gdhb@126.com (Y.X.); 13092320301@163.com (Y.X.)
[2] China Energy Science and Technology Research Institute Co., Ltd., Nanjing 210046, China
[3] State Key Laboratory of Clean and Efficient Coal-Fired Power Generation and Pollution Control, Nanjing 210046, China
* Correspondence: zzhong@seu.edu.cn

**Abstract:** In this study, the manganese-doped manganese–cobalt–iron spinel was prepared by the sol–gel self-combustion method, and its physical and chemical properties were analyzed by XRD (X-ray diffraction analysis), SEM (scanning electron microscope), and VSM (vibrating sample magnetometer). The mercury removal performance of simulated flue gas was tested on a fixed bed experimental device, and the effects of Mn doping amount, fuel addition amount, reaction temperature, and flue gas composition on its mercury removal capacity were studied. The results showed that the best synthesized product was when the doping amount of Mn was the molar ratio of 0.5, and the average mercury removal efficiency was 87.5% within 120 min. Among the fuel rich, stoichiometric ratio, and fuel lean systems, the stoichiometric ratio system is most conductive to product synthesis, and the mercury removal performance of the obtained product was the best. Moreover, the removal ability of $Hg^0$ was enhanced with the increase in temperature in the test temperature range, and both physical and chemical adsorption play key roles in the spinel adsorption of $Hg^0$ in the medium temperature range. The addition of $O_2$ can promote the removal of $Hg^0$ by adsorbent, but the continuous increase after the volume fraction reached 10% had little effect on the removal efficiency of $Hg^0$. While $SO_2$ inhibited the removal of mercury by adsorbent, the higher the volume fraction, the more obvious the inhibition. In addition, in an oxygen-free environment, the addition of a small amount of HCl can promote the removal of mercury by adsorbent, but the addition of more HCl does not have a better promotion effect. Compared with other reported adsorbents, the adsorbent has better mercury removal performance and magnetic properties, and has a strong recycling performance. The removal efficiency of mercury can always be maintained above 85% in five cycles.

**Keywords:** mercury removal; magnetic; manganese; cobalt; iron; spinel

## 1. Introduction

Mercury, as a highly toxic trace element in nature, has attracted wide attention from environmentalists all over the world due to its volatility and bioaccumulation [1–3]. It mainly exists in nature in the form of organic mercury, inorganic mercury, and metallic mercury [4]. In China, coal-fired power plants are the main source of mercury emissions, accounting for about 38% [5,6] of total mercury emissions. This situation has been highly valued by the Chinese government. The "Emission standard of air pollutants for thermal power plants" (GB13223-2011) issued in 2011 clearly stipulated that the mercury emission limit of coal-fired boilers was 30 μg/m³ or less for the first time [7]. In addition, "the Minamata Convention on Mercury" signed by 128 countries and regions including China has also formally entered into force on 16 August 2017 [8]. In order to implement the green and sustainable development strategy and "the Minamata Convention on Mercury", the development of high-efficiency and low-cost coal-fired flue gas mercury removal technology is urgently needed. Flue gas mercury is mainly composed of particulate mercury ($Hg^P$), oxidized mercury ($Hg^{2+}$), and elemental mercury ($Hg^0$). Among them, $Hg^P$ can be removed by dust removal equipment, and $Hg^{2+}$ is easily removed by wet desulfurization equipment, only $Hg^0$ is difficult to remove with existing control equipment [9–14]. At present, the more mainstream coal-fired flue gas mercury removal technology is still activated carbon injection (ACI) mercury removal technology, but this technology has disadvantages such as high price, poor recovery, and reduced fly ash quality, which limits its large-scale application [15–20]. Therefore, there is an urgent need to develop an efficient and easy-to-recover mercury removal adsorbent that can replace activated carbon.

In recent years, the research on non-carbon-based adsorbents can be divided into fly ash, mineral adsorbents, precious metals, metal oxides, and metal sulfides. Zhang et al. [17] studied the mercury removal performance of fly ash modified by $CaCl_2$, $CaBr_2$, and HBr, and the results showed that the mercury removal capacity of fly ash modified with different halogen compounds was significantly improved. Shi et al. [21] synthesized a new type of attapulgite catalyst ($CeO_2$/Atp (1:1)) that can maintain good catalytic activity in a wide temperature window, which can achieve a mercury removal efficiency of 97.75% at 200 °C. Cai et al. [15] studied the mercury removal performance of bentonite modified by KI and KBr, and the results showed that with the increase in the active material loading and the temperature (80–180 °C), the mercury removal efficiency increased, and compared with KBr modified, KI modified bentonite has better mercury removal performance. He et al. [22] synthesized Ce–Mn/Ti–PILC with a large specific surface area, which used a clay material similar to zeolite as the matrix. The results showed that under the HCl-free atmosphere, the 6%Ce–6%MnOx∥Ti–PILC catalyst could maintain a mercury removal efficiency above 90% in the range of 100–350 °C. After depositing silver nanoparticles on SBA–15, Xie et al. [23] prepared a silver-loaded SBA–15 adsorbent and found that mercury could form a silver amalgam with nano silver particles, and the mercury capture efficiency could reach 90% at 150 °C. Cimino et al. [24] developed two manganese-based catalysts for synergistic removal of NOx and Hg. The study found that the conversion of NOx and the removal of Hg were greatly affected by the type of carrier. $TiO_2$ is more suitable as a carrier for manganese oxide than $Al_2O_3$. Liu et al. [25] conducted a mercury adsorption performance test on the amorphous CoS synthesized by the liquid-phase precipitation method under the condition of 50 °C, and found that its adsorption capacity could reach 20.7 mg/g at a penetration rate of 25%. Li et al. [26] prepared nano ZnS with excellent mercury removal performance under high temperature conditions of 180 °C, and its adsorption capacity of $Hg^0$ at 50% transmission rate could reach 0.498 mg/g. Kong et al. [27] studied the change in mercury removal performance of $CuO/TiO_2$ and $CuS/TiO_2$ in the presence of $SO_2$ and $H_2O$, and the results showed that the mercury removal performance of $CuO/TiO_2$ was significantly reduced, while $CuS/TiO_2$ showed good resistance to $H_2O$ and $SO_2$. Liu et al. [28] also showed that the mercury adsorption performance of CuS was hardly affected by $SO_2$ and $H_2O$. Although these non-carbon-based adsorbents have good mercury removal capacity, they are generally not easy to recycle, while magnetic adsorbents, which are easy to separate from fly ash, can be recycled and reused well, which can significantly reduce the operation cost of power plants.

At present, two common magnetic materials, $\gamma$–$Fe_2O_3$ and $Fe_3O_4$, are mainly used to modify the adsorbents [2,7,12,29,30], and then the modified adsorbents are separated from the fly ash by the magnetic separation method to realize their recycling. Although spinels containing $MF_2O_4$ (M = Co, Mn, Cu, Zn, Fe, Ni, Mg, etc.) with high catalytic activity and strong magnetic response have been widely used in other fields [31,32], the research on mercury removal from coal-fired flue gas is relatively rare. Liao et al. [11] synthesized an Fe–Ti–Mn spinel with excellent mercury removal performance, which can be regenerated after washing with water and heating at high temperatures. Xiong et al. [14] also showed that the Fe–Ti–Mn spinel had a good adsorption performance for $Hg^0$. The Mn–Fe spinel synthesized by Dang Hao [13] has good mercury removal performance in the temperature range of 50–100 °C, and it can be regenerated after being washed by acidic

NaClO solution. After five cycles of adsorption, it can still maintain more than 95% of $Hg^0$ mercury removal efficiency. In addition, it has been reported that Co-doped iron oxides can improve the adsorption capacity of $Hg^0$ [33,34]. Considering that Co is a natural ferromagnetic element, this study plans to dope manganese ions into cobalt ferrite spinel to prepare a magnetic manganese–cobalt–iron spinel, in order to achieve efficient mercury removal and recyclability of the adsorbent.

In this study, the magnetic spinel $Mn_xCo_{(1-x)}Fe_2O_4$ (x = 0–1.0) was prepared by the sol–gel self-combustion synthesis method and characterized by XRD, SEM, and VSM. The adsorption performance of gas-phase zero-valent mercury ($Hg^0$) was investigated in a fixed-bed reaction system, and the effects of Mn doping amount, fuel addition amount, reaction temperature, and flue gas components on the mercury removal performance were discussed, and the regeneration performance of the screening of the sorbent was explored, aiming to provide theoretical guidance and research basis for the development of efficient and recyclable spinel ferrite adsorbents.

## 2. Materials and Methods

### 2.1. Materials

$(CH_3COO)_2Mn·4H_2O$, $(CH_3COO)_2Co·4H_2O$, $Fe(NO_3)_3·9H_2O$, and citric acid ($C_6H_8O_7$) as well as the above chemical reagents were analytically pure; high-purity deionized water was used in the experiment.

### 2.2. Sample Preparation

Magnetic Mn–Co–Fe spinel was synthesized by the sol–gel auto-combustion method. Taking the preparation of $Mn_{0.5}Co_{0.5}Fe_2O_4$ as an example, the specific preparation method is as follows. First, 3.68 g $(CH_3COO)_2Mn·4H_2O$, 3.74 g $(CH_3COO)_2Co·4H_2O$, 24.24 g $Fe(NO_3)_3·9H_2O$, and 9.61 g citric acid were dissolved in a proper amount of deionized water. The molar ratio of $Mn^{2+}$, $Co^{2+}$, and $Fe^{3+}$ was 1:1:4, and the ratio of the total oxidation value (O) of the oxidant (ferric nitrate) to the total reduction value (F) of the fuel (citric acid) was the stoichiometric ratio 1 (O/F = 1). After stirring for 30 min, the mixture was placed in a magnetic water-bath stirring pot at 70 °C for evaporation and stirring to remove water, and a wet gel was obtained. The obtained wet gel was dried and milled at 100 °C, and then put into a muffle furnace that had been preheated to 400 °C for the reaction. After the reaction was completed, the obtained product was calcined in a muffle furnace that had been preheated to 500 °C for 4 h. Finally, the obtained sample was placed in a drying dish for later use and recorded as 1M1C-1.

In the same way, $CoFe_2O_4$, $Mn_{0.25}Co_{0.75}Fe_2O_4$, $Mn_{0.75}Co_{0.25}Fe_2O_4$, and $MnFe_2O_4$ were prepared by changing the amount of Mn doping, which were denoted as CFO-1, 1M3C-1, 3M1C-1, and MFO-1 respectively. By changing the amount of fuel added, $Mn_{0.5}Co_{0.5}Fe_2O_4$ (O/F = 0.5), $Mn_{0.5}Co_{0.5}Fe_2O_4$ (O/F = 0.75), $Mn_{0.5}Co_{0.5}Fe_2O_4$ (O/F = 1.25), and $Mn_{0.5}Co_{0.5}Fe_2O_4$ (O/F = 1.5) were prepared as 1M1C-0.5, 1M1C-0.75, 1M1C-1.25, and 1M1C-1.5. respectively.

### 2.3. Characterization of Samples

A D8 ADVANCE X-ray diffractometer (Bruker, Germany) was used to detect and analyze the composition of the sample, and the sample was scanned at 5–90°; a Hitachi SU800 scanning electron microscope (Hitachi, Japan) was used to observe the microstructure of the product, and the particle size and particle size distribution of the sample were characterized by the image analysis method. The samples were analyzed for magnetic responsiveness and separability using a PPMS–9 (VSM) integrated physical property measurement system (Quantum Design, San Diego, CA, USA).

### 2.4. Experimental Device

The mercury removal performance of $Mn_xCo_{1-x}Fe_2O_4$ spinel adsorbents was tested by a fixed-bed experimental device, which includes a gas distribution section, a mercury generator, a fixed-bed reactor, a mercury analyzer, and a waste gas treatment. The schematic diagram of the experimental device is shown in Figure 1. The gas distribution was provided by a compressed gas cylinder. The gas was accurately controlled by a mass flow meter and the total flow rate of the gas was kept at 1 L/min. A total of 200 mL/min of $N_2$ was used as a carrier gas to be introduced into the mercury permeation tube, and gaseous $Hg^0$ was introduced into a gas mixing chamber. $N_2$ (balance gas), $O_2$, $SO_2$, and HCl gases from the gas distribution system and mercury vapor were fully mixed in the gas mixing chamber to simulate coal-fired flue gas. Before the experiment, the gas flow was switched to the bypass of the reaction tube, and the initial mercury concentration in the flue gas was measured by a VM3000 mercury detector (MI Company, Germany). After the mercury source was stable, the gas flow was switched to the main path, and the simulated flue gas containing mercury passed through the adsorbent for the mercury removal experiment. The mercury concentration in the flue gas at the outlet of the reaction tube was recorded in real time by the VM3000. The time of each test was 120 min. The tail gas discharged during the experiment was absorbed by activated carbon to prevent environmental pollution. Except for the quartz reaction tube, all connecting pipes are made of polytetrafluoroethylene, and the corresponding pipes were heated and controlled with heating belts to prevent mercury vapor from depositing on the inner wall of the pipes. During the experiment, the balance gas $N_2$ was 800 mL/min. When other gases are added, the corresponding balance gas should be reduced. The initial mass concentration of mercury ($Hg^0$) was constant at $(85 \pm 0.5)$ µg/m³, and the adsorbent dosage was 50 mg (passing through a 200-mesh sieve). The mercury removal efficiency ($\eta$) is defined as shown in Equation (1), where $c_{in}$ and $c_{out}$ represent the mass concentrations of Hg at the reactor inlet and outlet (µg/m³), respectively.

$$\eta = (1 - c_{out}/c_{in}) \times 100\% \tag{1}$$

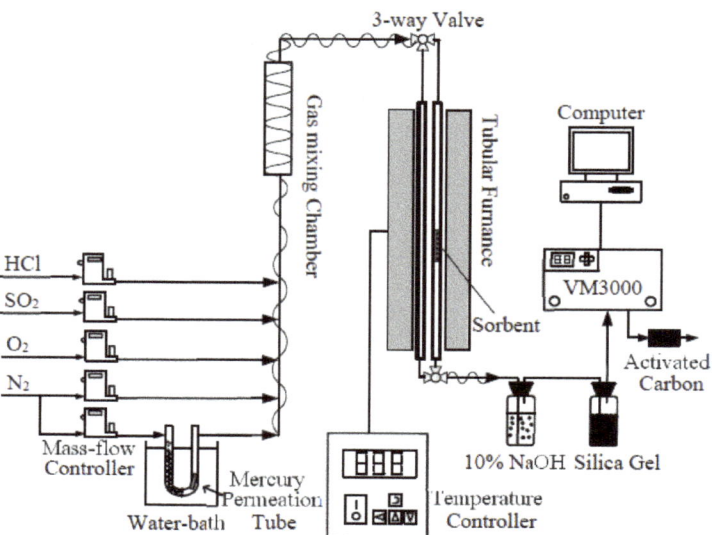

**Figure 1.** Sketch diagram of the mercury removal experimental apparatus.

## 3. Results and Discussion

### 3.1. Sample Characterization Analysis

#### 3.1.1. XRD Analysis

Figure 2 shows the XRD diffractogram of spinels prepared with different doping amounts of Mn. It can be seen from Figure 2 that the main diffraction peaks in all spinel XRD diffractograms were basically consistent with the standard diffractogram of maghemite ($\gamma$–$Fe_2O_3$) (JCPDS: 39–1346) and appeared at 18.28, 30.22, 35.48, 37.32, 43.16, 53.90, 57.10, and

62.70° corresponding to the diffraction planes (111), (220), (311), (222), (400), (422), (511), and (440), respectively. In addition, except for the characteristic peak of $Mn_2O_3$ (JCPDS: 41–1442) appearing at 55.22° and corresponding to the diffraction plane (440) in MFO-1, there were no characteristic peaks of cobalt and manganese oxides in the remaining spinels, which means that for CFO-1, 1M3C-1, 1M1C-1, and 3M1C-1, the cobalt and manganese ions had been completely doped into the spinel structure. As far as MFO-1 is concerned, manganese ions were not fully incorporated, and a small amount of 6.04% $Mn_2O_3$ phase impurities were generated. In addition, it can be seen that, except for 1M1C-1, the characteristic peaks of α–$Fe_2O_3$ (JCPDS: 89–0596) appearing at 33.02, 49.96 and 64.00° and corresponding to the diffraction planes (104), (024), and (300), respectively, appeared in the rest of the spinels. There were 8.74% and 9.20% α–$Fe_2O_3$ phase impurities in CFO-1 and 1M3C-1, respectively, 17.09% in 3M1C-1, and 37.96% in MFO-1.

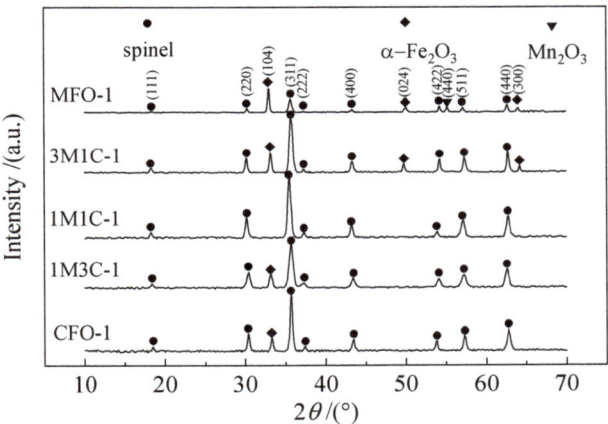

**Figure 2.** XRD patterns of the $Mn_xCo_{1-x}Fe_2O_4$ spinels.

Figure 3 shows the XRD diffractogram of the spinel prepared with different amounts of fuel. It can be seen from Figure 3 that the $Mn_{0.5}Co_{0.5}Fe_2O_4$ spinels prepared under different fuel addition amounts all belonged to a single-phase spinel structure, and no other impurity phases were generated. Compared with 1M1C-0.5 and 1M1C-1.5, the diffraction peak of 1M1C-1 was sharper and the peak intensity greater, which indicates that the crystallinity of 1M1C-1 was higher. According to the Scherrer formula shown in Equation (2), the average grain sizes of 1M1C-0.5, 1M1C-1, and 1M1C-1.5 were 17.67 nm, 21.21 nm, and 13.26 nm, respectively. It can be seen that the rich-burn and lean-burn systems are not conducive to grain growth. Only the stoichiometric ratio system is most conducive to product formation and grain growth, which is consistent with the results reported in the literature [35,36].

$$D = K\lambda / \beta \cos\theta \qquad (2)$$

where $D$ is the size of the crystal grain (nm); $K$ is the shape factor of the particle, generally 0.9; $\lambda$ is the X-ray wavelength used in the test, 0.1542 nm in this paper; $\beta$ is the maximum half-width of the diffraction peak; and $\theta$ is the diffraction angle corresponding to the X-ray diffraction peak.

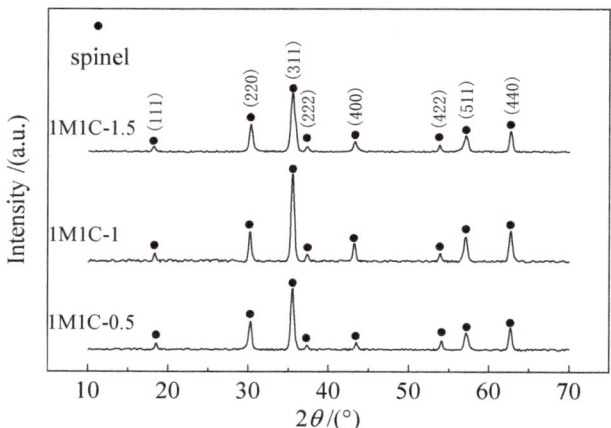

**Figure 3.** XRD patterns of spinels obtained with different fuel dosages.

3.1.2. SEM Analysis

Figure 4 shows the particle size distribution and SEM micrographies of the 1M1C-1, 3M1C-1, and MFO-1 spinels. We could see from Figure 4 that the particle size distribution of 1M1C-1 was relatively narrow, and the particle size was smaller. According to statistical calculations, the smallest particle size was 66 nm, the largest particle size was 1.52 µm, and the average particle size was 370 nm. Combined with the SEM micrographies, it can be seen that the particles were all spherical granular aggregates, conforming to the characteristics of an equiaxed crystal system, indicating that the synthesized product had a single-phase spinel structure. The minimum particle size of 3M1C-1 was 90 nm, the maximum particle size was 1.37 µm, and the average particle size was 381 nm. Compared to 1M1C-1, the particles of 3M1C-1 tended to be of medium size, and the average particle size increased slightly. Combined with the SEM micrographies, it can be seen that there are flaky particles in the product, which conform to the characteristics of a hexagonal crystal system. This shows that the impurity of $\alpha$–$Fe_2O_3$ appeared in the sample. At the same time, it is easy to learn that the minimum particle size of MFO-1 was 108 nm, the maximum particle size was 2.11 µm, and the average particle size was 388 nm. Compared with 3M1C-1, the particles of MFO-1 were more closed to larger particles. Combined with the SEM micrographies, it can be seen that the obvious secondary agglomeration phenomenon could be observed in the product, and relatively more flaky particles could be found.

3.1.3. VSM Analysis

Figure 5 shows the hysteresis loops of 1M1C-1, 3M1C-1, and MFO-1 spinels, which shows that 1M1C-1, 3M1C-1, and MFO-1 spinels all had superparamagnetism and could be spontaneously magnetized under the action of an external magnetic field. Magnetic agglomeration will not occur during the demagnetization process, and the specific saturation magnetization was 41 emu/g, 19.25 emu/g, and 10.43 emu/g, respectively, which is consistent with the analysis results of XRD and SEM. The decrease in specific saturation magnetization is due to the formation of $\alpha$–$Fe_2O_3$ impurity, and the higher the impurity ratio, the greater the decrease. According to the literature [2], these three spinels can be attracted by magnets, but it can also be seen from Figure 5 that the required external magnetic field to achieve the same specific saturation magnetization of 1M1C-1 was the smallest, followed by 3M1C-1, and MFO-1 was the largest. Taking 10 emu/g as an example, MFO-1 needs an external magnetic field of 9000 Oe, 3M1C-1 only needs 1750 Oe, and 1M1C-1 only needs 500 Oe. Therefore, from the perspective of engineering application, 1M1C-1 has better magnetic separation characteristics.

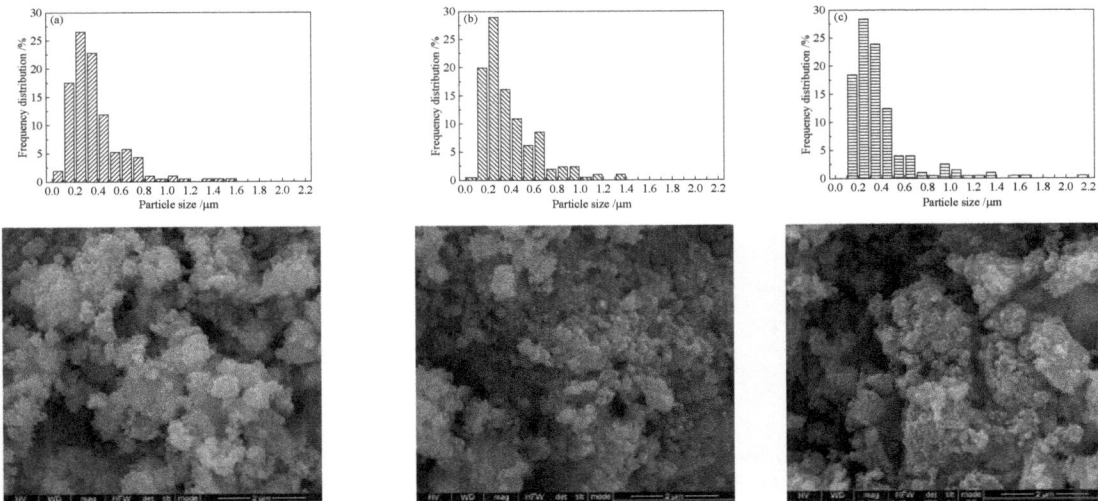

**Figure 4.** Particle size distributions and SEM micrographies of the spinel samples. (**a**) 1M1C-1; (**b**) 3M1C-1; (**c**) MFO-1.

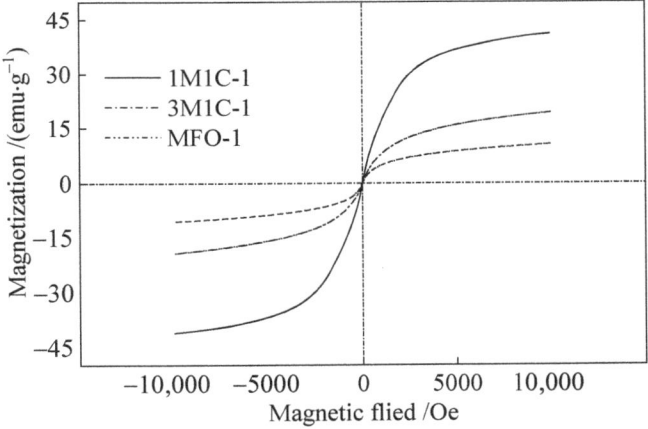

**Figure 5.** Hysteresis loop diagrams of three spinels.

### 3.2. Sorbent Mercury Removal Experiment

#### 3.2.1. Influence of Different Doping Amount of Mn on the Performance of Mercury Removal

In order to study the effect of Mn doping on mercury removal by spinel adsorbent, this research applied five types of spinels obtained in the experiment to a fixed bed mercury adsorption experiment to test their mercury removal performance. The reaction temperature in the experiment was 150 °C. The flue gas atmosphere was pure $N_2$. The experimental results are shown in Figure 6.

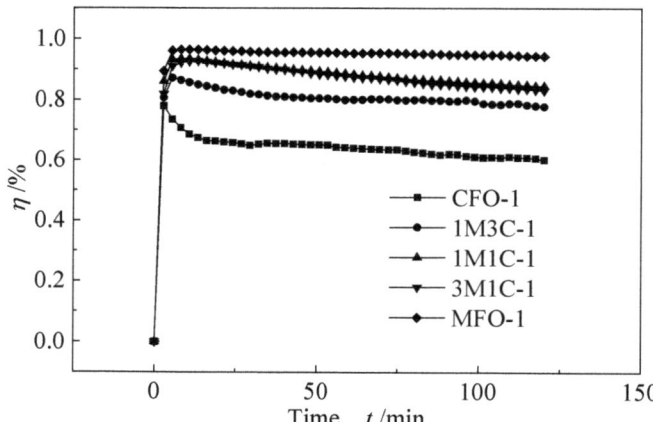

**Figure 6.** Effect of Mn doping content on the mercury removal efficiency of the adsorbent.

It can be seen from Figure 6 that the CFO-1 sample without Mn reached the highest mercury removal efficiency of 77.9% at the beginning, and then began to decline continuously, with an average mercury removal efficiency of 63.7% within 120 min. When the molar ratio of Mn was 0.25, the 1M3CFe sample initially reached a mercury removal efficiency of 80.6%, and a small increase occurred in a short period of time thereafter, reaching the highest mercury removal efficiency of 87.1%, and then began to slowly decrease. The average mercury removal efficiency during the test period was 80%, which shows that the doping of a small amount of Mn can greatly increase the mercury removal efficiency of the cobalt ferrite spinel on one hand, and on the other hand, can prolong the effective action time of the adsorbent. When the Mn doping amount is further increased to 0.5 by mole ratio, the initial, maximum, and average mercury removal efficiency of 1M1C-1 increased to 86%, 93.5%, and 87.5%, respectively, and the increase rates were 6.7%, 7.3%, and 9.4%, respectively. This shows that the downward trend of mercury removal efficiency had greatly slowed down, and the overall high-efficiency action time was further increased. When the molar ratio of Mn doping was 0.75, the mercury removal efficiency of 3M1C-1 dropped by 4.8%, 0.9%, and 1%, respectively, indicating that the formation of $\alpha$–$Fe_2O_3$ impurity makes the absolute content of high-efficiency mercury removal active ingredients in the same mass adsorbent. The decrease resulted in a slight decrease in the mercury removal efficiency when the Mn doping amount increased. When Mn was fully doped, the initial, highest and average mercury removal efficiency of the MFO-1 samples were 89.3%, 96.4%, and 94.2%, respectively, indicating that the mercury removal efficiency of MFO-1 dropped very slowly after rising to its highest, and almost remained unchanged. At the same time, it also showed that the Fe–Mn spinel had very good mercury removal performance. In the case of generating more $\alpha$–$Fe_2O_3$ impurities, it could still maintain a high mercury removal efficiency. It can be found that Mn-doped cobalt ferrite spinel had a good ability to remove $Hg^0$. As the Mn doping amount gradually increased from 0 to 0.5 in molar ratio, the mercury removal performance and high-efficiency action time gradually increased. However, when the doping amount reached a molar ratio of 0.75, the mercury removal efficiency of the adsorbent was slightly reduced, and the magnetic separation performance was also weakened. Even when the Mn was completely doped, the mercury removal efficiency of the adsorbent was again slightly improved. However, its magnetic separation performance was greatly reduced. In order to ensure the mercury removal performance and magnetic separation performance of the adsorbent at the same time, a molar ratio of 0.5 was taken as the best Mn doping amount.

### 3.2.2. The Effect of Different Fuel Addition on the Performance of Mercury Removal

In the case that the optimal doping amount of Mn was a 0.5 in molar ratio, in order to further study the influence of different fuel addition on the mercury removal performance of the prepared spinel, the mercury removal performance of the $Mn_{0.5}Co_{0.5}Fe_2O_4$ spinel prepared under the rich combustion, stoichiometric ratio, and lean-burn system was tested through fixed bed mercury adsorption experiments. The reaction temperature in the experiment was 150 °C, and the flue gas atmosphere was pure $N_2$. The experimental results are shown in Figure 7.

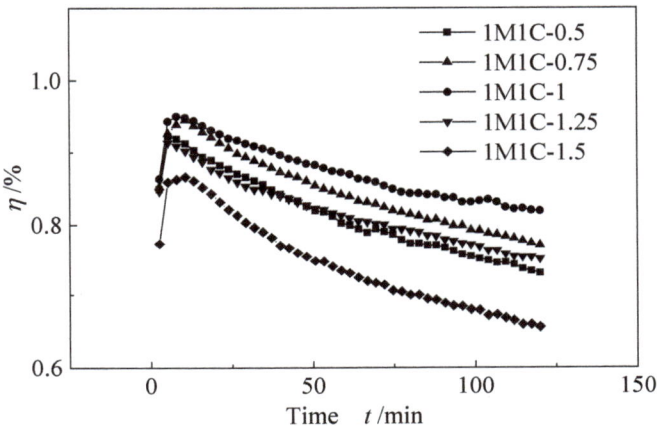

**Figure 7.** Effect of fuel addition content on the mercury removal efficiency of the adsorbent.

It can be seen from Figure 7 that when the Mn doping amount was 0.5 in molar ratio, the spinels prepared under different fuel addition amounts had relatively good mercury removal performance, and the relatively worst 1M1C-1.5 sample also had a 73.6% average mercury removal efficiency. Among them, the spinel prepared under the stoichiometric system had the best mercury removal performance, which is attributed to the fact that the stoichiometric system is most conducive to product formation and grain growth. For the rich combustion system, the oxidizer cannot provide the O required for the complete combustion of the fuel, and more O needs to be obtained from the surrounding environment. In the process of sample synthesis, $NO^{3-}$ cannot completely oxidize the surrounding organic matter, and the heat released during oxidation is provided to the unreacted remaining organic matter in addition to heating the reaction system to react with external oxygen, and the reaction has some issues such as high initial temperature and slow reaction rate. It also showed that the fuel-rich system is not conducive to the formation of products, which explains the reason for $\eta$(1M1C-1) > $\eta$(1M1C-0.75) > $\eta$(1M1C-0.5). For the lean-burn system, the factor restricting the synthesis of the sample is the fuel. The heat released during oxidation cannot make the system reach the most suitable temperature for product formation. Therefore, the lean-burn system is not conducive to product formation, which explains the reason for $\eta$(1M1C-1) > $\eta$(1M1C-1.25) > $\eta$(1M1C-1.5). It can be found that in a fuel-rich system, reducing the amount of fuel added can improve the mercury removal performance of the prepared spinel. Similarly, in the lean-burn system, increasing the amount of fuel added will also increase the average mercury removal efficiency of the prepared spinel. In other words, the closer the O/F ratio is to 1, the better the mercury removal performance of the prepared spinel. According to the XRD analysis results, the products obtained under different systems are all single-phase spinel structures, and the magnetic separation performance should be similar. Therefore, to ensure the best mercury removal performance, O/F = 1 is the best fuel addition. All subsequent studies on other influencing factors have taken 1M1C-1 as the research object.

3.2.3. The Effect of Reaction Temperature on the Performance of Mercury Removal

In order to study the influence of temperature on the mercury removal effect of magnetic manganese–cobalt–iron spinel, the temperature of the fixed-bed reactor was set to 100 °C, 125 °C, 150 °C, 175 °C, and 200 °C, and the mercury removal experiment was carried out. The research object selected in the experiment was 1M1C-1, and the flue gas atmosphere was pure $N_2$. Figure 8 shows the mercury removal efficiency curve of 1M1C-1 at different reaction temperatures. It can be seen from Figure 8 that the $Hg^0$ removal efficiency of 1M1C-1 gradually increases with the increase in temperature. When the reaction temperature is 200 °C, the mercury removal efficiency was the highest, which was 90.1%. Across the entire reaction temperature range, 1M1C-1 could maintain good mercury removal performance, and there was an average mercury removal efficiency of 82.3% at 100 °C, indicating that the reaction temperature had relatively little effect on its mercury removal performance. In addition, it can be found that when the reaction temperature was increased from 100 °C to 125 °C, the increase in mercury removal efficiency was relatively small. From 125 °C to 150 °C and then to 175 °C, the increase in mercury removal efficiency was relatively large. However, from 175 °C to 200 °C, the increase in mercury removal efficiency weakened again. This is because the mercury adsorption process is the result of the combined effect of physical adsorption and chemical adsorption [10,37]. When the temperature was low (100–125 °C), the chemical adsorption was not obvious, and the physical adsorption was dominant. The effect of mercury removal performance was small. When the temperature rose (125–175 °C), chemical adsorption began to take effect, and the mercury removal performance was relatively greatly improved. When the temperature continued to rise (175–200 °C), physical adsorption began to be suppressed, and mercury desorption began to appear on the surface. Therefore, the improvement trend of the mercury removal performance slowed down. Based on the above analysis, in the following study, in order to highlight the effect of flue gas components on the mercury removal performance, 150 °C was selected as the benchmark experimental condition.

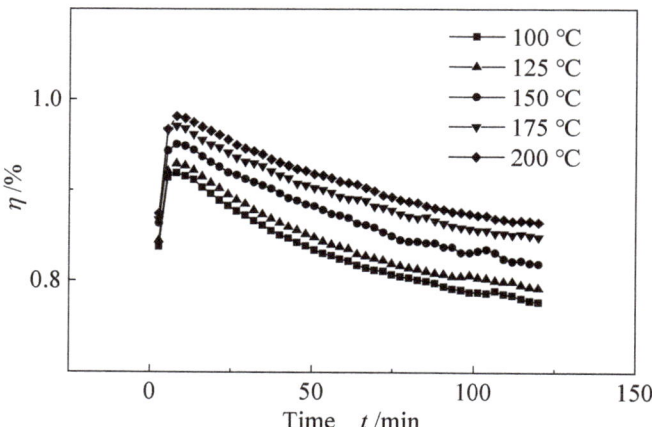

**Figure 8.** Effect of reaction temperature on the mercury removal efficiency of 1M1C-1.

3.2.4. Influence of Flue Gas Components on the Performance of Mercury Removal
Impact of $O_2$

Since different components ($O_2$, $SO_2$, HCl, etc.) in the flue gas will affect the mercury removal performance of the adsorbent, it is necessary to investigate the effect of different concentrations of $O_2$, $SO_2$, and HCl, on the mercury removal effect of the spinel. Figure 9 shows the effect of different volume fractions of $O_2$ on the removal of $Hg^0$ by 1M1C-1. It can be seen from the figure that without $O_2$, the mercury removal efficiency of 1M1C-1 showed a downward trend with the extension of the reaction time. When 5% $O_2$ was introduced into the atmosphere, the downward trend of mercury removal efficiency was obviously curbed. The average mercury removal efficiency was significantly increased to 91.1%, and the volume fraction of $O_2$ continued to increase to 10%. The downward trend was further slowed down, and the average mercury removal efficiency was also further increased to 94.2%, indicating that $O_2$ plays a positive role in the removal of $Hg^0$. It can provide abundant active oxygen, and supplement the lattice oxygen consumed in the adsorption process or chemically adsorbed oxygen, thereby promoting the $Hg^0$ removal process [6,33,37,38]. However, when the $O_2$ volume fraction was further increased to 15%, the $Hg^0$ removal efficiency was basically not improved, indicating that 10% $O_2$ is sufficient to regenerate the lattice oxygen or chemisorption oxygen consumed on the surface.

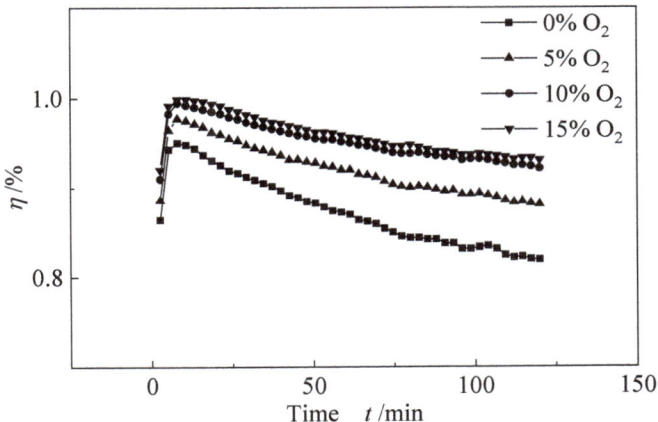

**Figure 9.** Effect of $O_2$ on the mercury removal efficiency of 1M1C-1.

Impact of $SO_2$

Figure 10 shows the effect of different volume fractions of $SO_2$ on the mercury removal performance of 1M1C-1. It is easy to see that after adding 0.04% $SO_2$ on the basis of pure $N_2$, the $Hg^0$ removal efficiency at the initial stage of adsorption had a slight increase, but after reaching the highest mercury removal efficiency, it dropped quickly, and the average mercury removal efficiency decreased compared to a pure $N_2$ atmosphere by 9%. The reason may be as shown in reactions (3)–(5) (O* is the surface active oxygen of the adsorbent) [11,39]. $SO_2$ molecules are more likely to occupy surface active sites. In the initial stage of adsorption, the preferentially adsorbed $SO_2$ molecules react with adsorbed $Hg^0$ to generate $HgSO_4$, which promotes the adsorption of $Hg^0$. However, the continuous introduction of $SO_2$ will occupy part of the active sites on one hand, and on the other hand, it may consume lattice oxygen or chemically adsorbed oxygen and react with metal oxides to form metal sulfates, occupy the surface, or block the pores, making the mercury removal efficiency decline faster. When the volume fraction of $SO_2$ continued to increase to 0.08% and 0.12%, the average mercury removal efficiency also continued to dropped to 74.5% and 70.7%. Therefore, different volume fractions of $SO_2$ have an inhibitory effect on 1M1C-1 mercury removal, and the higher the volume fraction, the more obvious the inhibitory effect. The reason may be the occurrence of competitive adsorption and side reactions [2,40].

$$SO_2(g) + 1M1CFe(surface) \to SO_2(ad) \qquad (3)$$

$$SO_2(ad) + O^* \to SO_3(ad) \qquad (4)$$

$$Hg^0(ad) + SO_3(ad) + O^* \to HgSO_4(ad) \qquad (5)$$

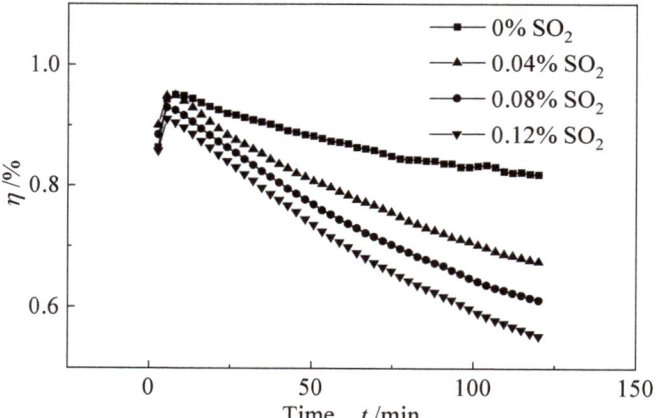

**Figure 10.** Effect of SO$_2$ on the mercury removal efficiency of 1M1C-1.

Compared with the side reactions, competitive adsorption accounts for a heavier proportion of the inhibition. Therefore, in order to further understand the competitive adsorption behavior of SO$_2$ on the surface of 1M1C-1, a mercury pre-adsorption experiment was carried out, as shown in Figure 11. In this experiment, first, we pre-adsorbed 1M1C-1 in Hg$^0$-containing N$_2$ for 120 min, then the mercury source was cut off and the total flue gas flow was kept constant. When the Hg$^0$ concentration in the flue gas was reduced to 0 μg/m$^3$, we added 0.12% SO$_2$ to the flue gas. It can be seen that the Hg$^0$ concentration in the flue gas rapidly increased to 20.1 μg/m$^3$, and then gradually decreased to 0 μg/m$^3$. This indicates that the introduction of SO$_2$ into the reaction system will cause the desorption of the weakly adsorbed mercury on the surface of the adsorbent, indicating that SO$_2$ and Hg$^0$ are competitively adsorbed during the removal of Hg$^0$ by 1M1C-1.

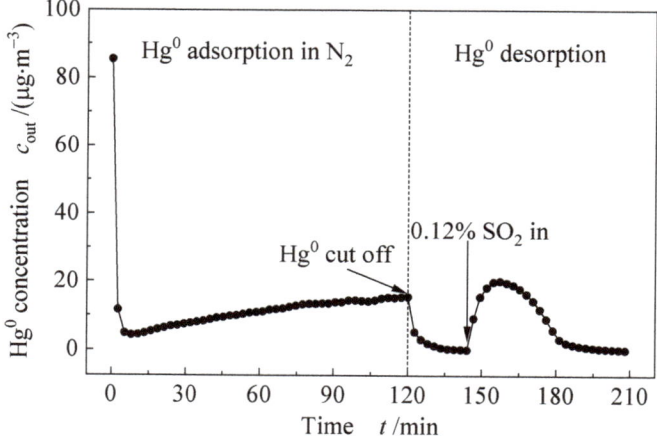

**Figure 11.** Effect of SO$_2$ on the desorption of mercury on 1M1C-1.

Impact of HCl

Figure 12 shows the effect of different volume fractions of HCl on the mercury removal performance of 1M1C-1. We could see that after 0.001% HCl was introduced into the reaction system, the average mercury removal efficiency increased by 3.4% to 89.3%, indicating that the addition of a small amount of HCl has a promoting effect on the mercury removal of 1M1C-1. Because it is difficult for $Hg^0$ to directly react with HCl under a $N_2$ atmosphere and lower temperature, it is speculated that $Mn^{4+}$ exists in the adsorbent and the reaction of Equations (6)–(8) occurs [3,29,41]. First, HCl is adsorbed on the surface of the adsorbent by separation and adsorption to form O–H bonds and Mn–Cl bonds, and the electrons of $Mn^{4+}$ become $Mn^{3+}$, and the electrons lost by chloride ions become active chlorine atoms, and then the active chlorine atoms and the adsorbed state $Hg^0$ reacted to generate HgCl, which in turn generated $HgCl_2$. When the HCl volume fraction increased to 0.003%, the mercury removal efficiency only increased to 90.7%, and when it increased to 0.005%, the mercury removal efficiency hardly increased. The higher the HCl volume fraction, the decline trend after the adsorbent reached the highest mercury removal efficiency became more obvious. This is because in an oxygen-free environment, when the concentration of HCl is high, the surface oxygen of the adsorbent is not enough to support the conversion of HCl and the remaining HCl will compete with $Hg^0$ for adsorption, covering some active sites on the surface [20], resulting in the overall mercury removal efficiency decreasing relatively quickly. Furthermore, the mercury removal performance improved slightly. Therefore, in an oxygen-free environment, adding a small amount of HCl can promote 1M1C-1 mercury removal, but adding more HCl cannot play a better role.

$$Mn^{4+}(s) + O^{2-}(ad) + HCl(g) \rightarrow Mn^{3+}(s) + OH^-(ad) + Cl(ad) \qquad (6)$$

$$Hg^0(ad) + Cl(ad) \rightarrow HgCl(ad) \qquad (7)$$

$$HgCl(ad) + Cl(ad) \rightarrow HgCl_2(ad) \qquad (8)$$

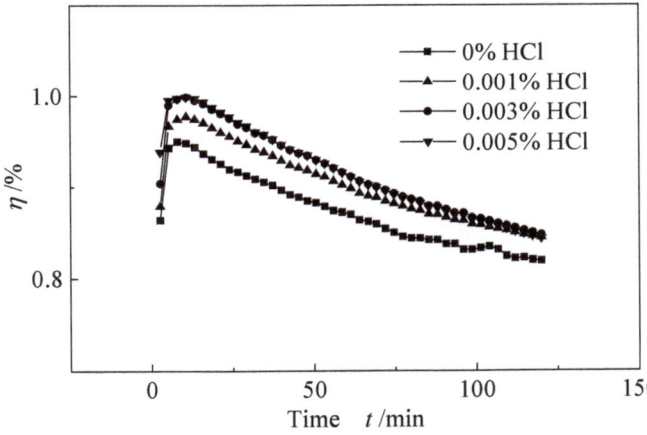

Figure 12. Effect of HCl on the mercury removal efficiency of 1M1C-1.

### 3.3. Comparison of Performance between Mn-Co-Fe Spinel and Other Adsorbents

Table 1 compares the $Hg^0$ capture performance and specific saturation magnetization of the 1M1C-1 adsorbent with those of other reported magnetic adsorbents. The capacity of the 1M1C-1 adsorbent for $Hg^0$ capture was 178.5 μg/g at 150 °C with the $Hg^0$ removal efficiency of 87.5%, and the specific saturation magnetization was 41 emu/g. This $Hg^0$ removal performance is basically at the best level among iron-based spinels, which was significantly higher than the Mn–Fe spinel and Fe–Ti–Mn spinel, and only slightly lower than the modified Fe–Ti spinel. It was also at a relatively good level among the magnetically modified natural mineral adsorbent, which was significantly higher than 1M1Atp and 0.2Fe1ATT, and only inferior to 5%CuMAtp. In addition, the specific saturation magnetization of 1M1C-1 was significantly better than the other adsorbents, which means that it will be easier to magnetically separate it. In general, 1M1C-1 had the best performance in capturing $Hg^0$ from flue gas and then recovering it by magnetic separation.

**Table 1.** Comparison of the adsorption capacity and specific saturation magnetization between 1M1C-1 and other adsorbents.

| Adsorbents | Adsorption Capacity (μg/g) | Temperature (°C) | Specific Saturation Magnetization (emu/g) | Reference |
|---|---|---|---|---|
| 1M1Atp | 70.15 | 150 | 29.45 | [2] |
| 0.2Fe1ATT | 56.7 | 150 | 29.5 | [12] |
| Mn–Fe spinel | 33 | 60 | 37.1 | [13] |
| Modified Fe-Ti spinel | 192.6 | 100 | 24.6 | [10] |
| 5%CuMAtp | 307.8 | 150 | 18.19 | [29] |
| Fe–Ti–Mn spinel | 75 | 60 | 29.6 | [11] |
| 1M1C-1 | 178.5 | 150 | 41 | this work |

*3.4. Regeneration and Reuse of Spent Mn–Co–Fe Spinel*

At present, the most commonly used adsorbent regeneration method is the direct thermal desorption method. According to the literature [13], for the actual flue gas of coal-fired power plants, the optimum temperature for thermal desorption regeneration of the Mn–Fe spinel cannot be lower than 500 °C. In light of the fact that the adsorbent finally screened in this study was calcined at 500 °C for 4 h, and the XRD results also showed that the sample crystal form was single and impurity-free, the adsorbent was regenerated at 500 °C by the direct thermal desorption method, and the mercury adsorption performance test was performed on the adsorbent obtained after regeneration, and the specific test results are shown in Figure 13. As can be seen from the figure, the mercury removal efficiency of the 1M1C-1 adsorbent can be maintained above 85% in the five cycles of regeneration. Therefore, the 1M1C-1 adsorbent had good recycling and regeneration performance.

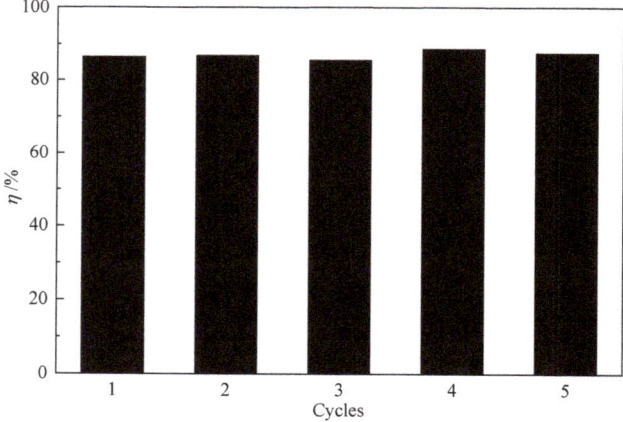

**Figure 13.** Regeneration performance of 1M1C-1 over five cycles.

## 4. Conclusions

In this study, a manganese-doped manganese–cobalt–iron spinel adsorbent was prepared by the sol–gel self-combustion method. The sample characterization results and mercury removal performance experiments showed that the synthesized product was the best when the molar ratio of Mn was 0.5. It had a single-phase spinel structure and good magnetic separation performance (41 emu/g), and the mercury removal performance was also relatively good. The average mercury removal efficiency within 120 min was

87.5%; and the Mn doping amount was maintained at 0.5. In the rich combustion, stoichiometric, and lean-burn system, the $Mn_{0.5}Co_{0.5}Fe_2O_4$ spinel was prepared, respectively. The results showed that the synthesized products were all single-phase spinel structures. However, the stoichiometric ratio system was most conducive to product synthesis, and the resulting product had the most excellent mercury removal performance. Within the experimental temperature, the mercury removal ability of 1M1C-1 was relatively good, and it gradually increased with the increase in temperature. However, physical adsorption was dominant at lower temperature, while physical adsorption was inhibited and mercury desorption occurred at a higher temperature. In this case, only moderate temperature, physical adsorption, and chemical adsorption work together to play an important role in the range. $O_2$ can promote the mercury removal performance of 1M1C-1. However, 10% $O_2$ was enough to regenerate the lattice oxygen or chemisorption oxygen consumed on the surface, and the higher $O_2$ volume fraction had little further effect on the mercury removal performance. The addition of different volume fractions of $SO_2$ has an inhibitory effect on the mercury removal of 1M1C-1, and the higher the volume fraction, the more obvious the inhibitory effect. The main reason is that $SO_2$ competes with $Hg^0$ for adsorption, and there may be side reactions that weaken the mercury removal capacity of the adsorbent. The addition of a small amount of HCl can promote 1M1C-1 mercury removal, but the addition of more HCl does not have a better promotion effect. This is mainly due to the insufficient surface oxygen of the adsorbent in an oxygen-free environment to support more HCl conversion, and the remaining HCl will instead begin competitive adsorption with $Hg^0$, covering part of the active sites on the surface, and weakening the active chlorine atom's promotion of mercury removal from the adsorbent. Compared with other reported adsorbents, the 1M1C-1 adsorbent had the best performance for $Hg^0$ capture from flue gas and then recovering it by magnetic separation. In addition, it also had a strong recycling performance, and the mercury removal efficiency could be maintained above 85% in five cycles after thermal desorption regeneration at 500 °C.

**Author Contributions:** This work presented here was carried out with collaboration among all authors. Methodology, J.H., Z.Z. and Y.X. (Yueyang Xu); validation, J.H. and Y.X. (Yuanqiang Xu); formal analysis, J.H.; investigation, J.H. and Y.X. (Yuanqiang Xu); data curation, J.H.; writing—original draft preparation, J.H.; writing—review and editing, Z.Z.; funding acquisition and supervision, Y.X. (Yueyang Xu). All authors have read and agreed to the published version of the manuscript.

**Funding:** The project was supported by the National Key Research and Development Program of China (2017YFC0210203) and Science and Technology Innovation Project of CHN Energy (GJNY-20-109).

**Institutional Review Board Statement:** Not applicable.

**Informed Consent Statement:** Not applicable.

**Data Availability Statement:** The data presented in this study are available on request from the corresponding author.

**Conflicts of Interest:** The authors declare no conflict of interest.

# References

1. Xu, Y.L.; Zhong, Q.; Liu, X.Y. Elemental mercury oxidation and adsorption on magnesite powder modified by Mn at low temperature. *J. Hazard. Mater.* **2015**, *283*, 252–259. [CrossRef] [PubMed]
2. Chen, H.; Huang, Y.J.; Dong, L.; Cao, J.H.; Xia, Z.P.; Qin, W.H. Study on the preparation of magnetic attapulgite and its mercury removal performance. *J. Fuel Chem. Technol.* **2018**, *46*, 1392–1400.
3. Yang, Y.J.; Liu, J.; Zang, B.K.; Liu, F. Density functional theory study on the heterogeneous reaction between $Hg^0$ and HCl over spinel-type $MnFe_2O_4$. *Chem. Eng. J.* **2017**, *308*, 897–903. [CrossRef]
4. Wang, J.W.; Xu, C.; Qin, W.; Zang, J.L.; Zang, X.L.; Dong, Y.J.; Cui, X.F. $Hg^0$ removal by palygorskite (PG) supported $MnO_x$ catalyst. *J. Fuel Chem. Technol.* **2020**, *48*, 1442–1451. [CrossRef]
5. Zhang, A.C.; Zheng, W.W.; Song, J.; Hu, S.; Liu, Z.C.; Xiang, J. Cobalt manganese oxides modified titania catalysts for oxidation of elemental mercury at low flue gas temperature. *Chem. Eng. J.* **2014**, *236*, 29–38. [CrossRef]
6. Liu, F.F.; Zhang, J.Y.; Zhao, Y.C.; Zheng, C.G. Mercury removal from flue gas by metal oxide-loaded attapulgite mineral sorbent. *Combust. Sci. Technol.* **2014**, *20*, 553–557.

7. Sun, Q.K.; Huang, Y.J.; Wang, L.; Guan, Z.W.; Li, M.; Zhou, J.; Wang, Y. Experimental study on mercury removal efficiencies of magnetic $Fe_3O_4$-Ag composite nanoparticles. *Chem. Ind. Eng. Prog.* **2017**, *36*, 1101–1106.
8. Ma, Y.P.; Xu, T.F.; Wang, J.D.; Shi, Y.R.; Wang, H.Y.; Xiong, F.G.; Xu, H.M.; Ma, Y.X.; Zhang, H.Z. Superior $Hg^0$ capture performance and $SO_2$ resistance of Co-Mn binary metal oxide-modified layered MCM-22 zeolite for $SO_2$-containing flue gas. *Environ. Sci. Pollut. Res. Int.* **2021**, *28*, 16447–16457. [CrossRef]
9. Chen, L.; Liu, S.Y.; Lv, W.Y.; Yang, K.; Li, Y. Effect of Manganese loading on zero valent mercury adsorption on magnetic iron oxides. *Environ. Eng.* **2019**, *37*, 131–137.
10. Zou, S.J.; Liao, Y.; Xiong, S.C.; Huang, N.; Geng, Y.; Yang, S.J. $H_2S$-modified Fe-Ti spinel: A recyclable magnetic sorbent for recovering gaseous elemental mercury from flue gas as a co-benefit of wet electrostatic precipitators. *Environ. Sci. Technol.* **2017**, *51*, 3426–3434. [CrossRef]
11. Liao, Y.; Xiong, S.C.; Dang, H.; Xiao, X.; Yang, S.J.; Wong, P.K. The centralized control of elemental mercury emission from the flue gas by a magnetic rengenerable Fe-Ti-Mn spinel. *J. Hazard. Mater.* **2015**, *299*, 740–746. [CrossRef] [PubMed]
12. Dong, L.; Huang, Y.J.; Chen, H.; Liu, L.Q.; Liu, C.Q.; Xu, L.G.; Zha, J.R.; Wang, Y.X.; Liu, H. Magnetic $\gamma$-$Fe_2O_3$-loaded attapulgite sorbent for $Hg^0$ removal in coal-fired flue gas. *Energy Fuels* **2019**, *33*, 7522–7533. [CrossRef]
13. Dang, H. *The Centralized Control of Elemental Mercury Emission from the Flue Gas Using Magnetic Mn-Fe Spinel*; Nanjing University of Science & Technology: Nanjing, China, 2017.
14. Xiong, S.C.; Xiao, X.; Huang, N.; Dang, H.; Liao, Y.; Zou, S.J.; Yang, S.J. Elemental mercury oxidation over Fe-Ti-Mn spinel: Performance, mechanism, and reaction kinetics. *Environ. Sci. Technol.* **2017**, *51*, 531–539. [CrossRef] [PubMed]
15. Cai, J.; Shen, B.X.; Li, Z.; Chen, J.H.; He, C. Removal of elemental mercury by clays impregnated with KI and KBr. *Chem. Eng. J.* **2014**, *241*, 19–27. [CrossRef]
16. Liu, H.; Yang, J.P.; Tian, C.; Zhao, Y.C.; Zhang, J.Y. Mercury removal from coal combustion flue gas by modified palygorskite adsorbents. *Appl. Clay Sci.* **2017**, *147*, 36–43. [CrossRef]
17. Zhang, Y.S.; Duan, W.; Liu, Z.; Cao, Y. Effects of modified fly ash on mercury adsorption ability in an entrained-flow reactor. *Fuel* **2014**, *128*, 274–280. [CrossRef]
18. Yang, Y.J.; Zhang, B.H.; Liu, J.; Wang, Z.; Miao, S. Mercury removal by recyclable and regenerable $Cu_xMn_{(3-x)}O_4$ spinel-type sorbents. *Combust. Sci. Technol.* **2017**, *23*, 511–515.
19. Dong, L.; Huang, Y.J.; Yuan, Q.; Cheng, H.Q.; Ding, S.Y.; Wang, S.; Duan, Y.F. Experimental study on the mercury removal from flue gas using manganese modified titanium-zirconium and titanium-tin composite oxide catalysts. *J. Fuel Chem. Technol.* **2020**, *48*, 741–751.
20. Wang, Z.; Yang, Y.J.; Liu, J.; Liu, F.; Yan, X.C. Experimental and theoretical insights into the effect of syngas components on $Hg^0$ removal over $CoMn_2O_4$ sorbent. *Ind. Eng. Chem. Res.* **2020**, *59*, 8078–8085. [CrossRef]
21. Shi, D.L.; Lu, Y.; Tang, Z.; Han, F.N.; Chen, R.Y.; Xu, Q. Removal of elemental mercury from simulated flue gas by cerium oxide modified attapulgite. *Korean J. Chem. Eng.* **2014**, *31*, 1405–1412. [CrossRef]
22. He, C.; Shen, B.X.; Chen, J.H.; Cai, J. Adsorption and oxidation of elemental mercury over Ce-$MnO_x$/Ti-PILCs. *Environ. Sci. Technol.* **2014**, *48*, 7891–7898. [CrossRef]
23. Xie, Y.J.; Yan, B.; Tian, C.; Liu, Y.X.; Liu, Q.X.; Zeng, H.B. Efficient removal of elemental mercury ($Hg^0$) by SBA-15-Ag adsorbents. *J. Mater. Chem. A* **2014**, *2*, 17730–17734. [CrossRef]
24. Cimino, S.; Scala, F. Removal of elemental mercury by $MnO_x$ catalysts supported on $TiO_2$ or $Al_2O_3$. *Ind. Eng. Chem. Res.* **2015**, *55*, 5133–5138. [CrossRef]
25. Liu, H.; You, Z.W.; Yang, S.; Liu, C.; Xie, X.F.; Xiang, K.S.; Wang, X.Y.; Yan, X. High-efficient adsorption and removal of elemental mercury from smelting flue gas by cobalt sulfide. *Environ. Sci. Pollut. Res.* **2019**, *26*, 6735–6744. [CrossRef] [PubMed]
26. Li, H.L.; Zhu, L.; Wang, J.; Li, L.Q.; Shih, K. Development of nano-sulfide sorbent for efficient removal of elemental mercury from coal combustion fuel gas. *Environ. Sci. Technol.* **2016**, *50*, 9551–9557. [CrossRef] [PubMed]
27. Kong, L.N.; Zou, S.J.; Mei, J.; Geng, Y.; Zhao, H.; Yang, S.J. Outstanding resistance of $H_2S$-modified $Cu/TiO_2$ to $SO_2$ for capturing gaseous $Hg^0$ from nonferrous metal smelting flue gas: Performance and reaction mechanism. *Environ. Sci. Technol.* **2018**, *52*, 10003–10010. [CrossRef]
28. Liu, W.; Xu, H.M.; Liao, Y.; Quan, Z.W.; Li, S.C.; Zhao, S.J.; Qu, Z.; Yan, N.Q. Recyclable CuS sorbent with large mercury adsorption capacity in the presence of $SO_2$ from non-ferrous metal smelting flue gas. *Fuel* **2019**, *235*, 847–854. [CrossRef]
29. Ding, S.Y.; Huang, Y.J.; Chen, H.; Dong, L.; Fan, C.H.; Hu, H.J.; Qi, E.B. Mercury removal performance of $CuCl_2$-modified magentic attapulgite. *Chem. Ind. Eng. Prog.* **2020**, *39*, 1187–1195.
30. Yang, J.P.; Zhao, Y.C.; Zhang, J.Y.; Zheng, C.G. Regenerable cobalt oxide loaded magnetosphere catalyst from fly ash for mercury removal in coal combustion flue gas. *Environ. Sci. Technol.* **2014**, *48*, 14837–14843. [CrossRef]
31. Zhang, Z.J.; Wang, Z.L.; Chakoumakos, B.C.; Yin, J.S. Temperature dependence of cation distribution and oxidation state in magnetic Mn-Fe ferrite nanocrystals. *J. Am. Chem. Soc.* **1998**, *120*, 1800–1804. [CrossRef]
32. Chandel, M.; Ghosh, B.K.; Moitra, D.; Patra, M.K.; Vadera, S.R.; Ghosh, N.N. Synthesis of various ferrite ($MFe_2O_4$) nanoparticles and their application as efficient and magnetically separable catalyst for biginelli reaction. *J. Nanosci. Nanotechnol.* **2018**, *18*, 2481–2492. [CrossRef] [PubMed]
33. Wang, Y.X.; Huang, Y.J.; Dong, L.; Yuan, Q.; Ding, S.Y.; Cheng, H.Q.; Wang, S.; Duan, Y.F. Experimental study on mercury removal of coal-fired flue gas over Co-doped iron-based oxide sorbent. *J. Fuel Chem. Technol.* **2020**, *48*, 785–794.

34. Shi, Y.J.; Deng, S.; Wang, H.M.; Huang, J.Y.; Li, Y.K.; Zhang, F.; Shu, X.Q. Fe and Co modified vanadium-titanium steel slag as sorbents for elemental mercury adsorption. *RSC. Adv.* **2016**, *6*, 15999–16009. [CrossRef]
35. Yue, Z.X.; Zhou, J.; Zhang, H.G.; Gui, Z.L.; Li, L.T. Auto-combustion behavior of nitrate-citrate gels and synthesis of ferrite nano-particles. *J. Chin. Ceram. Soc.* **1999**, *27*, 84–88.
36. Guo, M.Y.; Wang, Y.M.; Pan, Z.D.; Liu, S. Synthesis of nanocrystalline $(Co_{0.5}Cu_{0.5})(MnFe)O_4$ ceramic pigment via solution combustion technique. *J. Chin. Ceram. Soc.* **2015**, *43*, 411–417.
37. Yang, Y.J.; Liu, J.; Zhang, B.K.; Liu, F. Mechanistic studies of mercury adsorption and oxidation by oxygen over spinel-type $MnFe_2O_4$. *J. Hazard. Mater.* **2017**, *321*, 154–161. [CrossRef]
38. Xu, Y.; Luo, G.Q.; Pang, Q.C.; He, S.W.; Deng, F.F.; Xu, Y.Q.; Yao, H. Adsorption and catalytic oxidation of elemental mercury over regenerable magnetic Fe-Ce mixed oxides modified by non-thermal plasma treatment. *Chem. Eng. J.* **2019**, *358*, 1454–1463. [CrossRef]
39. Dong, L.; Huang, Y.J.; Liu, L.Q.; Liu, C.Q.; Xu, L.G.; Zha, J.R.; Chen, H.; Liu, H. Investigation of elemental mercury removal from coal-fired boiler flue gas over MIL101-Cr. *Energy Fuels* **2019**, *33*, 8864–8875. [CrossRef]
40. Zhang, Z.; Wu, J.; Li, B.; Xu, H.B.; Liu, D.J. Removal of elemental mercury from simulated flue gas by ZSM-5 modified with Mn-Fe mixed oxides. *Chem. Eng. J.* **2019**, *375*, 121946. [CrossRef]
41. Yang, S.J.; Yan, N.Q.; Guo, Y.F.; Wu, D.Q.; He, H.P.; Qu, Z.; Li, J.F.; Zhou, Q.; Jia, J.P. Gaseous elemental mercury capture from flue gas using magnetic nanosized $(Fe_{3-x}Mn_x)_{1-\delta}O_4$. *Environ. Sci. Technol.* **2011**, *45*, 1540–1546. [CrossRef]

Article

# The Crosslinker Matters: Vinylimidazole-Based Anion Exchange Polymer for Dispersive Solid-Phase Extraction of Phenolic Acids

Matthias Harder [1], Rania Bakry [1], Felix Lackner [1], Paul Mayer [1], Christoph Kappacher [1], Christoph Grießer [2], Sandro Neuner [3], Christian W. Huck [1], Günther K. Bonn [1,4] and Matthias Rainer [1,*]

[1] Institute of Analytical Chemistry and Radiochemistry, Leopold-Franzens University of Innsbruck, Innrain 80-82, A-6020 Innsbruck, Austria; matthias.harder@uibk.ac.at (M.H.); rania.bakry@uibk.ac.at (R.B.); felix.lackner@student.uibk.ac.at (F.L.); p.mayer@student.uibk.ac.at (P.M.); christoph.kappacher@uibk.ac.at (C.K.); christian.w.huck@uibk.ac.at (C.W.H.); guenther.bonn@uibk.ac.at (G.K.B.)

[2] Institute of Physical Chemistry, Leopold-Franzens University of Innsbruck, Innrain 52c, A-6020 Innsbruck, Austria; christoph.griesser@uibk.ac.at

[3] Department of General, Inorganic and Theoretical Chemistry, Leopold-Franzens University of Innsbruck, Innrain 80-82, A-6020 Innsbruck, Austria; sandro.neuner@uibk.ac.at

[4] ADSI-Austrian Drug Screening Institute, Innrain 66a, A-6020 Innsbruck, Austria

* Correspondence: m.rainer@uibk.ac.at; Tel.: +43-512-507-57307; Fax: +43-512-507-57399

**Abstract:** Crosslinkers are indispensable constituents for the preparation of SPE materials with ethylene glycol dimethacrylate (EGDMA) and divinylbenzene (DVB) among the most prominent representatives. A crosslinker that has not yet been used for the preparation of SPE sorbents is 3,3′-(hexane-1,6-diyl)bis(1-vinylimidazolium) bromide [$C_6$-bis-VIM] [Br]. In this study, we synthesized differently crosslinked vinylimidazole polymers with EGDMA, DVB and [$C_6$-bis-VIM] [Br] and evaluated their extraction efficiencies towards phenolic acids. Dispersive SPE experiments performed with the [$C_6$-bis-VIM] [Br] crosslinked polymers exhibited significantly higher extraction recoveries for the majority of analytes. Due to these promising results, the [$C_6$-bis-VIM] [Br] crosslinked polymer was optimized in terms of the monomer to crosslinker ratio and an efficient dispersive SPE protocol was developed, with maximum recoveries ranging from 84.1–92.5% and RSD values < 1%. The developed extraction procedure was also applied to cartridges resulting in recoveries between 97.2 and 98.5%, which were on average 5% higher than with the commercial anion exchange sorbent Oasis® MAX. Furthermore, the sorbent was regenerated showing a good reusability for the majority of analytes. In conclusion, this study clearly highlights the yet untapped potential of the crosslinker, [$C_6$-bis-VIM] [Br], with respect to the synthesis of efficient anion exchange polymers for SPE.

**Keywords:** crosslinker; phenolic acids; vinylimidazole; anion exchanger; co-polymer; solid-phase extraction; sustainable analytical sample preparation

## 1. Introduction

Naturally occurring phenolic acids belong to the substance class of phenolic compounds, which is one of the largest groups of secondary plant metabolites biosynthesized by vegetables, fruits, cocoa, teas and other plants [1]. Although the roles of phenolic acids in plants has not yet been completely clarified, they have been associated with various functions, including enzyme activity [2], nutrient uptake [3,4], allelopathy [5], as well as photosynthesis and protein synthesis [6]. Besides their important functions in plants, phenolic acids have been associated with the nutritional, organoleptic and antioxidative properties of foods. In this context, their content and profile have been intensively investigated by the food industry, mainly due to their effect on maturation, preservation and enzymatic browning [7,8]. Since phenolic acids are widespread in plant-based foods, they account for almost one third of dietary phenols. Consequently, there is a growing awareness and interest in the antioxidant behavior and human health benefits associated with these compounds [7,9]. Countless scientific studies have already examined their effects on different oxidative stress related diseases, including cancer [10,11], diabetes [12] and cardiovascular diseases [13]. Nowadays, it is generally recognized that the uptake of phenolic acids in conjunction to a cereal, fruit and vegetable rich diet strongly contributes to human health and reduced disease risk [14–16]. In this context, and due to the increasing industrial attention, efficient extraction and purification techniques regarding these compounds are in high demand [17].

Extraction techniques in analytical sample preparation are typically used for isolation, matrix simplification, pre-concentration or solvent exchange to facilitate a successful detection of analytes [18]. In this context, solid-phase extraction (SPE) has become an indispensable tool for both the laboratory and industrial fields [19] and even though SPE has been utilized for many years, it is still a dynamic field of science which is heavily researched especially in the field of miniaturization, automation, material science and eco-friendly analytical processes [18,20]. An alternative to conventional SPE is dispersive solid-phase extraction (DSPE) in which the sorbent is directly dispersed into the sample matrix. The close contact between the sorbent and analytes enhances both the sorption and elution process and therefore increases the efficiency of the overall extraction procedure. Its simplicity especially makes DSPE a valuable tool for analytical sample preparation [21,22].

Since the interaction mechanism and thus the quality of the extraction is highly dependent on the choice of sorbent, constant progress in the synthesis of novel materials has been and is still a driving force in the development of SPE [18,23]. The first emerging sorbents were silica- or carbon-based; however, their numerous disadvantages promoted the development of porous polymers to overcome the limitations presented by those materials [18,24]. Unlike silica-based sorbents, they are stable at virtually any pH value, contain no troublesome silanol groups and generally exhibit higher surface areas. In comparison to carbon-based sorbents, porous polymers are superior for recovering organic compounds from aqueous samples [24] and since their introduction, polymer-based materials have become an indispensable sorbent and have driven progress in the field of SPE. This is mainly due to their morphology and the possibility of incorporating various chemical functionalities into their porous framework, resulting in high retention capacities for different types of analytes and improved stability under different extraction conditions [18]. At present, a large selection of sorbents are available on the market, both as bulk material and in the form of packed cartridges, discs, pipette tips and 96-wellplates, enabling the use of SPE for a wide range of applications [25,26]. In this regard, polymer-based sorbents still count as one of the most important SPE materials with progressive research being conducted over the last decades [27–30].

The structure of polymerizable compounds and the polymerization process are primarily responsible for the physical and chemical properties of the obtained polymer [31]. There are several mechanisms for polymer synthesis with free-radical polymerization (FRP) being one of the most significant. Its robustness against impurities of the raw materials and the possibility to customize polymer properties by the choice of initiator and polymerization conditions are the main advantages over other processes. FRP is of enormous industrial importance, due to the large-scale production of vinyl polymers [32,33]. Furthermore, this method is also highly significant for the synthesis of efficient sorbents for SPE applications regarding scientific research. It is one of the most widely used techniques for the preparation of crosslinked polymers, both by academics and industrialists [34].

Incorporating compounds that have at least two free-radically polymerizable double bonds, so-called crosslinking agents or crosslinkers, enables the synthesis of polymers with three-dimensional networks. By changing the monovinyl to divinyl molar ratios, the degree of cross-linking can be altered to form weakly to highly crosslinked materials, which directly affects the properties of the obtained polymer. Many commercially available polymers owe their value to their specific three-dimensional structures, which makes crosslinking an important tool for polymer tuning [35]. Commonly used free-radically polymerizable crosslinking agents are divinylbenzene (DVB) or ethylene glycol dimethacrylate (EGDMA) [33]. DVB, for example, is an important crosslinker for commercial SPE sorbents, with Oasis® from Waters™ and Strata™-X from Phenomenex® being two of the most significant product lines regarding analytical sample preparation. EGDMA has become increasingly important in scientific research due to its higher polarity compared with DVB and the possibility to form hydrogen bonds with certain types of analytes [36–39]. A newly developed crosslinker that has been utilized for the synthesis of polymers is 3,3′-(hexane-1,6-diyl)bis(1-vinylimidazolium) bromide [$C_6$-bis-VIM] [Br]. Recent studies have shown that it can be used for the preparation of thermally and mechanically robust hydrogels [40], inverse opal microspheres [41], monolithic stationary phases [42] as well as anion-exchange membranes [43]; however, after a thorough search of the literature, it became apparent that [$C_6$-bis-VIM] [Br] has not yet been applied as a crosslinking agent for the preparation of SPE materials. Due to its structural properties, [$C_6$-bis-VIM] [Br] appeared to be well suited for the preparation of anion exchange sorbents and according to Wang et al., its synthesis is simple and feasible with high yields [40].

Phenolic acids can be isolated by different SPE materials, such as mixed-mode, ion-exchange, or reversed-phase sorbents. Yılmaz et al. reported the extraction of phenolic acids using different commercially available sorbents including Chromabond® C18, SAX and Oasis® HLB [44]. Furthermore, phenolic acids and flavonoids could be isolated and determined in honey by applying Bond Elut C18, Oasis® HLB, Strata-X and Amberlite XAD-2 [45]. Recent studies have shown that, in addition to commercially available sorbents, there is still great potential for the preparation of novel SPE materials to further improve the extraction of phenolic acids [46,47]. The aim of the presented study was to synthesize a novel [$C_6$-bis-VIM] [Br] crosslinked anion exchange polymer and subsequently develop an efficient extraction procedure for phenolic acids. Therefore, differently crosslinked polymers were prepared and compared with respect to their sorption capacity and analyte recoveries. The poly(n-VIM/$C_6$-bis-VIM) polymer was optimized regarding its monomer to crosslinker ratio and an efficient DSPE procedure was established. By adapting it to cartridges, the novel sorbent could be compared with the commercial anion exchange sorbent Oasis® MAX from Waters™. Furthermore, the reusability of the sorbent was investigated. Based on the obtained results, the potential of the crosslinker [$C_6$-bis-VIM] [Br] for the preparation of efficient and sustainable anion exchange materials was clearly demonstrated. We hope that this study is a promising step towards increasing the attractiveness of this substance for polymer science and sustainable SPE applications in the future.

## 2. Materials and Methods

Ferulic acid (FeraA, 99%), cinnamic acid (CinA, ≥99%), caffeic acid (CaffA, ≥98%), chlorogenic acid (ChlorA, ≥95%), 2,3-dihydroxybenzoic acid (DHB, 99%), 1-vinylimidazole (VIM, ≥99%), divinylbenzene (DVB, 80%), 2,2′-azobis(2-amidinopropane) dihydrochloride (AAPH, 97%), trifluoroacetic acid (TFA, 99%), diethyl ether ($Et_2O$, ≥99.9%), ethylene glycol dimethacrylate (EGDMA, 98%), and sodium phosphate monobasic monohydrate ($NaH_2PO_4 \cdot H_2O$, ≥99.0%), as well as basic, activated aluminum oxide ($Al_2O_3$, approx. 150 mesh, standard grade) were all purchased from Sigma Aldrich (St. Louis, MI, USA), an affiliate of Merck (Darmstadt, Germany). Sodium phosphate dibasic dehydrate ($Na_2HPO_4 \cdot 2H_2O$, ≥99.5%) and hydrochloric acid (HCl, 37% in $H_2O$) was obtained from Merck. The 1,6-dibromohexane (98%) was obtained from Acros Organics affiliate to Thermo Fisher Scientific (Waltham, MA, USA). The solvents methanol (MeOH, LC-MS grade), acetonitrile (ACN, LC-MS grade) and ethyl acetate (EtOAc, 99.5%) were purchased from Th. Geyer GmbH & Co. KG (Renningen, Germany) and 1-decanol (for synthesis) and 1-propanol (99.8%) from VWR™ (Radnor, PA, USA), and affiliate of Avantor™ (Radnor, PA, USA). Sodium hydroxide (NaOH, >99%), ammonium hydroxide solution (30% $NH_3$ in water) and formic acid (FA, ≥98%) were obtained from Carl Roth® (Karlsruhe, Germany). Empty SPE cartridges (1 mL) and matching polyethylene frits with 20 µm porosity were obtained from Supelco® (Bellefonte, PA, USA). Pre-packed Oasis® MAX cartridges (30 mg) were purchased from Waters™ (Milford, MA, USA).

Water was obtained from a Milli-Q water purification system from Merck Millipore (Burlington, MA, USA). The phenolic acid standard stock solution mixture contained 500 mg L$^{-1}$ of each analyte, including FerA, CinA, CaffA, ChlorA and DHB in 20 vol.% MeOH in water. Working standards were daily prepared by diluting the standard stock solution with 20 vol.% MeOH in water. All stock solutions were stored at 4 °C. Three phosphate buffers (0.1 M) with a pH 6.6, 7.0 and 7.7 were prepared in 5 vol.% MeOH in water by dissolving certain quantities of $Na_2HPO_4 \cdot 2H_2O$ and $NaH_2PO_4 \cdot H_2O$. The pH was adjusted by adding HCl or NaOH solution.

The DVB was extracted three times with 10% (w/v) NaOH solution and distilled under vacuum. The EGDMA was extracted with activated basic $Al_2O_3$ prior its use.

### 2.1. Instrumentation and Chromatographic Separation

HPLC analyses were performed on a 1100 Series HPLC Value System from Agilent (Santa Clara, CA, USA) equipped with a 1100 Series diode array detector (DAD). Analyte separation was performed with an Excel 5 C18 analytical column from ACE (Aberdeen, UK) with the dimensions 150 × 4.6 mm. The mobile phase consisted of 0.1 vol.% TFA in water (eluent A) and MeOH (eluent B). A gradient program with the following settings was applied (min/vol.% of eluent B): 0/10, 14/80, 14.1/99, 16/99, 16.1/10, 20/10. The injection temperature was set to 10 °C and the injection volume was 5 µL with a needle-wash in MeOH. The mobile-phase flow-rate was 700 µL min$^{-1}$ and the temperature of the column oven was set to 40 °C. The detection of analytes was performed at 218 nm.

DSPE experiments were conducted on a ThermoMixer® C and the evaporation of solvents was performed using a Concentrator plus, both from Eppendorf AG (Hamburg, Germany). Drying of the synthesized crosslinker and polymers was accomplished in a Vacutherm vacuum drying oven from Thermo Fisher Scientific. The pH value of the phosphate buffer solutions was checked with a pH meter SevenEasy™ from Mettler Toledo (Columbus, OH, USA). BET measurements were performed utilizing a NOVA 2000e Surface Area and Pore Size Analyzer from Quantachrome Instruments (Boynton Beach, FL, USA) and the software, Quantachrome NovaWin. Prior to the measurement, the sample was heated to 120 °C in vacuum for 30 min followed by $N_2$ adsorption at 77 K (five points from 0.05 to 0.30 p/p$_0$). Images of the polymer surface were made by scanning electron microscopy (SEM) on a JSM-6010LV from JEOL (Freising, Germany). Prior to measurements the samples were sputtered with gold and measurements were conducted at high vacuum with an excitation voltage of 15 kV. NMR spectra were recorded with a Avance DPX 300 MHz spectrometer from Bruker (Billerica, MA, USA) equipped with a 5 mm broadband probe at 25 °C. ATR-FTIR measurements were performed using a Spectrum 100 equipped with a Universal ATR Sampling Accessory with the software Spectrum (version 2.0.0.0) from PerkinElmer (Waltham, MA, USA). Prior to measurements, the samples were dried in a vacuum centrifuge for 24 h with the mode setting V-AQ. Six scans were carried out with a wavelength range of 4000–650 cm$^{-1}$ and a resolution of 4 cm$^{-1}$.

### 2.2. Method Validation

The HPLC-DAD method was validated according to the "Society of Toxicological and Forensic Chemistry" (GTFCh) to examine its applicability towards the analysis of phenolic acids [48]. A linear regression model was established to investigate the linearity and precision of the detection method. Analyte stabilities during the HPLC-DAD measurements and storage period were monitored. Furthermore, the limit of detection (LOD) and limit of quantification (LOQ) were determined.

#### 2.2.1. Linearity of Calibration

Two dilution series from 25 to 250 mg L$^{-1}$ and 250 to 500 mg L$^{-1}$ were prepared by diluting the standard stock solution with 20 vol.% MeOH in water. Three calibration sets for each dilution series were prepared and each standard concentration was measured three times.

#### 2.2.2. Repeatability

Two phenolic acid standards (50 and 500 mg L$^{-1}$) were measured using the same HPLC-DAD method and the same instrumentation for a period of 8 days. Each standard was therefore divided into three aliquots, stored at 4 °C and measured once a day.

#### 2.2.3. Processed Sample Stability

Stabilities of phenolic acids at two concentrations (50 and 500 mg L$^{-1}$) were investigated over a period of 19 h in the autosampler at 10 °C. Each standard was divided into three separate vials and measured every hour.

#### 2.2.4. Long-Term Stability

Phenolic acid stabilities at two concentrations (50 and 500 mg L$^{-1}$) were investigated for 14 days at 4 °C and 23 °C. Therefore, each standard was divided into six separate vials. Three aliquots were stored at 4 °C and three at 23 °C. All standards were measured once a day.

#### 2.2.5. LOD and LOQ

A calibration model ranging between 0.25 and 1.50 mg L$^{-1}$ was established by diluting the standard stock solution (500 mg L$^{-1}$) with 20 vol.% MeOH in water. Three identical dilution series were prepared, and each standard concentration was measured three times. The LOD and LOQ were calculated according to DIN 32645:2008-11 [49].

### 2.3. Synthesis of 3,3′-(Hexane-1,6-diyl)bis(1-vinylimidazolium) Bromide

The dicationic crosslinker, 3,3′-(hexane-1,6-diyl)bis(1-vinylimidazolium) bromide [C$_6$-bis-VIM] [Br], was synthesized according to Wang et al. [40]. For the implementation, 1-vinylimidazol (4.2 mL, 47 mmol) and 1,6-dibromohexane (3.2 mL, 21.2 mmol) were dissolved in 10 mL of MeOH. The mixture was stirred at 60 °C for 24 h. Subsequently, 100 mL EtOAc was added to precipitate the product. After cooling the mixture to 4 °C, the solid was filtered off, washed with 20 mL EtOAc three times, and dried under high vacuum to a constant weight. Yield: 7.9 g (86%) white, crystalline powder—$^1$H NMR (300 MHz, DMSO): δ 9.73 (s, 2H), 8.26 (s, 2H), 8.01 (s, 2H), 7.34 (dd, $J$ = 15.7, 8.8 Hz, 2H), 6.01 (d, $J$ = 15.6 Hz, 2H), 5.42 (d, $J$ = 8.7 Hz, 2H), 4.23 (t, $J$ = 7.2 Hz, 4H), 1.85 (s, 4H), 1.31 (s, 4H).

### 2.4. Synthesis of Crosslinked Vinylimidazole Polymers

The synthesis of vinylimidazole-based co-polymers was achieved according to a published article from the literature [42]. For the implementation of the free-radical polymerization procedure, certain quantities of VIM (0.77 mL, 8.50 mmol), 1-propanol (2.24 mL, 37.26 mmol), 1-decanol (0.48 mL, 3.05 mmol), water (1.20 mL, 66.72 mmol) and the crosslinker (1.85 mmol) were mixed in a sealable amber glass vial. Subsequently, the initiator AAPH (0.121 g, 0.45 mmol) was added and the mixture ultrasonified. After flushing with nitrogen for 10 min, the vial was placed into an oil bath at 70 °C for 24 h. In the next step, the obtained monolith was grinded using a mortar and pestle and washed with water, MeOH and Et$_2$O. Subsequently the white powder was dried in a vacuum drying oven at 30 °C (150 mbar) until a constant weight.

Three differently crosslinked polymers were synthesized with the crosslinkers [C$_6$-bis-VIM] [Br] (0.800 g, 1.85 mmol), EGDMA (0.35 mL, 1.85 mmol) and DVB (0.26 mL, 1.85 mmol) according to the previously described procedure, keeping the same monomer to crosslinker molar ratio of 4.59 to 1.

Furthermore, several poly(n-VIM/C$_6$-bis-VIM) polymers were prepared by changing the ratio between the monomer and crosslinker, by using the described polymerization procedure above. The molar sum of the monomer and crosslinker (10.35 mmol) was kept constant, whereas their molar ratio was varied between 1.15 to 1 and 13.80 to 1. All synthesized poly(n-VIM/$C_6$-bis-VIM) polymers are summarized in Table S1. At the monomer to crosslinker molar ratios below 2.30 to 1, 100 µL ACN was added to the mixture before sonification to avoid phase separation.

### 2.5. DSPE Protocol for Phenolic Acids

For the extraction of phenolic acids from aqueous samples, a four-step DSPE protocol was established including sorbent conditioning, sample loading, washing and analyte elution. In the first step, $10.0 \pm 0.2$ mg of solid sorbent was weighed into a 2 mL centrifuge tube. Subsequently, 0.5 mL of conditioning solution was added, and the mixture shaken on a thermomixer (5 min, 1200 rpm, 25 °C). After centrifugation (5 min, 14,000 rpm, 23 °C) the conditioning solution was discarded, and the polymer washed with 1 mL of 5 vol.% MeOH in a water solution by applying the same shaking and centrifugation settings as before. An amount of 1 mL of the sample was added, and the mixture shaken (15 min, 1200 rpm, 25 °C) and centrifuged (5 min, 14,000 rpm, 23 °C) with the supernatant analyzed by HPLC-DAD. The residual sorbent was washed with 1 mL of 5 vol.% MeOH in a water solution. The analyte elution was obtained by adding 1 mL of eluting solution and shaking (15 min, 1200 rpm, 25 °C). After centrifugation (5 min, 14,000 rpm, 23 °C) the supernatant was analyzed.

To maximize the phenolic acid recoveries, the conditioning and eluting solution were optimized. Furthermore, the effect of the extraction time on the adsorption equilibrium was investigated to minimize the duration of the extraction process. These DSPE experiments were performed with the poly(n-VIM/$C_6$-bis-VIM) sorbent with a monomer to crosslinker ratio of 11.50 to 1.

#### 2.5.1. Optimization of the Conditioning Solution

Five different conditioning solutions were applied including a solution of 5 vol.% MeOH in water containing 0.5% ($w/w$) LiCl, three phosphate buffers with different pH values (6.6, 7.0 and 7.7) and a 5 vol.% MeOH in water solution. All applied conditioning solutions are summarized in Table S2. The experiments were performed according to the four-step DSPE protocol in Section 2.5. The loading was performed with 1 mL of 500 mg $L^{-1}$ phenolic acid standard mix. After shaking and centrifugation the supernatant was analyzed.

#### 2.5.2. Optimization of the Eluting Solution

The elution was optimized by testing various eluting solutions containing different quantities of TFA, LiCl, ACN and water. All applied eluting solutions are provided in Table S3. DSPE experiments were carried out according to Section 2.5. The sorbent was conditioned with a 0.5 mL phosphate buffer (pH 7.0) and the loading was performed with 1 mL of 500 mg $L^{-1}$ phenolic acid standard mix. After washing, 1 mL of the eluting solution was added, the mixture shaken, centrifuged, and subsequently analyzed.

#### 2.5.3. Effect of the Extraction Time

To investigate the effect of the extraction time on the adsorption equilibrium, analyte extraction was performed for different time periods (1.0, 2.5, 5.0, 10.0 and 20.0 min). DSPE experiments were carried out according to Section 2.5. The conditioning was performed with a 0.5 mL phosphate buffer pH 7.0 and the sample consisted of 1 mL of 500 mg $L^{-1}$ phenolic acid standard mix. The obtained supernatant after loading was analyzed.

### 2.6. Effect of the Crosslinker on DSPE Efficiency

To compare the differently crosslinked polymers with each other, the extraction experiments were performed according to Section 2.5. As a solid sorbent, polymers with a monomer to crosslinker ratio of 4.59 to 1 were used. The conditioning was carried out with a 0.5 mL phosphate buffer pH 7.0 and loading with 1 mL of 500 mg $L^{-1}$ phenolic acid standard mix. The elution was performed with 1 mL 2 vol.% TFA, 1.5% ($w/w$) LiCl and 50 vol.% ACN in water. The supernatant after loading and the eluting solution were both analyzed.

*2.7. Effect of the Monomer to Crosslinker Ratio on DSPE Efficiency*

The effect of polymer composition on the extraction efficiency was investigated by performing DSPE experiments with poly(n-VIM/$C_6$-bis-VIM) polymers with different monomer to crosslinker ratios. DSPE experiments were carried out according to Section 2.5. The conditioning was performed with a 0.5 mL phosphate buffer pH 7.0 and loading with 1 mL of 500 mg $L^{-1}$ phenolic acid standard mix. The elution was carried out with 1 mL 2 vol.% TFA, 1.5% ($w/w$) LiCl and 50 vol.% ACN. The supernatant after loading and the eluting solution were analyzed.

*2.8. Sorption Capacity*

Different standard concentrations (10, 20, 50, 100, 250 and 500 mg $L^{-1}$) were applied during the sample loading to investigate the effect of the concentration on the extraction recovery. The DSPE experiments were performed according to Section 2.5 with the poly(n-VIM/$C_6$-bis-VIM) sorbent with a monomer to crosslinker ratio of 11.50 to 1. The elution was performed with 1 mL 2 vol.% TFA, 1.5% ($w/w$) LiCl and 50 vol.% ACN in water. The supernatant after loading and the eluting solution were subsequently analyzed.

*2.9. Polymer Reusability*

The reusability of the poly(n-VIM/$C_6$-bis-VIM) polymer with a monomer to crosslinker ratio of 11.50 to 1, was investigated by performing several DSPE cycles in succession. Firstly, 10.0 ± 0.2 mg of solid sorbent was weighed into a 2 mL centrifuge tube. Sorbent conditioning and sample loading were performed according to Section 2.5. with a 0.5 mL phosphate buffer pH 7.0 and 1 mL of 100 mg $L^{-1}$ phenolic acid standard mix. The sorbent was subsequently washed four times with 1 mL of 2 vol.% TFA, 1.5% ($w/w$) LiCl and 50 vol.% ACN in water and once with 1 mL of pure MeOH, with shaking (5 min, 1200 rpm, 25 °C) and centrifuging (5 min, 14,000 rpm, 23 °C) after every step. Finally, the polymer was dried at 60 °C and reused for the next DSPE cycle. The described procedure was carried out for a total of five times.

*2.10. Sorbent Comparison Study*

The poly(n-VIM/$C_6$-bis-VIM) polymers with a monomer to crosslinker ratio of 11.50 to 1 was compared with the commercially available anion exchange sorbent Oasis® MAX from Waters™. Therefore, the SPE cartridges were equipped with a polyethylene frit (20 μm) and 30.0 ± 0.2 mg of poly(n-VIM/$C_6$-bis-VIM) was added. The Oasis® MAX was purchased as pre-packed cartridges containing 30.0 mg of solid sorbent.

Experiments were conducted with an adjustable air pressure machine using a SPE procedure similar to the DSPE procedure described in Section 2.5. Firstly, the sorbent was conditioned with 0.5 mL of phosphate buffer (pH 7.0) and 1 mL 5 vol.% MeOH in water. Subsequently, 1 mL of phenolic acid standard mix (100 mg $L^{-1}$) was added. Then the sorbent was washed once with 1 mL 5 vol.% MeOH in water. Elution was accomplished by applying 1 mL of 1.5% ($w/w$) LiCl and 50 vol.% ACN in water.

In addition, the recommended SPE standard protocol for Oasis® MAX from Waters™ was also tested [50]. In this case, the conditioning was performed with 1 mL MeOH and 1 mL water, loading with 1 mL of phenolic acid standard mix (100 mg $L^{-1}$) and washing with 1 mL 5 vol.% $NH_3$ in water and 1 mL MeOH. The elution of analytes was performed with 2 vol.% FA in MeOH.

The solutions obtained after loading and elution were centrifuged (5 min, 14,000 rpm, 23 °C) and an aliquot of the supernatants was subsequently analyzed.

## 3. Results and Discussion

A validated HPLC-DAD method was established for the quantification of different phenolic acids from the aqueous samples. In order to assess the accuracy, reliability and repeatability of the obtained data, SPE experiments and HPLC-DAD measurements were all carried out at least in triplicate.

### 3.1. Method Validation

The HPLC-DAD method was successfully validated and included linearity of calibration, repeatability, processed sample stability, long-term stability as well as the calculation of LOD and LOQ.

Both calibration ranges (25–250 mg $L^{-1}$ and 250–500 mg $L^{-1}$) showed a linearity in the examined concentration range with coefficient of determinations ranging from 0.9995–0.9999 and 0.9959–0.9985, respectively. For the calibration model 25–250 mg $L^{-1}$, the maximum relative standard deviation (RSD) value was 2.2%, the minimum bias −5.4 and the maximum bias 6.1%. The RSD as well as minimum and maximum bias for the calibration model 250–500 mg $L^{-1}$, were 0.9, −3.3 and 2.0%, respectively. All results are summarized in Tables S4 and S5.

Repeatability of the applied HPLC-DAD method was investigated by calculating the interday and intraday RSD values over a period of eight days. All RSD values were far below 15%, demonstrating repeatability of the applied HPLC-DAD method. All results are given in Tables S6 and S7.

Processed sample stability results showed that the concentration did not significantly change during a measurement period of 19 h at 10 °C in the autosampler. The maximum deviations from the initial value were ranging between 1.1 and 2.4%, respectively.

The long-term stability was investigated and showed that all phenolic acids at a concentration of 50 and 500 mg $L^{-1}$ were stable for at least 14 days at 4 °C as well as 23 °C. The maximum deviations from the initial value at 4 °C were ranging from 0.9–1.8% and at 23 °C from 1.2–5.2%.

The LOD and LOQ were calculated according to DIN 32645:2008-11 and gave values ranging from 0.02–0.04 mg $L^{-1}$ and 0.08–0.12 mg $L^{-1}$, respectively. All results are given in Table S8.

### 3.2. DSPE Method Optimization

An efficient DSPE protocol for the extraction of phenolic acids was established by optimizing the extraction time, the conditioning and eluting solution. Subsequently, the EGDMA, DVB and [$C_6$-bis-VIM] [Br] crosslinked vinylimidazole polymers were compared regarding their phenolic acid extraction efficiency. It was observed that the type of crosslinker had a significant influence on the extraction recoveries. Furthermore, the influence of the monomer to crosslinker ratio of the poly(n-VIM/$C_6$-bis-VIM) polymer on the extraction efficiency was investigated.

It was shown that by using a phosphate buffer for the conditioning, a greater percentage of analytes could be bound to the polymer than by using an aqueous solution containing 0.5% LiCl ($w/w$) and 5 vol.% MeOH or a solution of 5 vol.% MeOH in water. All results are summarized in Table S2. In particular, the CaffA, FerA and CinA were strongly affected by the type of conditioning solution. This can probably be explained by their similar pKa values, which are significantly higher compared with ChlorA and DHB.

The amount of bound CaffA, FerA and CinA could be increased by 3.2, 6.2 and 7.7%, respectively, when using a phosphate puffer (pH 7.0) instead of 5 vol.% MeOH in water. The effect of conditioning pH on the bonding of phenolic acids was tested between 6.0 to 7.7. Higher pH values were avoided due to the instability of phenolic acids [51].

With increasing pH value of the phosphate buffer, the adsorption of CaffA, FerA and CinA slightly increased, whereby this trend was more pronounced for FerA and CinA than for CaffA; however, the difference between pH 7.0 and 7.7 was not very pronounced. Based on the knowledge that phenolic acids can degrade at basic pH, a phosphate buffer with pH 7.0 was selected as the conditioning solution for further DSPE experiments.

The adsorption equilibrium was investigated to minimize the duration of the extraction process during the DSPE experiments. All results are given in Table S9. As can be shown in Figure 1 the quantity of bound analytes did not significantly change after 10 min. We decided to use an extraction time of 15 min for following the DSPE experiments to be certain that a complete adsorption equilibrium was established.

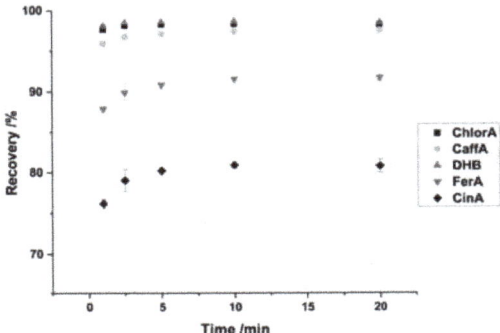

**Figure 1.** Analyte recovery results obtained from DSPE experiments with the poly(n-VIM/$C_6$-bis-VIM) polymer with a monomer to crosslinker ratio of 11.50 to 1, by applying different extraction times. Standard deviations are given as error bars.

In order to optimize the elution of analytes from the solid sorbent, different eluting conditions were applied. In the preliminary experiments, MeOH and ACN with and without different concentrations of TFA were tested. Furthermore, the addition of NaCl and LiCl was investigated. It was found that higher recoveries could be achieved with ACN compared to MeOH and LiCl compared to NaCl. By adding water to the organic solvent, the solubility of the LiCl was improved and the analyte recoveries could further be increased. Based on these results, six different eluting solutions containing various quantities of TFA, LiCl, ACN and water were tested. The highest recoveries for all analytes (except DHB) could be obtained with 2 vol.% TFA, 1.5% ($w/w$) LiCl and 50 vol.% ACN in water (Figure 2). All results are provided in Table S3.

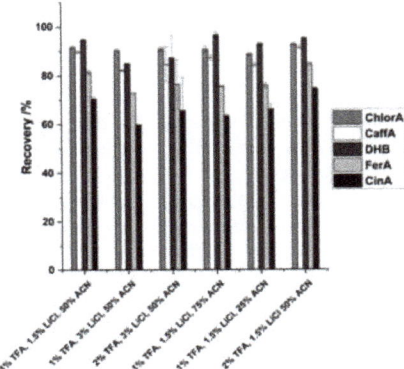

**Figure 2.** Analyte recovery results obtained from DSPE experiments with the poly(n-VIM/$C_6$-bis-VIM) polymer with a monomer to crosslinker ratio of 11.50 to 1, by applying different eluting solutions. Standard deviations are given as error bars.

## 3.3. Effect of the Crosslinker on DSPE Efficiency

The effect of differently crosslinked polymers on the extraction efficiency towards phenolic acids is displayed in Figure 3 and the results are summarized in Table S10. All solid sorbents exhibited different percentage recoveries indicating the significant influence of the crosslinker. All sorbents showed the same systematic increase in recoveries following the order: CinA < FerA < CaffA < ChlorA < DHB. This circumstance could be due to the common basic structure of all three polymers, which is due to the use of the same monomer VIM. The reason why the extraction recoveries followed this particular order can probably be explained by the different pKa values. The lower the pKa, the stronger the acid, which means that the substance dissociates more strongly in water. Therefore, DHB and ChlorA with the lowest pKa values should also exhibit the highest proportion of dissociated molecules in anionic form. Since the sorbent used is an anion exchanger, more analytes could be bound to the sorbent via ionic interactions compared to the other phenolic acids. This consequently also led to higher extraction recoveries. CaffA, FerA and CinA have similar pKa values, however they exhibited different extraction recoveries. A reason could be the different number of hydroxyl groups in their chemical structures. Occurring hydrogen bonds may increase the adsorption force between the phenolic acids and sorbent, which could explain the previously stated order of recoveries.

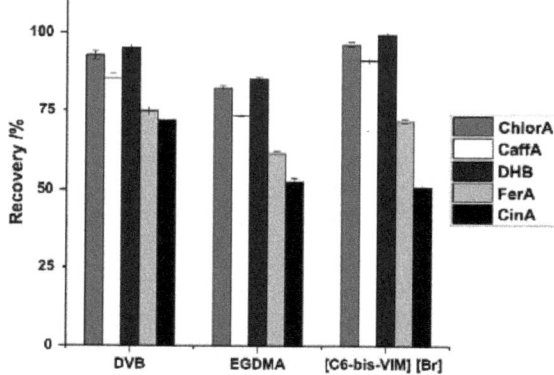

**Figure 3.** Analyte recovery results obtained from DSPE experiments with three differently crosslinked vinylimidazole polymers. Standard deviations are given as error bars.

The EGDMA crosslinked polymer exhibited the lowest extraction recoveries (52.4–85.3%) for all phenolic acids except the CinA. This can probably be explained by the absence of an aromatic moiety. Significantly better recoveries (71.9–95.1%) were obtained with the DVB crosslinked polymer. It was assumed that the EGDMA crosslinked polymer would be better suited for the extraction of phenolic acids due to its more polar structure and the possibility of forming hydrogen bonds. On the contrary, DVB has an aromatic ring structure, which enables the formation of π-π interactions with phenolic acids. This was probably the main reason for the higher recoveries with the DVB crosslinked polymer.

The polymer synthesized with [$C_6$-bis-VIM] [Br] showed promising recovery results especially for DHB, ChlorA and CaffA, ranging between 91.1 and 99.3%. In contrast, the recoveries for FerA (71.4%) and CinA (50.8%) were significantly lower. In comparison to the EGDMA crosslinked polymer, much higher recoveries were obtained for all analytes except the CinA. In detail, 10.2% higher recoveries were obtained for FerA, 13.8% for ChlorA, 14.0% for DHB and 17.9% for CaffA. Only the recovery for CinA was 1.7% lower. Compared to the sorbent synthesized with DVB, the crosslinked [$C_6$-bis-VIM] [Br] polymer showed significantly higher recoveries for ChlorA, CaffA and DHB. Slightly lower recoveries were found for FerA, and much lower recoveries were obtained for CinA. It is very likely that

FerA and CinA, due to their higher pKa values, were mainly present as uncharged species and therefore were more affected by the non-polar interactions of the DVB. In summary, the recovery with poly(n-VIM/$C_6$-bis-VIM) in comparison to the DVB crosslinked polymer for ChlorA was higher by 3.4%, for DHB by 4.2% and for CaffA by 5.6%. In contrast, the recoveries were 3.4% smaller for FerA and 21.1% for CinA.

Phenolic acid loading capacities (q) of the three differently crosslinked vinylimidazole polymers were calculated according to the following equation:

$$q = [(c_i - c_{eq}) * V]/m \quad (1)$$

where $c_i$ is the initial loading concentration, $c_{eq}$ the concentration at equilibrium, V the sample volume and m the dry mass of the applied polymer. From the summarized results in Table 1, it is evident that the type of crosslinker had a significant influence on the loading capacities of the different polymers. The EGDMA crosslinked sorbent exhibited the lowest capacities for all analytes except CinA. Polymers with the crosslinker DVB showed very high loading capacities for all phenolic acids. High capacities for ChlorA, CaffA and DHB were also obtained with the [$C_6$-bis-VIM] [Br] crosslinked polymer.

**Table 1.** Calculated loading capacities obtained from DSPE experiments with three differently crosslinked vinylimidazole polymers.

| Applied Crosslinker | ChlorA | CaffA | DHB | FerA | CinA |
|---|---|---|---|---|---|
| DVB | 49.2 ± 0.9 | 46.4 ± 0.7 | 49.4 ± 1.0 | 42.0 ± 0.4 | 40.2 ± 0.7 |
| EGDMA | 45.5 ± 0.7 | 41.9 ± 0.8 | 44.6 ± 0.5 | 37.1 ± 0.7 | 32.5 ± 0.5 |
| [$C_6$-bis-VIM] [Br] | 47.8 ± 0.3 | 45.9 ± 0.3 | 48.4 ± 0.3 | 38.5 ± 0.2 | 29.8 ± 0.2 |

Loading capacity: mean ± SD /mg g$^{-1}$ polymer; n = 3.

It is noteworthy that despite the lower loading capacities of poly(n-VIM/$C_6$-bis-VIM) compared to the DVB crosslinked polymer, higher recoveries could be achieved for DHB, ChlorA and CaffA. This circumstance can possibly be explained by the more complete elution when using the [$C_6$-bis-VIM] [Br] crosslinked polymer.

Furthermore, we investigated the effect of the molar ratio between VIM and [$C_6$-bis-VIM] [Br] on the extraction efficiency of phenolic acids. The monomer to crosslinker molar ratio was varied between 1.15 to 1.00 and 13.80 to 1.00. All applied molar ratios and results are summarized in Table S1. It was apparent that analyte recoveries were dependent on the monomer to crosslinker ratios (Figure 4), especially the recoveries of FerA and CinA that strongly increased with increasing monomer content, whereby a maximum was reached at a monomer to crosslinker ratio of 11.50 to 1. ChlorA and CaffA showed a similar behavior; however, the increase in recovery was not as pronounced. The DHB showed a different recovery pattern. With an increasing monomer content, the recoveries steadily decreased. In conclusion, the highest average recovery for the phenolic acids was obtained at a monomer to crosslinker ratio of 11.50 to 1 with the analyte recoveries ranging between 74.5 and 92.7%. Therefore, subsequent DSPE experiments were solely performed with this specific polymer.

By optimizing the VIM to crosslinker ratio, all loading capacities could consequently be significantly increased. The optimized [$C_6$-bis-VIM] [Br] crosslinked polymer with a molar ratio of 11.50 to 1 exhibited the highest average loading capacity for phenolic acids. Compared to the EGDMA and DVB crosslinked polymers, higher capacities for all analytes except the CinA were achieved with this specific composition.

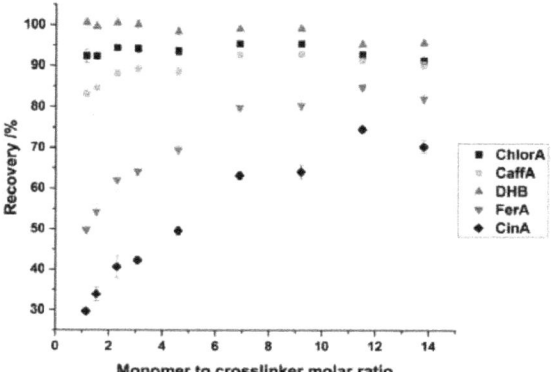

**Figure 4.** Analyte recovery results obtained from DSPE experiments with [$C_6$-bis-VIM] [Br] crosslinked vinylimidazole polymers with different monomer to crosslinker ratios. Standard deviations are given as error bars.

*3.4. Sorption Capacity*

DSPE experiments were performed with different loading concentrations (10–500 mg $L^1$) to investigate the influence of the standard concentration on the extraction efficiency. Results with the poly(n-VIM/$C_6$-bis-VIM) polymer with a monomer to crosslinker ratio of 11.50 to 1 are displayed in Figure 5 and given in Table S11.

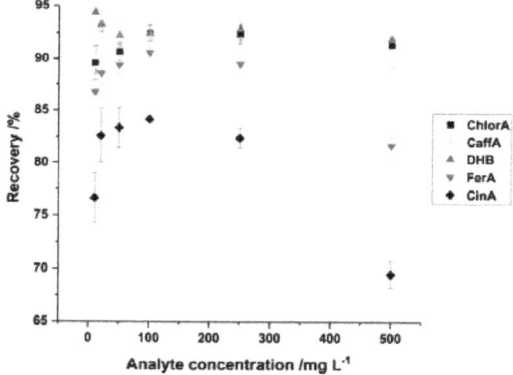

**Figure 5.** Extraction recoveries obtained from DSPE experiments with the poly(n-VIM/$C_6$-bis-VIM) polymer with a monomer to crosslinker ratio of 11.50 to 1, by applying different loading concentrations. Standard deviations are given as error bars.

It was observed that the concentration of phenolic acids in the sample had a direct influence on extraction recoveries. At a concentration of 10 mg $L^{-1}$ no phenolic acids were detected in the supernatant, indicating a quantitative binding; however, the highest recoveries could not be obtained in this particular case. It is possible that there existed a limited number of binding sites on the polymer that have very strong adsorption properties towards phenolic acids. Hence, a small proportion of analytes was strongly bound and was not eluted under the applied conditions. As a result, they remained attached to the polymer and significantly reduced the extraction recoveries. At analyte concentrations of 20, 50 and 100 mg $L^{-1}$, only CinA was detected in the supernatant and at 250 mg $L^{-1}$ all analytes were detected except for ChlorA. In contrast, at 500 mg $L^{-1}$ all analytes were visible in the chromatogram. The extraction recoveries obtained from the DSPE experiments with the highest sample concentration (500 mg L$^{-1}$) were limited by the loading capacity of the sorbent. Depending on the type of phenolic acid, 1.4 to 19.0% of the loaded concentration remained dissolved in the supernatant. These findings are in accordance with the results from Section 3.3, as the binding capacities were less than 50 mg g$^{-1}$ polymer, which indicated an overloading of the sorbent. Even though a high proportion of the bound analytes could be eluted from the polymer, the recoveries were still lower than at other concentration levels. The optimal concentration range for the extraction of phenolic acids with 10.0 ± 0.2 mg of sorbent was between 20 and 250 mg L$^{-1}$, with the highest average recovery observed at 100 mg L$^{-1}$.

*3.5. Polymer Reusability*

The reusability of the poly(n-VIM/C$_6$-bis-VIM) polymer with a monomer to crosslinker ratio of 11.50 to 1 was investigated by performing multiple DSPE cycles. Therefore, after each loading step the sorbent was regenerated, dried, and reused. In Figure 6, the quantity of bound phenolic acids after one to five DSPE cycles is displayed. Further data is given in Table S12. The applied polymer showed excellent reusability results for DHB with a maximum adsorption decrease of 0.6%. Good reusability results were also obtained for the ChlorA and CaffA with a maximum decrease in adsorption of 5.3 and 13.4%, respectively. A considerable decrease in the adsorption performance was observed for FerA and CinA with a maximum decrease of 30.7 and 50.3%, respectively. It is noteworthy that the quantity of bound ChlorA, CaffA, FerA and CinA mainly decreased after the first DSFE cycle and remained constant after the subsequent cycles. It is possible that the washing solution containing LiCl salt was responsible for this trend by influencing the binding properties of the sorbent. Since the FerA and CinA generally exhibited the weakest binding strength, they were probably the most affected by this effect. By optimizing the washing solution, it is likely to also result in a better reusability for FerA and CinA.

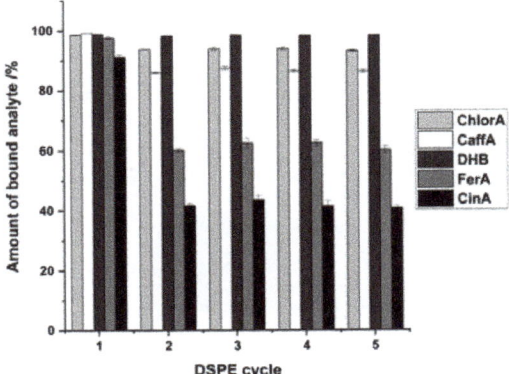

**Figure 6.** Binding results after performing multiple DSPE cycles with the poly(n-VIM/C$_6$-bis-VIM) polymer with a monomer to crosslinker ratio of 11.50 to 1. Standard deviations are given as error bars.

In summary, it was shown that the [C$_6$-bis-VIM] [Br] crosslinked polymer could be regenerated and reused for the majority of analytes making it a good candidate regarding sustainable analytical sample preparation.

*3.6. Sorbent Comparison Study*

The extraction efficiency of the novel poly(n-VIM/C$_6$-bis-VIM) polymer with a monomer to crosslinker ratio of 11.50 to 1 was evaluated by comparing it with Oasis® MAX from

Waters™. Oasis® MAX is a strong mixed-mode anion exchange and reversed-phase polymeric sorbent with a quaternary amine as the anion exchange function. Waters™ recommends it for the separation of acidic compounds [50]. Therefore, it was selected as a comparative polymer in this study.

Using the developed SPE protocol, extraction recoveries above 90% could be obtained with both sorbents. In addition, no phenolic acid residues were detected in the supernatants. Significantly higher extraction recoveries for all analytes were found with the synthesized sorbent in comparison with the Oasis® MAX. The poly(n-VIM/$C_6$-bis-VIM) polymer gave recovery results ranging between 97.2 and 98.5%, which were on average 5% higher than with the commercial solid sorbent. All results are given in Table 2.

**Table 2.** Recovery comparison of the developed poly(n-VIM/$C_6$-bis-VIM) polymer with a monomer to crosslinker ratio of 11.50 to 1 and the commercially available SPE sorbent Oasis® MAX.

| SPE Sorbent | ChlorA | CaffA | DHB | FerA | CinA |
|---|---|---|---|---|---|
| poly(n-VIM/$C_6$-bis-VIM) | 98.5 ± 3.3 | 97.2 ± 2.0 | 97.4 ± 3.1 | 97.4 ± 1.9 | 97.4 ± 1.7 |
| Oasis® MAX | 93.8 ± 0.6 | 92.5 ± 0.3 | 92.8 ± 0.3 | 92.3 ± 0.2 | 91.6 ± 0.2 |

Extraction recovery: mean ± SD /%; n = 3.

Since quantitative adsorption was achieved for both polymers, the differences in analyte recovery can only be due to differences in elution efficiency. Either the binding of phenolic acids with Oasis® MAX was too strong or the eluting conditions too weak to obtain higher recovery results.

Results with the recommended SPE standard protocol from Waters™ showed lower extraction recoveries for all phenolic acids compared to the developed protocol in this study. Furthermore, two new signals appeared in the chromatogram and the background noise increased considerably. It is very likely that the phenolic acids degraded during the washing step, due to the use of an ammonia solution and the instability of phenolic acids under alkaline conditions.

It has to be mentioned that the backpressure during the SPE experiments with the poly(n-VIM/$C_6$-bis-VIM) polymer was very high probably due to swelling effects. Therefore, the use of DSPE was clearly at an advantage in comparison to conventional SPE.

The developed DSPE and adapted SPE procedure with the poly(n-VIM/$C_6$-bis-VIM) polymer were compared with different solid-phase extraction methods and sorbents described in the literature (Table 3). It could be shown that the novel resin can compete with commercial sorbents as well as other synthesized SPE materials.

**Table 3.** Comparison of the developed DSPE and adapted SPE procedure with other solid-phase extraction methods for phenolic acids.

| Sorbent/ Amount | SPE Procedure | Sample Concentration/ mg L$^{-1}$ | Sample Volume/ mL | Recovery /% | | | | | Reference |
| | | | | ChlorA | CaffA | DHB | FerA | CinA | |
|---|---|---|---|---|---|---|---|---|---|
| C18/500 mg | SPE | 25 | - | - | 14.0 | - | 14.5 | - | [44] |
| SAX/200 mg | SPE | 25 | - | - | 29.9 | - | 30.5 | - | [44] |
| Oasis® HLB/200 mg | SPE | 25 | - | - | 102.5 | - | 102.7 | - | [44] |
| Magnetic MIP/ 10 mg | MSPE | 0.1 | 10 | 95–100 | 95–100 | - | 65–70 | - | [47] |
| Poly(n-VIM/EGDMA) /30 mg | SPE | 37.5–55 | 0.4 | 96.9 | 101.9 | 97.8 | - | - | [46] |
| Oasis® MAX/30 mg | SPE | 100 | 1 | 93.8 | 92.5 | 92.8 | 92.3 | 91.6 | This work |
| Poly(n-VIM/$C_6$-bis-VIM) /30 mg | SPE | 100 | 1 | 98.5 | 97.2 | 97.4 | 97.4 | 97.4 | This work |
| Poly(n-VIM/$C_6$-bis-VIM) /10 mg | DSPE | 100 | 1 | 92.5 | 92.0 | 92.5 | 90.6 | 84.1 | This work |

Extraction recoveries obtained from aqueous standard solutions.

### 3.7. Polymer Characterization

The poly(n-VIM/$C_6$-bis-VIM) polymer with a monomer to crosslinker ratio of 11.50 to 1 was investigated regarding its specific surface area and surface structure. The BET measurement of the sample gave a specific surface area of 49.3 m$^2$ g$^{-1}$. The adsorption and desorption isotherm are provided in Figure S1. Additionally, the pore size distribution calculated from the desorption branch (20 points from 0.95 to 0.3 p/p$_0$) of the isotherm is given in Figure S2. The results showed that the pore size distribution ranges between 25 and 50 Å with a maximum at 35 Å. The SEM measurements provided an image of the surface structure of the co-polymer, which is given in Figure S3. The surface is composed of many small, round, spherical components that were all about the same size. No macroscopic structural elements were observed. The recorded ATR-FTIR spectra (Figure S4) confirmed the presence of the functional groups incorporated by the monomer and the crosslinker. The following signals in the spectra of poly(n-VIM/C$_6$-bis-VIM) were observed and assigned: 3083 cm$^{-1}$ N-H stretching; 2932 cm$^{-1}$ C-H stretching (aromatic); 2859 cm$^{-1}$ C-H stretching (aliphatic); 1553 cm$^{-1}$ C-N stretching; 1414 cm$^{-1}$ C-H stretching (ring); 1284 cm$^{-1}$ C-H in plane bending; 1227 cm$^{-1}$ C-H in plane bending and C-N stretching; 1156 cm$^{-1}$ C-N (aliphatic); 1083 cm$^{-1}$ C-H in plane bending (ring) and C-C stretching; 908 cm$^{-1}$ C=C bending and stretching; 818 cm$^{-1}$ C-H bending (ring); 740 cm$^{-1}$ C-H bending (ring); and 662 cm$^{-1}$ N-C stretching (ring). Especially the signal at 3083 cm$^{-1}$ is of importance as it is the characteristic frequency of imidazolium salts. Furthermore, a signal at 1176 cm$^{-1}$ was observed in the spectra of the crosslinker, which was also found in the spectra of the polymer (1156 cm$^{-1}$) and not in the spectra of the VIM. The obtained results are consistent with the structural analysis of Lippert et al. and the IR spectrum table from Merck [52,53].

## 4. Conclusions

The synthesis and optimization of new polymers is an important branch of research that has been intensively pursued for years. This has allowed SPE to be continuously improved and advanced, making it by far the most important method in analytical sample preparation. Nowadays, there is a wide range of different polymers available and every year new, even more efficient sorbents are developed and manufactured.

The preparation of novel SPE materials is feasible in particular due to the huge variety of existing monomers and crosslinkers. In this study we focused on the effect of the type of crosslinker and the ratio between the monomer and crosslinker on the extraction efficiency toward phenolic acids. We synthesized and used [C$_6$-bis-VIM] [Br] as a crosslinker for the preparation of porous polymers. Subsequently, sorbents with the commonly used crosslinkers EGDMA and DVB were prepared and compared with the poly(n-VIM/C$_6$-bis-VIM) polymer. It was found that the type of crosslinker had a significant influence on the adsorption properties of the polymer. This finding strongly emphasizes the importance of selecting suitable crosslinkers for the preparation of efficient SPE materials. It was observed that polymers prepared with the commonly used crosslinkers EGDMA, or DVB exhibited lower extraction recoveries for the majority of phenolic acids.

The developed DSPE procedure for the extraction of phenolic acids represents an efficient analytical sample preparation method. It is simple and fast with maximum recoveries ranging between 84.1 and 92.5%. It was also shown that the DSPE procedure gave reproducible results, as the RSD values were very low. In addition, the developed extraction procedure was adapted to cartridges. No residual phenolic acids could be found in the supernatant and the extraction recoveries were ranging between 97.2 and 98.5%. It was demonstrated that the poly(n-VIM/C$_6$-bis-VIM) polymer could also compete with the commercially available sorbent Oasis® MAX from Waters™. Significantly, higher extraction recoveries for all analyzed phenolic acids were obtained, which were on average 5% higher than with Oasis® MAX.

Furthermore, it was demonstrated that the poly(n-VIM/C$_6$-bis-VIM) polymer could be regenerated and reused for the majority of analytes. Therefore, the novel sorbent is a good contender for its use in sustainable analytical sample preparation. This aspect is of great significance, as the minimization of resources is becoming more and more important in the development of new sample preparation methods.

Due to the ionic properties of [C$_6$-bis-VIM] [Br], it is well suited for the preparation of anion exchange sorbents. Due to its physicochemical properties, [C$_6$-bis-VIM] [Br] is not only an excellent constituent for the preparation of polymers but also because its synthesis is facile and inexpensive with high yields. Therefore, [C$_6$-bis-VIM] [Br] could present a good alternative to conventional crosslinkers such as EGDMA and DVB, which exhibited different properties when incorporated into vinylimidazole polymers. In conclusion, this study clearly highlights the yet untapped potential of the crosslinker [C$_6$-bis-VIM] [Br] with respect to the synthesis of new anion exchange polymers for their application in the field of SPE.

**Supplementary Materials:** The following supporting information can be downloaded at: https://www.mdpi.com/article/10.3390/separations9030072/s1, Figure S1. BET adsorption and desorption isotherm; Figure S2. Pore size distribution; Figure S3. SEM image; Figure S4. ATR-FTIR spectra; Table S1. Optimization of the [C$_6$-bis-VIM] [Br] crosslinked polymer; Table S2. Conditioning solution; Table S3. Eluting solution; Table S4. Calibration results 25–250 mg L$^{-1}$; Table S5. Calibration results 250–500 mg L$^{-1}$; Table S6. Repeatability 50 mg L$^{-1}$; Table S7. Repeatability 500 mg L$^{-1}$; Table S8. LOD and LOQ; Table S9. Adsorption time profiles; Table S10. Differently crosslinked polymers; Table S11. Sorption capacity; Table S12. Reusability.

**Author Contributions:** The study was supervised and reviewed by M.R., R.B. and G.K.B. and was largely visualized by M.H., M.R. and R.B. ATR-FTIR measurements and corresponding data evaluation were carried out by C.K. and C.W.H. BET measurements and interpretation of the obtained data were performed by C.G. NMR measurements and the associated data evaluation were carried out by S.N. Synthesis of the crosslinker and preparation of different polymers were carried out by P.M., F.L. and M.H. HPLC analysis as well as data evaluation were performed by F.L. and M.H. The manuscript was drafted by M.H. All authors have read and agreed to the published version of the manuscript.

**Funding:** This research received no external funding.

**Data Availability Statement:** Not applicable.

**Acknowledgments:** We would like to thank Mag. Martina Tribus (Institute of Mineralogy and Petrography, University of Innsbruck) for the SEM measurements. C. Grießer thanks the Austrian ACS Catalysis pubs.acs.org/acscatalysis Research Article https://doi.org/10.1021/acscatal.1c00415 ACS Catal. 2021, 11, 4920-4928 4926 Research Promotion Agency (FFG) for funding by the project number 870523.

**Conflicts of Interest:** The authors declare no conflict of interest.

# References

1. Kabera, J. Plant secondary metabolites: Biosynthesis, classification, function and pharmacological properties. *J. Pharm. Pharmacol.* **2014**, *2*, 377–392.
2. Marchiosi, R.; Dos Santos, W.D.; Constantin, R.P.; De Lima, R.B.; Soares, A.R.; Finger-Teixeira, A.; Mota, T.R.; de Oliveira, D.M.; Foletto-Felipe, M.D.P.; Abrahão, J.; et al. Biosynthesis and metabolic actions of simple phenolic acids in plants. *Phytochem. Rev.* **2020**, *19*, 865–906. [CrossRef]
3. Pospíšil, F.; Šindelářová, M. The effect of phenolic acids on metabolism and nutrient uptake of roots. In *Structure and Function of Plant Roots, Proceedings of the 2nd International Symposium, Bratislava, Czechoslovakia, 1–5 September 1980*; Brouwer, R., Gašparíková, O., Kolek, J., Loughman, B.C., Eds.; Springer: Dordrecht, The Netherlands, 1981; pp. 253–257.
4. Brouwer, R.; Gašparíková, O.; Kolek, J.; Loughman, B.C. Structure and Function of Plant Roots. In Proceedings of the 2nd International Symposium, Bratislava, Czechoslovakia, 1–5 September 1980; Springer: Dordrecht, The Netherlands, 1981.
5. Blum, U. Plant-Plant Allelopathic Interactions II. In *Laboratory Bioassays for Water-Soluble Compounds with an Emphasis on Phenolic Acids*; Springer International Publishing: Cham, Switzerland, 2014.
6. Mersie, W.; Singh, M. Phenolic acids affect photosynthesis and protein synthesis by isolated leaf cells of velvet-leaf. *J. Chem. Ecol.* **1993**, *19*, 1293–1301. [CrossRef] [PubMed]
7. Ramawat, K.G.; Mérillon, J.-M. *Natural Products: Phytochemistry, Botany and Metabolism of Alkaloids, Phenolics and Terpenes*; Springer: Heidelberg, Germany, 2013.
8. Shahidi, F.; Naczk, M. *Phenolics in Food and Nutraceuticals*, 2nd ed.; CRC Press: Hoboken, NJ, USA, 2003.
9. Robbins, R.J. Phenolic Acids in Foods: An Overview of Analytical Methodology. *J. Agric. Food Chem.* **2003**, *51*, 2866–2887. [CrossRef]

10. Huang, W.-Y.; Cai, Y.-Z.; Zhang, Y. Natural Phenolic Compounds From Medicinal Herbs and Dietary Plants: Potential Use for Cancer Prevention. *Nutr. Cancer* 2009, *62*, 1–20. [CrossRef]
11. Anantharaju, P.G.; Gowda, P.C.; Vimalambike, M.G.; Madhunapantula, S.V. An overview on the role of dietary phenolics for the treatment of cancers. *Nutr. J.* 2016, *15*, 99. [CrossRef]
12. Vinayagam, R.; Jayachandran, M.; Xu, B. Antidiabetic Effects of Simple Phenolic Acids: A Comprehensive Review. *Phytother. Res.* 2015, *30*, 184–199. [CrossRef]
13. Ali, S.S.; Ahmad, W.A.N.W.; Budin, S.B.; Zainalabidin, S. Implication of dietary phenolic acids on inflammation in cardiovascular disease. *Rev. Cardiovasc. Med.* 2020, *21*, 225–240. [CrossRef]
14. Kumar, N.; Goel, N. Phenolic acids: Natural versatile molecules with promising therapeutic applications. *Biotechnol. Rep.* 2019, *24*, e00370. [CrossRef]
15. Călinoiu, L.F.; Vodnar, D.C. Whole Grains and Phenolic Acids: A Review on Bioactivity, Functionality, Health Benefits and Bioavailability. *Nutrients* 2018, *10*, 1615. [CrossRef]
16. Rashmi, H.B.; Negi, P.S. Phenolic acids from vegetables: A review on processing stability and health benefits. *Food Res. Int.* 2020, *136*, 109298. [CrossRef] [PubMed]
17. Saibabu, V.; Fatima, Z.; Khan, L.A.; Hameed, S. Therapeutic Potential of Dietary Phenolic Acids. *Adv. Pharmacol. Sci.* 2015, *2015*, 1–10. [CrossRef] [PubMed]
18. Poole, C.F. *Solid-Phase Extraction*; Elsevier: Amsterdam, The Netherlands, 2020.
19. Simpson, N.J.K. *Solid-Phase Extraction: Principles, Techniques, and Applications*; Dekker: New York, NY, USA, 2000.
20. Pawliszyn, J. *Handbook of Solid Phase Microextraction*; Elsevier: Chennai, India; Oxford, UK, 2011.
21. Socas-Rodríguez, B.; Herrera-Herrera, A.V.; Asensio-Ramos, M.; Hernández-Borges, J. Dispersive Solid-Phase Extraction. In *Analytical Separation Science*; Anderson, J.L., Berthod, A., Pino, V., Stalcup, A., Eds.; Wiley-VCH: Weinheim, Germany, 2015; pp. 1525–1570.
22. Chisvert, A.; Cárdenas, S.; Lucena, R. Dispersive micro-solid phase extraction. *TrAC Trends Anal. Chem.* 2019, *112*, 226–233. [CrossRef]
23. Thurman, E.M.; Mills, M.S. *Solid Phase Extraction: Principles and Practice*; Wiley: New York, NY, USA, 1998.
24. Fritz, J.S. *Analytical Solid-Phase Extraction*; Wiley-VCH: New York, NY, USA, 1999.
25. Majors, R. New designs and formats in solid-phase extraction sample preparation. *LCGC N. Am.* 2001, *19*, 678–687.
26. Faraji, M.; Yamini, Y.; Gholami, M. Recent Advances and Trends in Applications of Solid-Phase Extraction Techniques in Food and Environmental Analysis. *Chromatographia* 2019, *82*, 1207–1249. [CrossRef]
27. Fontanals, N.; Marcé, R.M.; Borrull, F. Materials for Solid-Phase Extraction of Organic Compounds. *Separations* 2019, *6*, 56. [CrossRef]
28. Ścigalski, P.; Kosobucki, P. Recent Materials Developed for Dispersive Solid Phase Extraction. *Molecules* 2020, *25*, 4869. [CrossRef]
29. Płotka-Wasylka, J.; Marć, M.; Szczepańska, N.; Namieśnik, J. New Polymeric Materials for Solid Phase Extraction. *Crit. Rev. Anal. Chem.* 2017, *47*, 373–383. [CrossRef]
30. Nollet, L.M.; Lambropoulou, D.A. *Chromatographic Analysis of the Environment: Mass Spectrometry Based Approaches*, 4th ed.; CRC Press: Boca Raton, FL, USA, 2017.
31. Alb, A.M.; Reed, W.F. *Monitoring Polymerization Reactions: From Fundamentals to Applications*; John Wiley & Sons Inc.: Hoboken, NJ, USA, 2014.
32. Mishra, M.K.; Yagci, Y. *Handbook of Vinyl Polymers: Radical Polymerization, Process, and Technology*, 2nd ed.; Taylor & Francis: Boca Raton, FL, USA, 2009.
33. Matyjaszewski, K. *Handbook of Radical Polymerization*, 1st ed.; John Wiley & Sons, Inc.: Hoboken, NJ, USA, 2002.
34. Gutiérrez, T.J. *Reactive and Functional Polymers Volume Two: Modification Reactions, Compatibility and Blends*, 1st ed.; Springer International Publishing: Cham, Switzerland, 2020.
35. Saldivar-Guerra, E.; Vivaldo-Lima, E. *Handbook of Polymer Synthesis, Characterization, and Processing*, 1st ed.; Wiley: Hoboken, NJ, USA, 2013.
36. Zhang, T.; Zhou, F.; Huang, J.; Man, R. Ethylene glycol dimethacrylate modified hyper-cross-linked resins: Porogen effect on pore structure and adsorption performance. *Chem. Eng. J.* 2018, *339*, 278–287. [CrossRef]
37. Meischl, F.; Losso, K.; Kirchler, C.G.; Stuppner, S.E.; Huck, C.W.; Rainer, M. Synthesis and Application of Histidine-Modified Poly(Glycidyl Methacrylate/Ethylene Glycol Dimethacrylate) Sorbent for Isolation of Caffeine from Black and Green Tea Samples. *Chromatographia* 2018, *81*, 1467–1474. [CrossRef]
38. Meischl, F.; Schemeth, D.; Harder, M.; Köpfle, N.; Tessadri, R.; Rainer, M. Synthesis and evaluation of a novel molecularly imprinted polymer for the selective isolation of acetylsalicylic acid from aqueous solutions. *J. Environ. Chem. Eng.* 2016, *4*, 4083–4090. [CrossRef]
39. Schemeth, D.; Kappacher, C.; Rainer, M.; Thalinger, R.; Bonn, G.K. Comprehensive evaluation of imidazole-based polymers for the enrichment of selected non-steroidal anti-inflammatory drugs. *Talanta* 2016, *153*, 177–185. [CrossRef] [PubMed]
40. Wang, C.; Guan, X.; Yuan, Y.; Wu, Y.; Tan, S. Polyacrylamide crosslinked by bis-vinylimidazolium bromide for high elastic and stable hydrogels. *RSC Adv.* 2019, *9*, 27640–27645. [CrossRef]

41. Cui, J.; Zhu, W.; Gao, N.; Li, J.; Yang, H.; Jiang, Y.; Seidel, P.; Ravoo, B.J.; Li, G. Inverse Opal Spheres Based on Polyionic Liquids as Functional Microspheres with Tunable Optical Properties and Molecular Recognition Capabilities. *Angew. Chem. Int. Ed.* **2014**, *53*, 3844–3848. [CrossRef] [PubMed]
42. Murauer, A.; Bakry, R.; Partl, G.; Huck, C.W.; Ganzera, M. Optimization of an innovative vinylimidazole-based monolithic stationary phase and its use for pressured capillary electrochromatography. *J. Pharm. Biomed. Anal.* **2019**, *162*, 117–123. [CrossRef]
43. Kim, S.-Y.; Kim, W.; Choi, S.-H. Anion-Exchange Membrane with Poly(3,3'-(hexyl) bis(1-vinylimidazolium) bromide)/PVC Composites Prepared by Inter-polymerization. *Eur. J. Eng. Res. Sci.* **2019**, *4*, 116–120. [CrossRef]
44. Yilmaz, P.K.; Kolak, U. SPE-HPLC Determination of Chlorogenic and Phenolic Acids in Coffee. *J. Chromatogr. Sci.* **2017**, *55*, 712–718. [CrossRef]
45. Michalkiewicz, A.; Biesaga, M.; Pyrzynska, K. Solid-phase extraction procedure for determination of phenolic acids and some flavonols in honey. *J. Chromatogr. A* **2008**, *1187*, 18–24. [CrossRef]
46. Schemeth, D.; Noel, J.-C.; Jakschitz, T.; Rainer, M.; Tessadri, R.; Huck, C.W.; Bonn, G.K. Poly(N-vinylimidazole/ethylene glycol dimethacrylate) for the purification and isolation of phenolic acids. *Anal. Chim. Acta* **2015**, *885*, 199–206. [CrossRef]
47. Tashakkori, P.; Erdem, P.; Bozkurt, S.S. Molecularly Imprinted Polymer Based on Magnetic Ionic Liquid for Solid Phase Extraction of Phenolic Acids. *J. Liq. Chromatogr. Relat. Technol.* **2017**, *40*, 657–666. [CrossRef]
48. Peters, F.T.; Hartung, M.; Herbold, M.; Daldrup, T.; Mußhoff, F. Appendix B to the GTFCh guidelines for quality assurance in forensic-toxicological analyses. Requirements for the validation of analytical methods. *Toxichem Krimtech* **2009**, *76*, 185–208.
49. DIN 32645. *Chemische Analytik—Nachweis-, Erfassungs-und Bestimmungsgrenze unter Wiederholbedingungen—Begriffe, Verfahren, Auswertung*; Beuth Verlag GmbH: Berlin, Germany, 2008.
50. Waters Corporation. Simplifying Solid-Phase Extraction. Available online: https://www.waters.com/webassets/cms/library/docs/720001692en.pdf (accessed on 8 February 2022).
51. Friedman, M.; Jürgens, H.S. Effect of pH on the Stability of Plant Phenolic Compounds. *J. Agric. Food Chem.* **2000**, *48*, 2101–2110. [CrossRef] [PubMed]
52. Lippert, J.L.; Robertson, J.A.; Havens, J.R.; Tan, J.S. Structural studies of poly(N-vinylimidazole) complexes by infrared and Raman spectroscopy. *Macromolecules* **1985**, *18*, 63–67. [CrossRef]
53. Merck, I.R. Spectrum Table & Chart: IR Spectrum Table by Frequency Range. Available online: https://www.sigmaaldrich.com/AT/en/technical-documents/technical-article/analytical-chemistry/photometry-and-reflectometry/ir-spectrum-table. (accessed on 25 February 2022).

Article

# Effective Solid Phase Extraction of Toxic Pyrrolizidine Alkaloids from Honey with Reusable Organosilyl-Sulfonated Halloysite Nanotubes

Tobias Schlappack [1,†], Nina Weidacher [1,†], Christian W. Huck [1], Günther K. Bonn [1,2] and Matthias Rainer [1,*]

1 Institute of Analytical Chemistry and Radiochemistry, Leopold-Franzens-University Innsbruck, Innrain 80/82, A-6020 Innsbruck, Austria
2 Austrian Drug Screening Institute–ADSI, Innrain 66a, A-6020 Innsbruck, Austria
* Correspondence: m.rainer@uibk.ac.at; Tel.: +43-512-507-57307
† These authors contributed equally to this work.

**Abstract:** Pyrrolizidine alkaloids are plant secondary metabolites that have recently attracted attention as toxic contaminants in various foods and feeds as they are often harvested by accident. Furthermore, they prove themselves as hard to analyze due to their wide structural range and low concentration levels. However, even low concentrations show toxic behavior in the form of chronic liver diseases and possible carcinogenicity. Since sample preparation for this compound group is in need of more green and sustainable alternatives, modified halloysite nanotubes present an interesting approach. Based on the successful use of sulfonated halloysite nanotubes as inexpensive, easy-to-produce cation exchangers for solid phase extraction in our last work, this study deals with the further modification of the raw nanotubes and their performance in the solid phase extraction of pyrrolizidine alkaloids. Conducting already published syntheses of two organosilyl-sulfonated halloysite nanotubes, namely HNT-PhSO$_3$H and HNT-MPTMS-SO$_3$H, both materials were used as novel materials in solid phase extraction. After the optimization of the extraction protocol, extractions of aqueous pyrrolizidine alkaloid mixtures showed promising results with recoveries ranging from 78.3% to 101.3%. Therefore, spiked honey samples were extracted with an adjusted protocol. The mercaptopropyl-sulfonated halloysite nanotubes revealed satisfying loading efficiencies and recoveries. Validation was then performed, which displayed acceptable performance for the presented method. In addition, reusability studies using HNT-MPTMS-SO$_3$H for solid phase extraction of an aqueous pyrrolizidine alkaloid mixture demonstrated excellent results over six cycles with no trend of recovery reduction or material depletion. Therefore, organosilyl-sulfonated halloysite nanotubes display a green, efficient and low-cost alternative to polymeric support in solid phase extraction of toxic pyrrolizidine alkaloids from complex honey matrix.

**Keywords:** halloysite nanotubes; organosilyl-sulfonated halloysite nanotubes; solid phase extraction; pyrrolizidine alkaloids; honey

## 1. Introduction

Clays are well known to humanity and have been used in a wide range of applications for a long time. However, as humankind continues to evolve, more possibilities for these compounds arise through different methods for controlling their morphological characteristics [1]. One type of material derived from these natural clays is halloysite. Halloysite nanotubes, also called nanosized tubular halloysite or halloysite nanoclay is the naturally most occurring halloysite [2]. Large deposits were found in Australia, the United States, China, New Zealand, Mexico and Brazil [2,3]. Morphologically, halloysite nanotubes are two-layered aluminosilicates with a hollow, tubular structure, which is formed through the rolling of 15 to 20 aluminosilicate layers, similar to carbon nanotubes [2–6]. The external surface consists of silicone dioxide groups, whereas the inner surface is composed of aluminum oxide groups [7]. It is a member of the kaolin group and shares chemical similarity with kaolinite [2,8,9]. However, monolayers of water molecules separate the unit layers in halloysites in contrast to the layers in kaolinite. Hydrated forms share a sum formula of $Al_2(OH)_4Si_2O_5 \cdot nH_2O$. Halloysite–(10 Å) presents the hydrated form with n = 2 and one layer of water, whereas halloysite–(7 Å) is the name of the dry mineral with n = 0. The angstrom term defines the $d_{001}$–value of the respective mineral. A conversion from the hydrated to the dry state can be achieved through a mild temperature and/or vacuum. The dimensions of halloysite nanotubes vary from the submicron scale to several microns in length, 30 to 190 nm external diameter and 10 to 100 nm internal diameter [2,5]. However, the morphology and porosity of these halloysites can strongly vary between origins and are influenced through acidic or basic treatment [10,11]. Furthermore, platy and spheroidal morphologies were also observed next to tubular structures, which is the dominant form. This tubular structure is caused by a mismatch between adjacent silicone dioxide and aluminum oxide layers [7]. In the past decade, these tubes became the main focus of many studies and showed promising characteristics [2]. As they are non-toxic, inexpensive, biocompatible, possess high specific surface areas and show different possibilities for inner-outer surface chemistry, the first applications in polymer filling, catalysis, nanoencapsulation and wastewater treatment have already been reported [1,2,5,8,9,12,13]. All these properties make halloysite nanotubes a strong competitor against expensive carbon nanotubes [9], even though carbon nanotubes are currently still in frequent use [14,15]. In addition, the aluminosilicate structure of halloysites with their external siloxane surface gives the possibility of chemically modifying hydroxyl groups on said surface [2,7,8,16,17]. In addition to this, high natural cation exchange capacities of $30-50 \times 10^{-2}$ mol kg$^{-1}$ suggests favorable ion exchange properties [3]. This opens the possibility for the use of highly selective, chemically modified halloysite nanotubes to extract compounds of interest. In our last work, we presented a method using one-pot synthesized sulfonated halloysite nanotubes (HNT-SO$_3$H) for the selective solid phase extraction of toxic pyrrolizidine alkaloids from honey [18]. These pyrrolizidine alkaloids present a hepatotoxic and possible cancerogenic group of compounds with over 660 known, highly diverse structures present in over 6000 plant species [19–24]. As they are often harvested by accident they are a highly important topic for food safety, since regular intake of even small amounts can cause chronic liver diseases [22–29]. Furthermore, the transportation of pollen from pyrrolizidine alkaloid-containing plants through bees can contaminate pollen products such as honey as well [19,30]. Highly sensitive methods are needed, since maximum limits of pyrrolizidine alkaloids in food are in the magnitude of micrograms per kilogram [31]. Therefore, working groups and official institutions use solid phase extraction with reversed phase or cation exchange interactions to enrich the alkaloids and reduce the matrix effect before analysis [18,21,30,32–36]. Especially reversed phase methods show lower selectivity in general and can therefore be problematic in terms of interfering compounds and matrix effects. Hence, we want to expand the possibilities for the selective extraction of pyrrolizidine alkaloids from complex matrices and to further reduce the matrix effect. In this study, we synthesized two organosilyl-sulfonated halloysite nanotubes according to previously published studies to selectively extract toxic pyrrolizidine alkaloids from a honey sample [16,17]. Apart from our previously published work, according to our knowledge there has been no application of modified halloysites as solid phase extraction material so far. However, different modifications on the halloysites could provide even better results in solid phase extractions. Furthermore, in addition to the impressive analytical performance of modified halloysite nanotubes, exploring the applicability of these materials for solid-phase extraction is another important building block in making analytical chemistry more environmentally friendly and sustainable, and further eliminating extraction methods using polymer resins.

## 2. Experimental

### 2.1. Materials and Methods

#### 2.1.1. Reagents and Standards

Acetonitrile (for LC-MS; ≥99.95%) and methanol (for LC-MS; ≥99.95%) were obtained from Chemsolute® (Th. Geyer, Renningen, Germany). Dichloromethane (for HPLC, 100%) was ordered from VWR (VWR International, Radnor, USA). Chlorosulfonic acid (purum, >98.0% (T)) and dimethyl sulfoxide (p.a.; ACS: >99.9% (GC)) were bought from Fluka AG (Honeywell International Inc., Morristown, NJ, USA). Thiourea (p.a.) was purchase from Merck (Merck KGaA, Darmstadt, Germany) for the determination of dead time. Ammonium formate (≥95%) and formic acid (ROTIPURAN® ≥98%, p.a., ACS) were purchased from Carl Roth (Carl Roth GmbH + Co. KG, Karlsruhe, Germany). Phyproof® reference substances were purchased from PhytoLab (PhytoLab GmbH & Co. KG, Vestenbergsgreuth, Germany), namely heliotrine (minimum 85% (HPLC), lycopsamine (minimum 85% (HPLC)), and senecionine (minimum 85% (HPLC)). The pyrrolizidine alkaloid reference substances were dissolved in acetonitrile in the first step. The appropriate final concentration was achieved by dilution with Milli-Q™ water. Caffeine (ReagentPlus®, minimum 99%), halloysite nanoclay, monocrotaline (≥98%), anhydrous toluene (99.8%), triethoxyphenlysilane (≥98%) and 3-(mercaptopropyl)trimethoxysilane (95%) were ordered from Sigma-Aldrich (Sigma-Aldrich, St. Louis, USA). Purified water was collected from a Merck Millipore Milli-Q™ Reference Ultrapure Water Purification System. Empty 1 mL polypropylene SPE cartridges with prefiltration (polyethylene, 20 mm porosity) were acquired from Sigma-Aldrich.

#### 2.1.2. Honey Sample

Honey from Bergland-Honig (Bergland-Honig GmbH, Urban, Austria) was purchased from a nearby convenience store to analyze a spiked field sample. The honey was prepared as described in the following chapter, without the addition of analytes. The analyte-free real sample was analyzed for pyrrolizidine alkaloids prior to the experiments using the developed UHPLC-MS/MS method to prevent any possible bias.

#### 2.1.3. Sample Preparation

A slightly modified method, compared to our previous study [18], was used to prepare the honey sample. For this purpose, 4 g of bee honey was completely dissolved in 40 mL of 0.05 M formic acid (FA) in a falcon tube by shaking. Centrifugation was then carried out at 14,000 rpm for 10 min. Subsequently, an aqueous standard solution of the pyrrolizidine alkaloids heliotrine, lycopsamine, monocrotaline, and senecionine was then added to the sample to get a final concentration of 12.5 µg L$^{-1}$ per pyrrolizidine alkaloid. These pyrrolizidine alkaloids were selected because they represent four of the six main structures of this class of compounds. The spiking amount was based on the current European Union regulation for pyrrolizidine levels in honey. This states that a maximum level of 500 µg kg$^{-1}$ of pyrrolizidine alkaloids in food supplements, pollen and pollen-based products must not be exceeded [31].

#### 2.1.4. Synthesis of Organosilyl-Sulfonated Halloysite Nanotubes

Synthesis of both alkyl-sulfonated halloysite nanotube materials was achieved according to a similar protocol published by Silva et al. [17] and Peixoto et al. [16]. The scales of the reactions were altered if needed. In the following subsections, the general synthesis steps of both materials will be described more thoroughly. Before using both synthesized materials for solid phase extraction, grinding was performed to achieve a more uniform distribution of particles.

#### HNT-PhSO$_3$H

In the first step of the synthesis of HNT-PhSO$_3$H, 2 g of dry halloysite nanotubes were suspended in 100 mL anhydrous toluene and 1.5 mmol triethoxyphenylsilane (PhTES).

The mixture was then refluxed for 24 h under constant stirring and nitrogen atmosphere. Subsequently, the product HNT-PhTES was centrifuged, washed with four 10 mL portions of toluene and dried at 100 °C for 24 h.

For the chlorosulfonation in the second step, 2 g of the previously synthesized HNT-PhTES were added to 30 mL of dichloromethane. Afterwards, 2.4 mL of chlorosulfonic acid were added dropwise to the mixture, which was then refluxed at 50 °C for 6 h. The resulting HNT-PhSO$_3$H material was then centrifuged, washed with four 10 mL portions of methanol and dried at 100 °C for 24 h. The synthesis steps are shown in Figure 1.

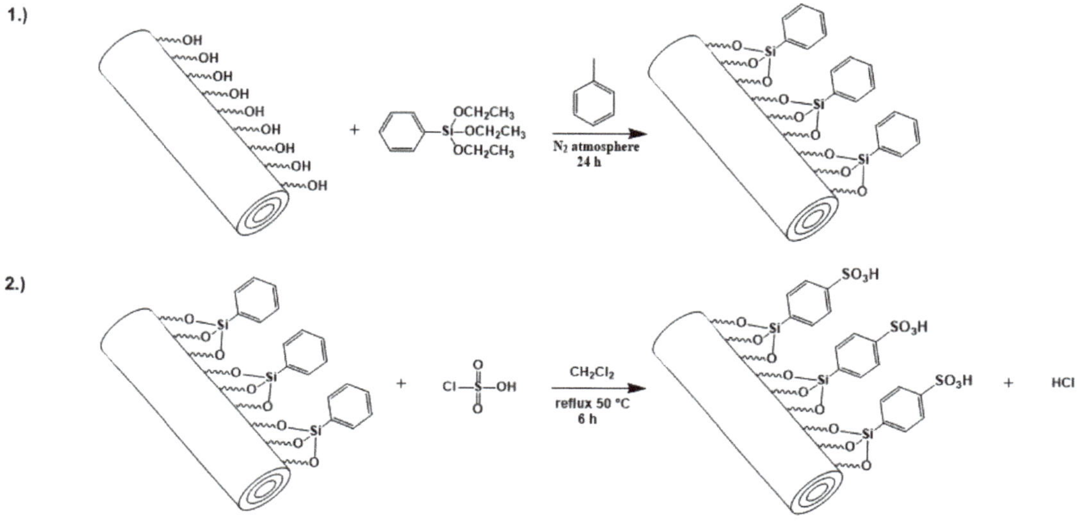

**Figure 1.** Two-step synthesis of HNT-PhSO$_3$H consisting of organosilylation (1.) and sulfonation step (2.) according to Silva et al. and Peixoto et al. [16,17].

HNT-MPTMS-SO$_3$H

Prior to the synthesis, 2 g of halloysite nanotubes were dried at 100 °C for 1 h. Subsequently, the organosilylation of the dried halloysite nanotubes was achieved through the dropwise addition of 1.13 mL (6 mmol) (3-mercaptopropyl)trimethoxysilane to 2 g of HNTs suspended in 100 mL anhydrous toluene. The reaction mixture was refluxed under constant stirring and nitrogen atmosphere for 24 h. This was followed by centrifugation and washing of the resulting HNT-MPTMS with four 10 mL portions of toluene and drying at 100 °C for 24 h.

In the second step of the synthesis, 2 g of HNT-MPTMS were suspended in 30 mL of dichloromethane through constant magnetic stirring while cooling the mixture with the help of an ice bath for 10 min. This was followed by the addition of 3.4 mL of chlorosulfonic acid through a constant pressure dropping funnel. The reaction mixture was then stirred for 4 h at room temperature with an equipped Dimroth condenser. The final HNT-MPTMS-SO$_3$H was then centrifuged, washed with four 10 mL portions of methanol and dried at 100 °C. Figure 2 shows the respective synthesis steps.

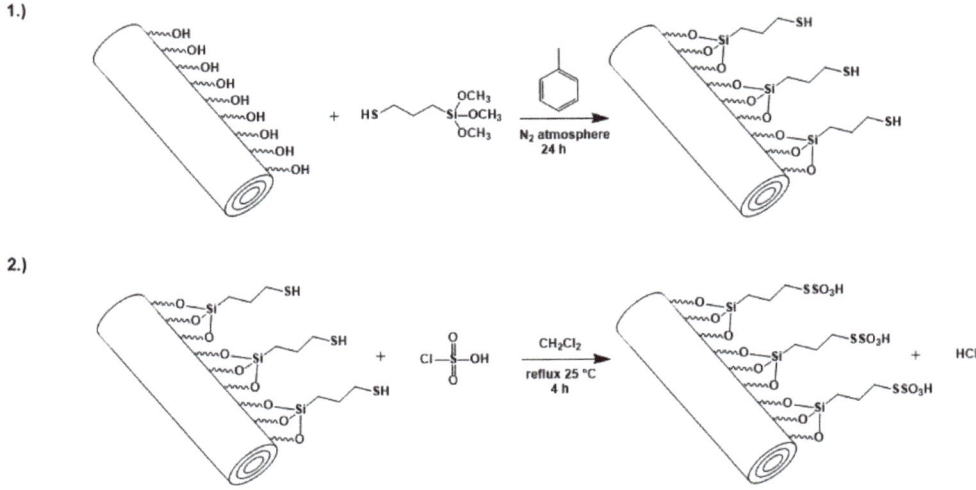

**Figure 2.** Two-step synthesis of HNT-MPTMS-SO$_3$H consisting of organosilylation (**1.**) and sulfonation step (**2.**) according to Silva et al. and Peixoto et al. [16,17].

### 2.1.5. FT-ATR Analysis

To verify the structural change in the modified halloysite nanotubes, FT-ATR analysis was performed with a PerkinElmer Spectrum 100 FT-IR Spectrometer (PerkinElmer Inc., Waltham, MA, USA) equipped with a universal ATR sampling accessory. Measurements were recorded in the range of 650 cm$^{-1}$ to 4000 cm$^{-1}$.

### 2.1.6. Solid Phase Extraction

Solid phase extraction protocols were optimized using an aquatic 10 µg L$^{-1}$ standard mixture of four pyrrolizidine alkaloids, namely heliotrine, lycopsamine, monocrotaline and senecionine as they present four of the six main structures of the pyrrolizidine alkaloid compound group. A schematic workflow starting from sample preparation up to UHPLC-MS/MS analysis can be seen in Figure 3.

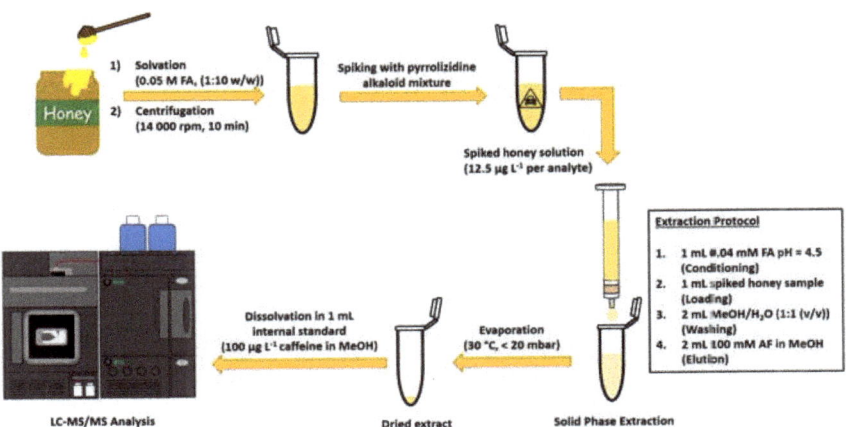

**Figure 3.** Schematic display of the experimental workflow using organosilyl-sulfonated halloysite nanotubes as material for solid phase extraction of toxic pyrrolizidine alkaloids in honey matrix.

Pyrrolizidine Alkaloid Standard Mixture Extraction

Empty, pre-fritted 1 mL solid phase extraction tubes were filled with 50 mg of ground organosilyl-sulfonated halloysites and covered with an additional 0.2 µm polyethylene (PE) frit. A force meter was used to press all cartridges with a weight of 2.5 kg to ensure uniform packing and reproducibility. For conditioning, 1 mL of 0.04 mM formic acid solution with a pH of 4.5 was used. Before the conditioning solvent reached the level of the first PE frit, 1 mL of aquatic pyrrolizidine alkaloid standard mixture was added onto the material. In order to decrease the pH of the sample, leading to the protonation of the nitrogen atom of the pyrrolizidine alkaloids, the standard mixture was diluted with conditioning solvent. However, since no effect was visible, this step was not implemented in the final extraction protocol. After the complete passing of the sample through the solid phase elution was performed with 2 mL of 100 mM ammonium formate in methanol. All steps were performed with the help of pressured air at a drop rate of three to four drops per minute. The residues of the sample after loading as well as the eluates of the solid phase extraction were then dried at 30 °C under vacuum and resuspended in 1 mL 100 µg L$^{-1}$ caffeine solution in dimethylsulfoxide/methanol 1/1 (v/v).

Spiked Honey Sample

Solid phase extraction of spiked honey sample and blank honey matrix was achieved through the identical protocol, which was used for the pyrrolizidine standard mixture. Furthermore, an additional washing step consisting of two portions of methanol/water 1/1 (v/v) was implemented. This was conducted to further reduce matrix effects originating from competing substances of the honey matrix. In our last work, this was not possible due to the weaker binding of the pyrrolizidine alkaloids to the solid material [18].

- Validation

To present the effectiveness and competitiveness of the synthesized solid materials for solid phase extraction, validation was performed for HNT-MPTMS-SO$_3$H. All parameters were determined according to found literature [37–41] and only slightly modified.

Specificity

Specificity was ensured through the use of highly selective multi reaction monitoring mode. Furthermore, blank solvent and honey matrix were analyzed with the presented UHPLC-MS/MS method to prove high specificity.

Linearity

Linearity was determined through measurement of three independently prepared matrix match calibrations with eight concentration steps ranging from 0 µg L$^{-1}$ to 14 µg L$^{-1}$ (0, 16, 32, 48, 64, 80, 96 and 112% of target concentration, respectively). The sample preparation protocol and UHPLC-MS/MS method was chosen identically to the final procedure.

Bias

Bias was calculated according to the following formula [39].

$$\text{Bias }[\%] = \frac{\bar{x} - \mu}{\mu} \cdot 100\% \quad (1)$$

where $\bar{x}$ represents the averaged value of each concentration step and $\mu$ represents the reference value.

LOD and LOQ

Limit of detection and limit of quantification were calculated according to DIN 32645 [37], using the following formulas.

$$LOD = s_{x_0} \cdot t_{f,\alpha} \sqrt{\frac{1}{m} + \frac{1}{n} + \frac{\bar{x}^2}{Q_x}} \quad (2)$$

$$LOD = s_{x_0} \cdot t_{f,\alpha} \sqrt{\frac{1}{m} + \frac{1}{n} + \frac{((k \cdot LOD) - \bar{x})^2}{Q_x}} \quad (3)$$

Parameters in the previous two equations are $s_{x_0}$, which describes the procedure standard deviation of the regression curve, and $t$, which represents the student factor with $P = 95\%$ used one-sided for the $LOD$ calculation and two-sided for $LOQ$ calculation. Additionally, $m$ is the number of replicate measurements and $n$ displays the number of calibration points. $\bar{x}$ describes the arithmetic mean of all calibration concentrations, while $Q_x$ describes the term $\sum_{i=1}^{n}(x_i - \bar{x})^2$. Furthermore, $k \cdot LOD$ was used as approximation, since the $LOQ$ is usually calculated iteratively, for which $k = 3$ was chosen.

Loading Efficiency

Loading efficiency was determined through the measurement of ten independent residues of the extracted spiked samples after solid phase extraction in triplicate.

Recovery

Recovery was assessed by performing ten independent solid phase extractions of spiked honey sample (12.5 µg L$^{-1}$ of each pyrrolizidine alkaloid) and comparing the measured analyte and internal standard areas with a matrix matched calibration curve. Each sample was measured in triplicate. Standard deviation of each triplicate measurement was calculated according to the formula based on the formula for random sampling of the data set.

Repeatability

Repeatability was determined from ten independent experiments at 100% of the target concentration. Experiments were executed by the same operator on the same day. Subsequently, ten solid phase extractions were independently performed with a honey sample spiked with 12.5 µg L$^{-1}$ of each of the four pyrrolizidine alkaloids. Each eluate of the solid phase extraction was measured in triplicate. Repeatability was then calculated as RSD in % with the standard deviation formula based on random sampling of the data set.

Matrix Effect

Matrix effect was calculated through the comparison of slopes between a matrix matched calibration curve and a calibration curve of the same levels (0–14 µg L$^{-1}$) in pure solvent with 200 mM ammonium formate, since it is also present in this amount in the matrix extracts, according to the following formula [42].

$$ME[\%] = \frac{k_{MM}}{k_{Solvent}} \cdot 100\% \quad (4)$$

With $k_{MM}$ representing the slope of the matrix matched calibration and $k_{Solvent}$ representing the slope of the calibration in pure solvent. Each calibration curve was measured in triplicate.

Autosampler Stability

Autosampler stability was tested through the measurement of three matrix matched calibration samples at low, medium and high concentration levels (4, 8 and 12 µg L$^{-1}$, respectively) in triplicate after every 6 h in a range of 48 h. Subsequently, the fraction of areas multiplied with the concentration of the internal standard were calculated for each measurement. Freshly prepared solutions (e.g., 0 h) were used as 100% to normalize the consecutive averaged triplicate measurements.

Reusability Study

As environmentally friendly and sustainable chemistry becomes more and more important in today's world, the HNT-MPTMS-SO$_3$H material was checked for its multiple usability. Solid phase extraction of the aqueous 10.0 µg L$^{-1}$ pyrrolizidine standard solution was therefore performed as earlier specified. In order to cleanse the solid material for re-use and prevent carryover from preceding extractions, we conducted our previously published protocol for the reusability study [18]. As in general, the first washing step was carried out with two extra portions of 1 mL each of the methanolic 100 mM ammonium formate solution. Subsequently, the material was washed twice with 1 mL methanol each. To verify the efficiency of the washing steps, UHPLC-MS/MS analysis of the washing steps was carried out, which revealed no traces of pyrrolizidine alkaloids. Subsequently, the solid in the cartridge was pre-dried with compressed air for 3 min before being completely dried at 50 °C for 10 min [18]. This procedure was performed before each new replicate cycle, so that a total of six SPE cycles could be performed with the same cartridge. After that, the experiments were stopped as the evidence for reusability could be clearly presented.

2.1.7. UHPLC-MS/MS Analysis

A Waters Acquity Premier liquid chromatograph with a Waters TQD triple quadrupole detector was used for UHPLC-MS/MS analysis. The sheath and auxiliary gas was nitrogen, whereas argon represented the collision gas. A Thermo Fisher Hypersil Gold™ C18 Selectivity column measuring 150 × 2.1 mm with a particle size of 1.9 mm, identical to official analysis protocols, was used [30,32]. The column oven was set to 50 °C, while the temperature of the autosampler was fixed at 25 °C. The injection volume was adjusted to 1 µL using 10% acetonitrile in water as wash solution. Analysis was performed in binary gradient mode with 0.1% FA in $H_2O$ (A) and acetonitrile (B) at a flow rate of 0.4 mL min$^{-1}$, similar to our previous work [18]. The following solvent composition was applied during analysis: 0.0–1.0 min (5% B), 1.0–7.5 min (5–50% B), 7.5–7.6 min (50–100% B), 7.6–8.3 min (100% B), 8.3–8.4 min (100–5% B), and 8.4–13.0 min (5% B). Waters MassLynx was used for acquisition. Multiple reaction monitoring (MRM) method tuning and generation was performed in positive electrospray (ES) mode using Waters' IntelliStart software. The measurements were carried out using the tuning method of monocrotaline, as it is a representative of the compound class. MRM methods for each analyte were created under solvent flow with initial gradient solvent composition. Three transitions per analyte, except for the structurally smaller internal standard caffeine, were selected for quantification via their total ion chromatogram (TIC). In Table 1, the transitions of the four pyrrolizidine alkaloids and caffeine (IS) are shown with the corresponding retention times, cone and collision voltages. A measurement of a 1 mg L$^{-1}$ thiourea solution revealed the dead time of the used method. To protect the mass spectrometer from unwanted contaminants such as highly polar substances, e.g., salts and saccharides, as well as strong nonpolar compounds, the MS acquisition was initiated after 1.5 min and stopped after 7.5 min. Data analysis was carried out using TargetLynx.

Table 1. Retention times and tandem MS parameters of the analyte transitions observed in MRM mode.

| Compound | $R_t$/min | Precursor Ion $[M + H]^+$ | Product Ions | Cone Voltage/V | Collision Voltage/V |
| --- | --- | --- | --- | --- | --- |
| Monocrotaline | 2.0 | 326 | 94; 120; 194 | 58 | 42; 32; 34 |
| Lycopsamine | 3.3 | 300 | 94; 138; 156 | 50 | 32; 22; 28 |
| Caffeine (IS) | 4.0 | 195 | 41; 138 | 32 | 40; 18 |
| Heliotrine | 4.5 | 314 | 94; 138; 156 | 44 | 44; 22; 30 |
| Senecionine | 5.1 | 336 | 93(.5); 93(.9); 120 | 58 | 48; 34; 38 |

## 3. Results and Discussion

*3.1. Synthesis of Organosilyl-Sulfonated Halloysite Nanotubes*

MIR-ATR Analysis

FT-ATR analyses of both synthesized, modified halloysite nanotubes showed newly appearing signals in the spectrum compared to the unmodified halloysite nanoclay. These are in accordance with the results obtained by Silva et al. [17].

*3.2. Solid Phase Extraction*

3.2.1. SPE with Unmodified Halloysite Nanotubes

To verify that modification of halloysite nanotubes is necessary to improve their properties as cation-exchange material for solid phase extraction, unmodified halloysite nanoclay was tested. Therefore, raw, unmodified halloysite nanotubes were used as solid phase extraction material for the extraction of an aqueous 10 µg L$^{-1}$ pyrrolizidine alkaloid mixture. Furthermore, the same procedure used for the modified materials was performed. However, solid phase extractions could not be executed as the backpressure of the unmodified material was too high to pass the solvent through the packed cartridge in a practical manner. As a result of this, structural modification of the raw halloysite nanotubes is necessary to perform solid phase extractions at a reasonable backpressure to ensure a well-working sample preparation method. These results are identical to the solid phase extractions of unmodified halloysite nanotubes from our last work [18]

3.2.2. Pyrrolizidine Alkaloid Standard Mixture
HNT-PhSO$_3$H

To validate the performance of the synthesized modified halloysite nanotubes, solid phase extractions were performed with an aqueous 10.0 µg L$^{-1}$ pyrrolizidine alkaloid mixture, according to the previously described extraction protocol. Table 2 shows the averaged recoveries of the four pyrrolizidine alkaloids after performing solid phase extraction using HNT-PhSO$_3$H. Hence, high recoveries of at least 81.5% for monocrotaline and up to 99.8% for heliotrine could be observed. Batch to batch recoveries are displayed in Figure 4. Subsequently, an analysis of the breakthrough showed only noise which displays high loading efficiency of the pyrrolizidine alkaloids on the solid material.

**Table 2.** Recoveries of an aqueous 10.0 µg L$^{-1}$ pyrrolizidine alkaloid standard mixture after solid phase extraction using HNT-PhSO$_3$H and HNT-MPTMS-SO$_3$H as solid materials.

| Recovery ± SD /% (N = 10; HNT-PhSO$_3$H) | | | |
|---|---|---|---|
| Monocrotaline | Lycopsamine | Heliotrine | Senecionine |
| 81.5 ± 3.4 | 98.3 ± 7.3 | 99.8 ± 5.1 | 92.1 ± 6.8 |
| Recovery ± SD/% (N = 10; HNT-MPTMS-SO$_3$H) | | | |
| Monocrotaline | Lycopsamine | Heliotrine | Senecionine |
| 78.3 ± 5.8 | 101.3 ± 5.9 | 99.1 ± 2.6 | 81.3 ± 5.7 |

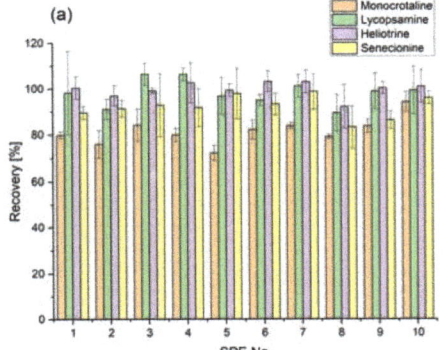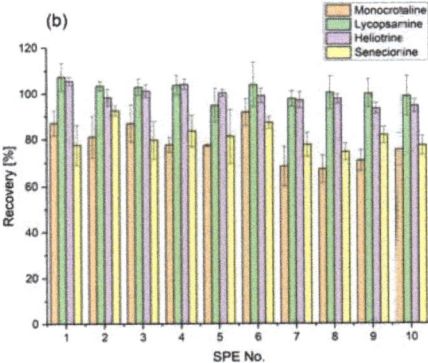

**Figure 4.** Solid phase extraction of an aqueous 10.0 µg L$^{-1}$ pyrrolizidine alkaloid standard mixture using HNT-PhSO$_3$H (**a**) and HNT-MPTMS-SO$_3$H (**b**) as solid phase.

HNT-MPTMS-SO$_3$H

The second modified organosilyl-sulfonated halloysite nanotubes were also investigated for their extraction performance using an aqueous 10.0 µg L$^{-1}$ pyrrolizidine alkaloid mixture. Control measurements of the breakthrough have shown to be identical to the previously described HNT-PHSO$_3$H material, as only noise could be observed in the UHPLC-MS/MS analysis. Similar recoveries compared to HNT-PhSO$_3$H could be achieved. The highest average recovery was obtained for lycopsamine with 101.3% and the lowest recovery for monocrotaline with 78.3%, as visible in Table 2. Batch to batch recoveries for ten performed solid phase extractions and a direct comparison to the previously presented material are displayed in Figure 4.

### 3.2.3. Spiked Honey Sample

As both synthesized materials, HNT-PhSO$_3$H and HNT-MPTMS-SO$_3$H, revealed satisfying performance during the solid phase extraction of the aqueous 10.0 μg L$^{-1}$ pyrrolizidine alkaloid standard, further extractions of spiked honey samples, with the adjusted pyrrolizidine alkaloid limit for honey were conducted.

HNT-PhSO$_3$H

Solid phase extraction for the spiked honey sample was executed as described earlier. Therefore, the same extraction protocol was performed for the aqueous pyrrolizidine alkaloid mixture, with additional washing steps implemented. Table 3 includes the obtained averaged recoveries and matrix effects for solid phase extractions with HNT-PhSO$_3$H as solid material. When comparing these results with the extractions of the aqueous sample mixture, an obvious decrease in recovery for all analytes is visible. Senecionine and heliotrine still show acceptable values, however, monocrotaline and lycopsamine could not be sufficiently recovered, as visible in Figure 5. After solid phase extraction, the breakthrough partly showed large residues of these compounds. Structural properties such as the phenyl group supporting van der Waals and π-π interactions with interfering compounds from the honey matrix could be a pillar of this hypothesis. However, the reduction in the matrix effect could be achieved to a large extent, as the strongest matrix effect was obtained for lycopsamine with only +1.8%.

**Table 3.** Recoveries and matrix effect of a 12.5 μg L$^{-1}$ spiked honey sample after solid phase extraction with HNT-PhSO$_3$H as solid material.

| Recovery ± SD/% (N = 10) | | | |
|---|---|---|---|
| Monocrotaline | Lycopsamine | Heliotrine | Senecionine |
| 62.5 ± 4.4 | 35.8 ± 2.0 | 75.9 ± 3.2 | 86.8 ± 8.0 |
| Matrix Effect/% | | | |
| Monocrotaline | Lycopsamine | Heliotrine | Senecionine |
| 98.3 | 101.8 | 100.1 | 93.3 |

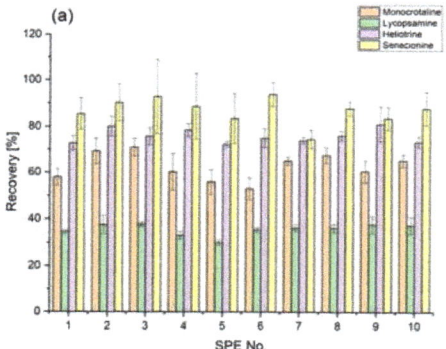 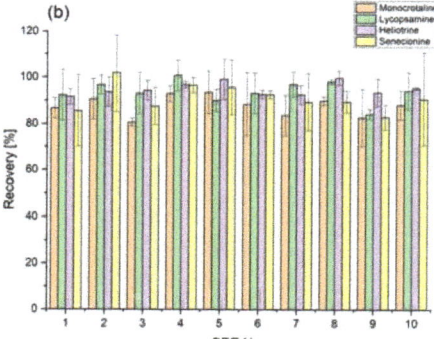

**Figure 5.** Solid phase extractions of a 12.5 μg L$^{-1}$ spiked honey sample with HNT-PhSO$_3$H (**a**) and HNT-MPTMS-SO$_3$H (**b**) as solid phase.

HNT-MPTMS-SO$_3$H

When comparing analyte recoveries in the eluate using HNT-MPTMS-SO$_3$H with the previously described recoveries of HNT-PhSO$_3$H (Figure 5) as solid, it becomes apparent that significantly higher recoveries can be achieved by using HNT-MPTMS-SO$_3$H. Average values present high recoveries up to 95.0% for heliotrine when using HNT-MPTMS-SO$_3$H. In comparison, HNT-PhSO$_3$H as solid material reveals recoveries of up to 86.8% for senecionine, while lycopsamine shows only poor recoveries of 35.8%. For this reason, validation was only performed for HNT-MPTMS-SO$_3$H.

- Validation

    Specificity

    Pyrrolizidine alkaloids were detected in highly specific multiple reaction monitoring mode. Hence, the combination of nominal precursor and product ions have to be given to fulfil detection criteria. Therefore, high specificity should be given theoretically. However, measurements of blank solvents and blank matrix were executed with the previously presented UHPLC-MS/MS method, which revealed no observable pyrrolizidine alkaloid signals.

    Linearity

    Coefficients of determination show satisfying $R^2$ values up to 0.992 and therefore display good linearity of the method. The calibration curves of the selected four analytes can be found in the supplementary information (Figure S1).

    Bias

    Bias values were calculated according to Equation (1). Table 4 shows the bias values of a low-, medium- and high-concentrated matrix-matched calibration solution. The obtained values are within the limit of ±15% and can therefore be seen as acceptable [39].

**Table 4.** Validation parameters of the presented method using HNT-MPTMS-SO$_3$H and 12.5 µg L$^{-1}$ spiked honey sample.

| Concentration level/µg L$^{-1}$ | Bias/% | | | |
|---|---|---|---|---|
| | Monocrotaline | Lycopsamine | Heliotrine | Senecionine |
| 4 | ±5.95 | ±2.07 | ±1.36 | ±10.0 |
| 8 | ±11.3 | ±2.47 | ±7.36 | ±2.85 |
| 12 | ±5.07 | ±0.03(4) | ±3.67 | ±1.89 |
| Limit of detection and Limit of quantification/µg L$^{-1}$ | | | | |
| | Monocrotaline | Lycopsamine | Heliotrine | Senecionine |
| LOD | 1.0 | 0.7 | 0.6 | 1.2 |
| LOQ | 3.2 | 2.3 | 1.9 | 3.6 |
| Recovery ± SD/% (N = 10) | | | | |
| Monocrotaline | Lycopsamine | Heliotrine | Senecionine | |
| 87.8 ± 7.0 | 94.0 ± 6.0 | 95.0 ± 3.9 | 91.3 ± 9.9 | |
| Repeatability RSD/% (N = 10) | | | | |
| Monocrotaline | Lycopsamine | Heliotrine | Senecionine | |
| 5.0 | 4.9 | 2.9 | 6.2 | |
| Matrix Effect/% | | | | |
| Monocrotaline | Lycopsamine | Heliotrine | Senecionine | |
| 106.6 | 104.1 | 110.1 | 102.2 | |

LOD and LOQ

In Table 4, the respective limits of detection and quantification, calculated according to DIN 32645 [37] Equations (2) and (3), are displayed, respectively. The limit of detection values for lycopsamine and heliotrine are in the ng L$^{-1}$ range, whereas monocrotaline and senecionine show values in the low µg L$^{-1}$ area. Therefore, both parameters demonstrate the suitability of the presented method for highly precise analysis of pyrrolizidine alkaloids in honey matrix.

Loading Efficiency

Measurements of the breakthrough from spiked honey samples showed only noise for all analytes at the respective retention times after solid phase extraction with HNT-MPTMS-SO$_3$H. Therefore, pyrrolizidine alkaloid levels can only be smaller than the respective limits of detection from Table 4. Hence, the loading step of the sample onto HNT-MPTMS-SO$_3$H for solid phase extraction can be seen as highly efficient and selective as complete adsorption of the compounds of interest takes place in a complex matrix such as honey.

Recovery

Table 4 shows the obtained averaged recoveries of all target analytes in the eluate from ten performed solid phase extractions. When comparing these results with the extraction of the aqueous pyrrolizidine alkaloid mixture, which was presented in a previous section, averaged recovery values for monocrotaline and senecionine present slightly better results, whereas lycopsamine and heliotrine show a small decrease. However, when standard deviation is included, the recovery ranges overlap. Therefore, no decrease in recovery is visible when performing solid phase extractions with a highly complex matrix such as honey. Looking at publications that use solid phase extractions as sample preparation for pyrrolizidine alkaloid analysis (Table 5), it can be seen that the recoveries of the method presented here are competitive [43–47].

**Table 5.** Recovery rates of pyrrolizidine alkaloids in different matrix in previously published solid phase extraction methods.

| Sample Matrix | Recovery Range/% | Solid Phase Material | Literature |
| --- | --- | --- | --- |
| Gynura procumbens | 21.8–99.4 | PCX | [43] |
| Gynura procumbens | 21.6–96.1 | SCX | [43] |
| Gynura procumbens | 57.8–101.9 | C18 | [43] |
| Gastrodia elata | 77.6–101.4 * | MCX | [44] |
| Atractylodes japonica | 85.2–101.9 * | MCX | [44] |
| Leonurus japonicus | 93.3–112.7 * | MCX | [44] |
| Glycyrrhiza uralensis | 73.8–98.1 * | MCX | [44] |
| Chrysanthemum morifolium | 70.6–103.5 * | MCX | [44] |
| Tussilago farfara | 73.1–111.4 * | MCX | [45] |
| Lithospermi erythrorhzion | 72.3–118.3 * | MCX | [45] |
| Tussilago farfara | 92.5–103.5 | MIP | [46] |
| Herbal teas (fennel, mixed tea and rooibos) | 72–122 | C18 | [47] |
| Honey (cornflower and lavender) | 66–96 | C18 | [47] |

* recoveries from medium spike levels.

Furthermore, the presented method also competes with sample preparation methods other than solid phase extraction in terms of analyte recovery, as the values already reported here are within the recovery ranges of the published methods [42,48–50].

Repeatability

Residual standard deviation values describing the repeatability of the presented method are displayed in Table 4. RSD values ranging from 2.9% up to 6.2% could be obtained. This can be seen as acceptable results, since cartridges were self-packed and therefore variation in packing is inevitable.

Matrix Effect

Matrix effects for all analyzed compounds of interest, which were calculated according to Equation (4), are displayed in Table 4. Values ranging from 102.2% to 110.1% could be obtained, showing slight ion enhancement for each compound of interest. According to the literature, the matrix effect of this method can be classified as soft [51,52]. Hence, matrix-matched calibration would not be necessary for the newly presented method, as matrix effects can be sufficiently reduced [42]. Furthermore, the novel method shows a significantly better reduction in matrix effects than in our last work, where we observed medium matrix effects for most analyzed compounds [18]. This can be reasoned due to the additional washing step, which is now possible due to the stronger binding of the analytes onto the solid material.

Autosampler Stability

Figure 6 displays the averaged triplicate measurements of each concentration level every six hours. The stability of the low concentrated matrix-matched standard, close to the limit of quantification, showed no deviations higher than 20%, with the exception of the 6 h measurement of senecionine (±79.8%) [39]. However, stability values of this sample before and after these durations suggest a random instrument error for this triplicate measurement, as they present in-range values. Subsequently, no deviations more than ±15% were observed for the medium and high concentration levels (8 and 12 µg L$^{-1}$) [39]. Therefore, acceptable stability of the samples in the autosampler is given.

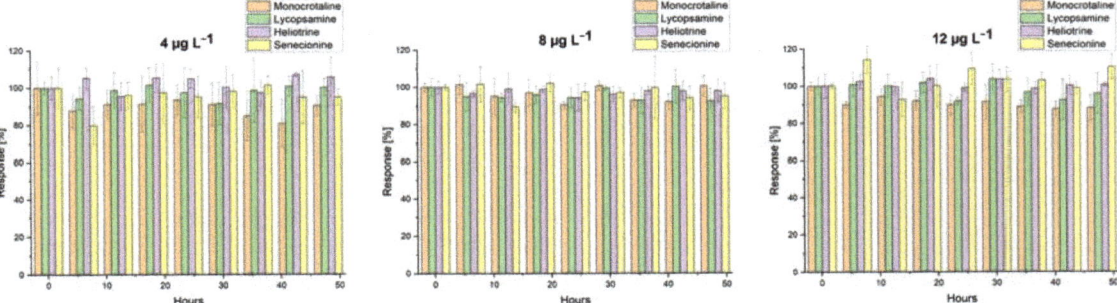

**Figure 6.** Autosampler stability of matrix-matched calibration standards with low, medium and high concentrations of pyrrolizidine alkaloids.

### 3.2.4. Reusability Study

Figure 7 displays the recoveries of all four compounds of interest after six SPE cycles, performed according to the previously described reusability protocol. Hence, the HNT-MPTMS-SO$_3$H material can be reused multiple times, as no significant decrease in recovery can be observed when taking the error bars into account. Subsequently, Table 6 displays the recovery consistency over all six performed solid phase extraction cycles with no signs of loss in performance or destruction of the solid material. After six SPE cycles, the test was terminated, since the reusability could be fully demonstrated. However, since no deficits in recovery rates or phase stability are apparent, use over six cycles is highly likely.

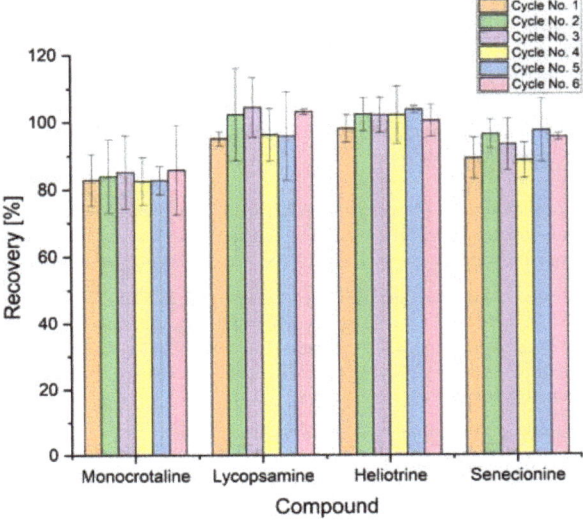

**Figure 7.** Recoveries of an aqueous 10.0 µg L$^{-1}$ pyrrolizidine alkaloid mixture after six consecutive solid phase extractions with HNT-MPTMS-SO$_3$H.

Table 6. Averaged recoveries over six cycles of solid phase extraction using the same HNT-MPTMS-SO$_3$H solid material.

| SPE Cycle No. | Recovery ± SD/% | | | |
|---|---|---|---|---|
| | Monocrotaline | Lycopsamine | Heliotrine | Senecionine |
| 1 | 82.9 ± 7.6 | 95.1 ± 2.0 | 98.1 ± 4.1 | 89.2 ± 6.2 |
| 2 | 84.0 ± 10.9 | 102.3 ± 13.8 | 102.3 ± 5.0 | 96.3 ± 4.3 |
| 3 | 85.2 ± 10.9 | 104.5 ± 8.9 | 102.1 ± 5.3 | 93.2 ± 7.7 |
| 4 | 82.5 ± 7.0 | 96.3 ± 7.8 | 102.0 ± 8.6 | 88.5 ± 5.2 |
| 5 | 82.7 ± 4.2 | 95.9 ± 13.2 | 103.7 ± 1.1 | 97.4 ± 9.5 |
| 6 | 85.8 ± 13.3 | 103.1 ± 0.8 | 100.4 ± 4.8 | 95.5 ± 1.1 |

## 4. Conclusions

Two modified halloysite nanotubes were tested for their performance in the selective solid phase extraction of toxic pyrrolizidine alkaloids as alternative candidates to polymeric resins. Both materials showed satisfying results in the extraction of an aqueous pyrrolizidine alkaloid mixture containing four of the six main structures of the pyrrolizidine alkaloid group. In addition, solid phase extraction of spiked honey samples was performed with both materials considering the respective maximum pyrrolizidine alkaloid level. While the HNT-PhSO$_3$H material showed a significant decrease for some analytes, yet good matrix reduction properties, the use of the HNT-MPTMS-SO$_3$H material does not lead to a decrease in recoveries when compared to the aqueous analyte mixture. Therefore, the latter was validated and displayed acceptable results. Matrix effects could be reduced down to a maximum of 10.1% when compared to our last work. In addition, repeatability presents satisfying values for self-packed solid phase tubes. To make a further step towards green chemistry, the solid material was tested for reusability and showed excellent performance over six solid phase extraction cycles with no decrease in recovery or depletion of the solid material. Comparing the recoveries obtained with the results of solid phase extractions using commercial SCX cartridges performed in our last work with the same pyrrolizidine alkaloids, concentrations and matrix, the significantly better performance of the mercaptopropyl-sulfonated halloysite material is again evident. Furthermore, halloysite nanotubes can once again be presented as an economical and environmentally friendly resource due to their massive natural occurrence and resulting low cost. Hence, a green, sustainable and novel solid material for selective solid phase extraction of toxic pyrrolizidine alkaloids from honey can be presented to further develop additional environmentally friendlier alternatives compared to usual used polymeric substrates in solid phase extraction.

**Supplementary Materials:** The following supporting information can be downloaded at: https://www.mdpi.com/article/10.3390/separations9100270/s1, Figure S1: Calibration curves of the four analytes, namely monocrotaline, lycopsamine, heliotrine and senecionine.

**Author Contributions:** Conceptualization, T.S. and M.R.; data curation, T.S. and N.W.; formal analysis, T.S. and N.W.; investigation, T.S. and N.W.; methodology, T.S. and N.W.; project administration, M.R.; supervision, M.R.; validation, T.S. and N.W.; visualization, T.S.; writing—original draft, T.S.; writing—review & editing, C.W.H., G.K.B. and M.R. All authors have read and agreed to the published version of the manuscript.

**Funding:** This research received no funding.

**Conflicts of Interest:** There are no conflict to declare.

## References

1. Lazzara, G.; Cavallaro, G.; Panchal, A.; Fakhrullin, R.; Stavitskaya, A.; Vinokurov, V.; Lvov, Y. An Assembly of Organic-Inorganic Composites Using Halloysite Clay Nanotubes. *Curr. Opin. Colloid Interface Sci.* **2018**, *35*, 42–50. [CrossRef]
2. Yuan, P.; Tan, D.; Annabi-Bergaya, F. Properties and Applications of Halloysite Nanotubes: Recent Research Advances and Future Prospects. *Appl. Clay Sci.* **2015**, *112–113*, 75–93. [CrossRef]

3. Lvov, Y.; Wang, W.; Zhang, L.; Fakhrullin, R. Halloysite Clay Nanotubes for Loading and Sustained Release of Functional Compounds. *Adv. Mater.* **2016**, *28*, 1227–1250. [CrossRef]
4. El-Sheikhy, R. Exploring of New Natural Saudi Nanoparticles: Investigation and Characterization. *Sci. Rep.* **2020**, *10*, 21557. [CrossRef]
5. Nicholson, J.; Weisman, J.; Boyer, C.; Wilson, C.; Mills, D. Dry Sintered Metal Coating of Halloysite Nanotubes. *Appl. Sci.* **2016**, *6*, 265. [CrossRef]
6. Riahi-Madvaar, R.; Taher, M.A.; Fazelirad, H. Synthesis and Characterization of Magnetic Halloysite-Iron Oxide Nanocomposite and Its Application for Naphthol Green B Removal. *Appl. Clay Sci.* **2017**, *137*, 101–106. [CrossRef]
7. Vergaro, V.; Abdullayev, E.; Lvov, Y.M.; Zeitoun, A.; Cingolani, R.; Rinaldi, R.; Leporatti, S. Cytocompatibility and Uptake of Halloysite Clay Nanotubes. *Biomacromolecules* **2010**, *11*, 820–826. [CrossRef]
8. Almasri, D.A.; Saleh, N.B.; Atieh, M.A.; McKay, G.; Ahzi, S. Adsorption of Phosphate on Iron Oxide Doped Halloysite Nanotubes. *Sci. Rep.* **2019**, *9*, 3232. [CrossRef]
9. Chen, X.-G.; Li, R.-C.; Zhang, A.-B.; Lyu, S.-S.; Liu, S.-T.; Yan, K.-K.; Duan, W.; Ye, Y. Preparation of Hollow Iron/Halloysite Nanocomposites with Enhanced Electromagnetic Performances. *R. Soc. Open Sci.* **2018**, *5*, 171657. [CrossRef]
10. Joo, Y.; Sim, J.H.; Jeon, Y.; Lee, S.U.; Sohn, D. Opening and Blocking the Inner-Pores of Halloysite. *Chem. Commun.* **2013**, *49*, 4519. [CrossRef]
11. White, R.D.; Bavykin, D.V.; Walsh, F.C. The Stability of Halloysite Nanotubes in Acidic and Alkaline Aqueous Suspensions. *Nanotechnology* **2012**, *23*, 065705. [CrossRef]
12. Fizir, M.; Dramou, P.; Dahiru, N.S.; Ruya, W.; Huang, T.; He, H. Halloysite Nanotubes in Analytical Sciences and in Drug Delivery: A Review. *Microchim. Acta* **2018**, *185*, 389. [CrossRef]
13. Rawtani, D.; Pandey, G.; Tharmavaram, M.; Pathak, P.; Akkireddy, S.; Agrawal, Y.K. Development of a Novel 'Nanocarrier' System Based on Halloysite Nanotubes to Overcome the Complexation of Ciprofloxacin with Iron: An in Vitro Approach. *Appl. Clay Sci.* **2017**, *150*, 293–302. [CrossRef]
14. Tabani, H.; Khodaei, K.; Moghaddam, A.Z.; Alexovič, M.; Movahed, S.K.; Zare, F.D.; Dabiri, M. Introduction of Graphene-Periodic Mesoporous Silica as a New Sorbent for Removal: Experiment and Simulation. *Res. Chem. Intermed.* **2019**, *45*, 1795–1813. [CrossRef]
15. Moghaddam, A.Z.; Bameri, A.E.; Ganjali, M.R.; Alexovič, M.; Jazi, M.E.; Tabani, H. A Low-Voltage Electro-Membrane Extraction for Quantification of Imatinib and Sunitinib in Biological Fluids. *Bioanalysis* **2021**, *13*, 1401–1413. [CrossRef]
16. Peixoto, A.F.; Fernandes, A.C.; Pereira, C.; Pires, J.; Freire, C. Physicochemical Characterization of Organosilylated Halloysite Clay Nanotubes. *Microporous Mesoporous Mater.* **2016**, *219*, 145–154. [CrossRef]
17. Silva, S.M.; Peixoto, A.F.; Freire, C. HSO3-Functionalized Halloysite Nanotubes: New Acid Catalysts for Esterification of Free Fatty Acid Mixture as Hybrid Feedstock Model for Biodiesel Production. *Appl. Catal. A Gen.* **2018**, *568*, 221–230. [CrossRef]
18. Schlappack, T.; Rainer, M.; Weinberger, N.; Bonn, G.K. Sulfonated Halloysite Nanotubes as a Novel Cation Exchange Material for Solid Phase Extraction of Toxic Pyrrolizidine Alkaloids. *Anal. Methods* **2022**, *14*, 2689–2697. [CrossRef]
19. *Fragen und Antworten zu Pyrrolizidinalkaloiden in Lebensmitteln*; Bundesinstitut für Risikobewertung: Berlin, Germany, 2020; Volume 15, pp. 1–7.
20. *Analytik und Toxizität von Pyrrolizidinalkaloiden Sowie Eine Einschätzung des Gesundheitlichen Risikos Durch Deren Vorkommen in Honig—Stellungnahme Nr. 038/2011 des BfR vom 11 August 2011, Ergänzt am 21 Januar 2013*; Bundesinstitut für Risikobewertung: Berlin, Germany, 2013; pp. 1–37.
21. Crews, C.; Berthiller, F.; Krska, R. Update on Analytical Methods for Toxic Pyrrolizidine Alkaloids. *Anal. Bioanal. Chem.* **2010**, *396*, 327–338. [CrossRef]
22. Ma, C.; Liu, Y.; Zhu, L.; Ji, H.; Song, X.; Guo, H.; Yi, T. Determination and Regulation of Hepatotoxic Pyrrolizidine Alkaloids in Food: A Critical Review of Recent Research. *Food Chem. Toxicol.* **2018**, *119*, 50–60. [CrossRef]
23. Schrenk, D.; Gao, L.; Lin, G.; Mahony, C.; Mulder, P.P.J.; Peijnenburg, A.; Pfuhler, S.; Rietjens, I.M.C.M.; Rutz, L.; Steinhoff, B.; et al. Pyrrolizidine Alkaloids in Food and Phytomedicine: Occurrence, Exposure, Toxicity, Mechanisms, and Risk Assessment—A Review. *Food Chem. Toxicol.* **2020**, *136*, 111107. [CrossRef]
24. Wiedenfeld, H.; Edgar, J. Toxicity of Pyrrolizidine Alkaloids to Humans and Ruminants. *Phytochem. Rev.* **2011**, *10*, 137–151. [CrossRef]
25. Brugnerotto, P.; Seraglio, S.K.T.; Schulz, M.; Gonzaga, L.V.; Fett, R.; Costa, A.C.O. Pyrrolizidine Alkaloids and Beehive Products: A Review. *Food Chem.* **2021**, *342*, 128384. [CrossRef]
26. *Pyrrolizidinalkaloide in Kräutertees und Tees—Stellungnahme 018/2013 des BfR vom 5 Juli 2013*; Bundesinstitut für Risikobewertung: Berlin, Germany, 2013; pp. 1–31.
27. Casado, N.; Morante-Zarcero, S.; Sierra, I. The Concerning Food Safety Issue of Pyrrolizidine Alkaloids: An Overview. *Trends Food Sci. Technol.* **2022**, *120*, 123–139. [CrossRef]
28. PAK und Pyrrolizidinalkaloide in Getrockneten Kräutern und Gewürzen., Bundesministerium für Arbeit, Soziales, Gesundheit und Konsumentenschutz & Österreichische Agentur für Gesundheit und Ernährungssicherheit GmbH, Vienna, Austria. 2019, pp. 1–4. Available online: www.ages.at (accessed on 12 September 2022).
29. Wiedenfeld, H. Plants Containing Pyrrolizidine Alkaloids: Toxicity and Problems. *Food Addit. Contam. Part A* **2011**, *28*, 282–292. [CrossRef]

30. *Bestimmung von Pyrrolizidinalkaloiden (PA) in Honig Mittels SPE-LC-MS/MS: Methodenbeschreibung*; Bundesinstitut für Risikobewertung: Berlin, Germany, 2013; Volume 1.0, pp. 1–17.
31. European Commission. *COMMISSION REGULATION (EU) 2020/2040 of 11 December 2020 Amending Regulation (EC) No 1881/2006 as Regards Maximum Levels of Pyrrolizidine Alkaloids in Certain Foodstuffs*; Official Journal of the European Union; 2022; p. 5. Available online: eur-lex.europa.eu (accessed on 12 September 2022).
32. *Bestimmung von Pyrrolizidinalkaloiden (PA) in Pflanzenmaterial Mittels SPE-LC-MS/MS: Methodenbeschreibung*; Bundesinstitut für Risikobewertung: Berlin, Germany, 2014; Volume 2.0, pp. 1–17.
33. Chung, S.W.C.; Lam, C.-H. Development of an Analytical Method for Analyzing Pyrrolizidine Alkaloids in Different Groups of Food by UPLC-MS/MS. *J. Agric. Food Chem.* **2018**, *66*, 3009–3018. [CrossRef]
34. Griffin, C.T.; Danaher, M.; Elliott, C.T.; Glenn Kennedy, D.; Furey, A. Detection of Pyrrolizidine Alkaloids in Commercial Honey Using Liquid Chromatography–Ion Trap Mass Spectrometry. *Food Chem.* **2013**, *136*, 1577–1583. [CrossRef]
35. He, Y.; Zhu, L.; Ma, J.; Wong, L.; Zhao, Z.; Ye, Y.; Fu, P.P.; Lin, G. Comprehensive Investigation and Risk Study on Pyrrolizidine Alkaloid Contamination in Chinese Retail Honey. *Environ. Pollut.* **2020**, *267*, 115542. [CrossRef]
36. Kowalczyk, E.; Sieradzki, Z.; Kwiatek, K. Determination of Pyrrolizidine Alkaloids in Honey with Sensitive Gas Chromatography-Mass Spectrometry Method. *Food Anal. Methods* **2018**, *11*, 1345–1355. [CrossRef]
37. *DIN 32645*; Deutsches Institut für Normierung e.V. Beuth Verlag GmbH: Berlin, Germany, 2008.
38. Nasiri, A.; Jahani, R.; Mokhtari, S.; Yazdanpanah, H.; Daraei, B.; Faizi, M.; Kobarfard, F. Overview, Consequences, and Strategies for Overcoming Matrix Effects in LC-MS Analysis: A Critical Review. *Analyst* **2021**, *146*, 6049–6063. [CrossRef]
39. Peters, F.T.; Hartung, M.; Herbold, M.; Schmitt, G.; Daldrup, T. Anforderungen an Die Validierung von Analysenmethoden. *Toxichem Krimtech* **2009**, *76*, 185.
40. Shabir Ghulam, A. Step-by-Step Analytical Methods Validation and Protocol in the Quality System Compliance Industry. *J. Valid. Technol.* **2005**, *10*, 314–325.
41. Shabir Ghulam, A. A Practical Approach to Validation of HPLC Methods under Current Good Manufacturing Practices. *J. Valid. Technol.* **2004**, *10*, 210–218.
42. Izcara, S.; Casado, N.; Morante-Zarcero, S.; Pérez-Quintanilla, D.; Sierra, I. Miniaturized and Modified QuEChERS Method with Mesostructured Silica as Clean-up Sorbent for Pyrrolizidine Alkaloids Determination in Aromatic Herbs. *Food Chem.* **2022**, *380*, 132189. [CrossRef]
43. Ji, Y.-B.; Wang, Y.-S.; Fu, T.-T.; Ma, S.-Q.; Qi, Y.-D.; Si, J.-Y.; Sun, D.-A.; Liao, Y.-H. Quantitative Analysis of Pyrrolizidine Alkaloids in *Gynura Procumbens* by Liquid Chromatography–Tandem Quadrupole Mass Spectrometry after Enrichment by PCX Solid-Phase Extraction. *Int. J. Environ. Anal. Chem.* **2019**, *99*, 1090–1102. [CrossRef]
44. Jeong, S.H.; Choi, E.Y.; Kim, J.; Lee, C.; Kang, J.; Cho, S.; Ko, K.Y. LC-ESI-MS/MS Simultaneous Analysis Method Coupled with Cation-Exchange Solid-Phase Extraction for Determination of Pyrrolizidine Alkaloids on Five Kinds of Herbal Medicines. *J. AOAC Int.* **2021**, *104*, 1514–1525. [CrossRef]
45. Ko, K.Y.; Jeong, S.H.; Choi, E.Y.; Lee, K.; Hong, Y.; Kang, I.H.; Cho, S.; Lee, C. A LC–ESI–MS/MS Analysis Procedure Coupled with Solid Phase Extraction and MeOH Extraction Method for Determination of Pyrrolizidine Alkaloids in Tussilago Farfara and Lithospermi Erythrorhzion. *Appl. Biol. Chem.* **2021**, *64*, 53. [CrossRef]
46. Luo, Z.; Chen, G.; Li, X.; Wang, L.; Shu, H.; Cui, X.; Chang, C.; Zeng, A.; Fu, Q. Molecularly Imprinted Polymer Solid-phase Microextraction Coupled with Ultra High Performance Liquid Chromatography and Tandem Mass Spectrometry for Rapid Analysis of Pyrrolizidine Alkaloids in Herbal Medicine. *J. Sep. Sci.* **2019**, *42*, 3352–3362. [CrossRef]
47. Bodi, D.; Ronczka, S.; Gottschalk, C.; Behr, N.; Skibba, A.; Wagner, M.; Lahrssen-Wiederholt, M.; Preiss-Weigert, A.; These, A. Determination of Pyrrolizidine Alkaloids in Tea, Herbal Drugs and Honey. *Food Addit. Contam. Part A* **2014**, *31*, 1886–1895. [CrossRef]
48. Rizzo, S.; Celano, R.; Campone, L.; Rastrelli, L.; Piccinelli, A.L. Salting-out Assisted Liquid-Liquid Extraction for the Rapid and Simple Simultaneous Analysis of Pyrrolizidine Alkaloids and Related N-Oxides in Honey and Pollen. *J. Food Compos. Anal.* **2022**, *108*, 104457. [CrossRef]
49. Celano, R.; Piccinelli, A.L.; Campone, L.; Russo, M.; Rastrelli, L. Determination of Selected Pyrrolizidine Alkaloids in Honey by Dispersive Liquid–Liquid Microextraction and Ultrahigh-Performance Liquid Chromatography–Tandem Mass Spectrometry. *J. Agric. Food Chem.* **2019**, *67*, 8689–8699. [CrossRef]
50. Martinello, M.; Borin, A.; Stella, R.; Bovo, D.; Biancotto, G.; Gallina, A.; Mutinelli, F. Development and Validation of a QuEChERS Method Coupled to Liquid Chromatography and High Resolution Mass Spectrometry to Determine Pyrrolizidine and Tropane Alkaloids in Honey. *Food Chem.* **2017**, *234*, 295–302. [CrossRef]
51. Ferrer Amate, C.; Unterluggauer, H.; Fischer, R.J.; Fernández-Alba, A.R.; Masselter, S. Development and Validation of a LC–MS/MS Method for the Simultaneous Determination of Aflatoxins, Dyes and Pesticides in Spices. *Anal. Bioanal. Chem* **2010**, *397*, 93–107. [CrossRef]
52. Kaczyński, P.; Łozowicka, B. A Novel Approach for Fast and Simple Determination Pyrrolizidine Alkaloids in Herbs by Ultrasound-Assisted Dispersive Solid Phase Extraction Method Coupled to Liquid Chromatography–Tandem Mass Spectrometry. *J. Pharm. Biomed. Anal.* **2020**, *187*, 113351. [CrossRef]

Article

# Simultaneous Detection of Chlorzoxazone and Paracetamol Using a Greener Reverse-Phase HPTLC-UV Method

Ahmed I. Foudah [1], Faiyaz Shakeel [2], Mohammed H. Alqarni [1], Tariq M. Aljarba [1], Sultan Alshehri [2] and Prawez Alam [1,*]

1 Department of Pharmacognosy, College of Pharmacy, Prince Sattam Bin Abdulaziz University, Al-Kharj 11942, Saudi Arabia
2 Department of Pharmaceutics, College of Pharmacy, King Saud University, Riyadh 11451, Saudi Arabia
* Correspondence: p.alam@psau.edu.sa or prawez_pharma@yahoo.com

**Abstract:** In the literature, greener/eco-friendly analytical techniques for simultaneous estimation of chlorzoxazone (CZN) and paracetamol (PCT) are scarce. As a consequence, greener reverse-phase high-performance thin-layer chromatography with ultraviolet (HPTLC-UV) detection was developed and validated for simultaneous estimation of CZN and PCT in commercial capsules and tablets. The greenness of the proposed HPTLC-UV technique was assessed quantitatively by utilizing the "Analytical GREENness (AGREE)" methodology. For simultaneous estimation of CZN and PCT, the greener HPTLC-UV technique was linear in the 40–1600 ng band$^{-1}$ and 30–1600 ng band$^{-1}$ ranges, respectively. Furthermore, the suggested HPTLC-UV methodology proved sensitive, accurate, precise, and robust for simultaneous detection of CZN and PCT. The assay of CZN in marketed capsules and tablets was found to be 99.01 ± 1.53 and 100.87 ± 1.61%, respectively, using the suggested HPTLC-UV method. The assay of PCT in commercial capsules and tablets was found to be 98.31 ± 1.38 and 101.21 ± 1.67%, respectively. The AGREE index for the greener HPTLC-UV technique was found to be 0.79, suggesting an excellent greenness profile for the proposed HPTLC-UV technique. These results and data suggested the suitability of the greener HPTLC-UV methodology for simultaneous estimation of CZN and PCT in commercial formulations.

**Keywords:** chlorzoxazone; greener HPTLC; paracetamol; simultaneous detection; validation

## 1. Introduction

Combined dosage forms are widely used due to their tolerability, synergistic effects, and patient's acceptability [1]. Paracetamol (PCT) (chemical name: 4-hydroxyacetanilide; molecular structure: Figure 1A) is a commonly used nonsteroidal anti-inflammatory, antipyretic, and analgesic drug [2,3]. It is marketed in the form of several dosage forms [3]. Chlorzoxazone (CZN) (chemical name: 5-chloro-2-benzoxazolinone; molecular structure: Figure 1B) is used as a muscle relaxant [4,5]. The combination of PCT and CZN (tablets and capsules) is commonly used as a muscle relaxant [1,4]. Although PCT has been recommended as the safest anti-inflammatory medicine, gastrointestinal bleeding, hypertension, and hepatotoxicity are common hazardous/toxic effects of PCT in higher doses [6]. The common overdose/hazardous effects of CZN are nausea vomiting, diarrhea, drowsiness, and dizziness [7]. Therefore, it is crucial to conduct a qualitative and quantitative investigation of PCT and CZN in commercialized dosage forms.

A literature survey demonstrated several analytical methods for simultaneous detection of CZN and PCT in marketed dosage forms and biological samples. For simultaneous detection of CZN and PCT in combined dosage forms, several ultraviolet (UV)-based spectrometry methods have been reported [1,8–12]. For simultaneous detection of CZN and PCT in various commercially available combined pharmaceutical preparations, a number of "high-performance liquid chromatography (HPLC)" methods are also reported [13–20].

Ultra-high-performance liquid chromatography (UHPLC) and supercritical fluid chromatography methods have also been utilized for simultaneous detection of CZN and PCT in commercially available combined dosage forms [21,22]. UHPLC and liquid chromatography tandem mass-spectrometry methods have also been used for simultaneous detection of CZN and PCT in human plasma samples [23,24]. For simultaneous detection of CZN and PCT in various commercially available combined pharmaceutical preparations, various "high-performance thin-layer chromatography (HPTLC)" approaches have also been used [25–28]. A capillary liquid chromatography technique has also been applied for concurrent detection of CZN and PCT in commercialized tablets [29]. Recently, we reported a single greener HPTLC method for simultaneous detection of caffeine and PCT in combined dosage forms [30]. However, no greener HPTLC methods have been reported for simultaneous detection of CZN and PCT in combined pharmaceutical preparations. Various analytical approaches have been recommended in published papers on simultaneous detection of CZN and PCT in combination dosage forms. Meanwhile, the greenness index in any of the pharmaceutical assay literature has not been determined. Furthermore, greener HPTLC methods have not been used for simultaneous detection of CZN and PCT in combined dosage forms. For the determination of the greenness index, several quantitative and qualitative analytical approaches have been documented [31–35]. Only the "Analytical GREENness (AGREE)" methodology takes all twelve green analytical chemistry (GAC) components into account when calculating the greenness index [33]. As a result, the greenness index of a suggested HPTLC-UV method has been determined using the AGREE analytical approach [33].

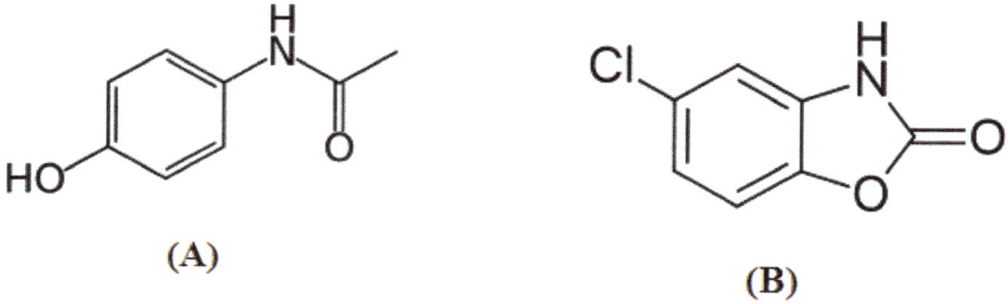

**Figure 1.** Molecular structures of (**A**) paracetamol (PCT) and (**B**) chlorzoxazone (CZN).

The goal of the current study was to design and validate a reverse-phase HPTLC-UV approach for rapid, accurate, and environmentally friendly simultaneous detection of CZN and PCT in combination dosage forms. The greener HPTLC-UV methodology for simultaneous detection of CZN and PCT was verified using "The International Council for Harmonization (ICH)" Q2-R1 protocols [36].

## 2. Materials and Methods

### 2.1. Materials

CZN and PCT working standards were procured from Sigma Aldrich (St. Louis, MO, USA). HPLC-grade ethanol and methanol were procured from E-Merck (Darmstadt, Germany). Ultra-pure water was procured using a Milli-Q unit. Marketed capsules (Relaxon containing 250 mg of CZN and 300 mg of PCT) and marketed tablets (Myodol having 250 mg of CZN and 325 mg of PCT) were purchased from pharmacy shops in Riyadh (Saudi Arabia) and New Delhi (India), respectively. Every other substance, including solvents, was of analytical grade.

### 2.2. Instrumentation and Chromatographic Analysis

Simultaneous detection of CZN and PCT in their pure forms, as well as in marketed capsules and tablets, was carried out using an "HPTLC CAMAG TLC system (CAMAG,

Muttenz, Switzerland)". The obtained samples were applied as 6 mm bands using a "CAMAG Automatic TLC Sampler 4 (ATS4) Sample Applicator (CAMAG, Geneva, Switzerland)". A "CAMAG microliter Syringe (Hamilton, Bonaduz, Switzerland)" was connected to the sample applicator. TLC plates were glass plates (plate size: 10 × 20 cm) pre-coated with RP silica gel (particle size: 5 µm) 60F254S plates. A constant application rate of 150 nL s$^{-1}$ was established for the simultaneous detection of CZN and PCT. The TLC plates were prepared in a "CAMAG automated developing chamber 2 (ADC2) (CAMAG, Muttenz, Switzerland)" in a linear ascending mode at an 80 mm distance. A binary ethanol/water (70:30, v v$^{-1}$) mixture was used as the mobile phase. The preparation chamber was filled with the vapors of the mobile phase for 30 min at 22 °C. CZN and PCT were measured at a wavelength of 268 nm. Scan speed was adjusted at 20 mm s$^{-1}$, and the slit size was set to 4 × 0.45 mm$^2$. For each measurement, either three or six replicates were used. "WinCAT's (version 1.4.3.6336, CAMAG, Muttenz, Switzerland)" was the program utilized for data processing and analysis.

*2.3. Calibration Curves and Quality Control (QC) Samples for CZN and PCT*

The required quantities of CZN and PCT were dispensed separately in the designated volumes of the mobile phase to create individual stock solutions, each of which contained 100 µg mL$^{-1}$ of the drug. By serial dilution of the CZN or PCT stock solution using the mobile phase, concentrations in the 40–1600 ng band$^{-1}$ range for CZN and the 30–1600 ng band$^{-1}$ range for PCT were produced. On RP-TLC plates, 200 µL of each concentration of CZN and PCT were spotted. Each CZN and PCT concentration's spot area was noted. By plotting the concentrations of both medications against the observed spot area in six replicates (n = 6), calibration curves for CZN and PCT were created. Three distinct QC samples were freshly created for the evaluation of several validation parameters.

*2.4. Sample Preparations for Simultaneous Detection of CZN and PCT in Marketed Tablets and Capsules*

Ten marketed tablets (each having 250 mg of CZN and 325 mg of PCT) were taken, and the average weight was noted. For the simultaneous detection of CZN and PCT in commercial capsules, ten capsules (each having 250 mg of CZN and 300 mg of PCT) were taken, and average weight was noted. The contents of the capsules were taken out from the capsule shell. The contents of the tablets or capsules were powdered after being roughly crushed. One hundred milliliters of methanol was used to dissolve a portion of the powder from each brand. Ten milliliters of methanol was used to dilute around 1 mL of this solution for each brand of tablet or capsule once more. To get rid of any undissolved excipients, the produced tablets and capsule solutions were filtered and sonicated for ten minutes. The produced solutions were utilized to detect CZN and PCT in the marketed tablets and capsules using the suggested HPTLC-UV approach.

*2.5. Validation Parameters*

The proposed HPTLC-UV method for simultaneous detection of CZN and PCT was validated for a number of parameters using the ICH-Q2-R1 protocols [36]. The linearity range for CZN and PCT was established by plotting their concentrations versus the measured spot areas. The linearity of the greener HPTLC-UV method was determined in the 40–1600 ng band$^{-1}$ range for CZN and the 30–1600 ng band$^{-1}$ range for PCT in six replicates (n = 6).

The determination of retardation factor ($R_f$), asymmetry factor (As), and theoretical plates number per meter (N m$^{-1}$) were used to assess the parameters for system acceptability for the suggested HPTLC-UV method for simultaneous detection of CZN and PCT. Using their published equations, $R_f$, As, and N m$^{-1}$ values were calculated [34].

The accuracy of the suggested HPTLC-UV approach for simultaneous detection of CZN and PCT was evaluated using the percentage of recovery. For CZN and PCT, three QC levels—lower QC (LQC; 100 ng band$^{-1}$), middle QC (MQC; 400 ng band$^{-1}$), and high

QC (HQC; 1600 ng band$^{-1}$)—were used to test the accuracy of the proposed HPTLC-UV method. At each QC level, the percentage of recovery for CZN and PCT was evaluated across six replicates (n = 6).

For CZN and PCT, the proposed HPTLC-UV method's intra- and inter-assay precision was assessed. Examining intra-assay variation for CZN and PCT involved quantifying newly made CZN and PCT solutions at LQC, MQC, and HQC on the same day in six replicates (n = 6). Inter-assay variability for CZN and PCT was examined using quantitation of newly generated solutions at LQC, MQC, and HQC on three consecutive days in six replicates (n = 6).

By making a few small, deliberate adjustments to the mobile phase mixture, the robustness for CZN and PCT was assessed for the greener HPTLC-UV method. The greener mobile phase for CZN and PCT, ethanol/water (70:30, v v$^{-1}$), was changed to ethanol/water (72:28, v v$^{-1}$) and ethanol/water (68:32, v v$^{-1}$), and the differences in chromatographic response and $R_f$ values were noted in six replicates (n = 6).

The sensitivity of the proposed HPTLC-UV methodology for CZN and PCT was evaluated as "limit of detection (LOD) and limit of quantification (LOQ)" using a "standard deviation" method. The "LOD and LOQ" values for CZN and PCT were determined using their published equations (n = 6) [36].

To assess the specificity of the proposed HPTLC-UV method, the $R_f$ values and UV-absorption spectra of standard CZN and PCT were compared to those of CZN and PCT in marketed tablets and capsules.

*2.6. Application of a Greener HPTLC-UV Method in Simultaneous Detection of CZN and PCT in Commercial Tablets and Capsules*

The acquired marketed tablet and capsule solutions were spotted onto RP-TLC plates for the HPTLC-UV technique, and the chromatographic responses were recorded using the same experimental conditions used for the simultaneous detection of standards CZN and PCT in three replicates (n = 3). Using the calibration curves for CZN and PCT, the amounts of CZN and PCT in the marketed tablets and capsules were calculated.

*2.7. Greenness Evaluation Using AGREE Methodology*

The AGREE methodology [33] was used to determine the greenness index for the greener HPTLC-UV method for simultaneous detection of CZN and PCT. In this methodology, different greenness scores (between 0.0 and 1.0) are assigned to twelve different components of GAC, and the average of twelve is finally taken. Twelve different components considered by the AGREE methodology are sample treatment, sample amount, device positioning, steps for sample preparation, automation of device, derivatization, amount of waste, analysis throughput, energy consumption, source of reagents/solvents, toxicity, and operator's safety [33]. AGREE indices (0.0–1.0) for the greener HPTLC-UV method were assessed using "AGREE: The Analytical Greenness Calculator (version 0.5, Gdansk University of Technology, Gdansk, Poland, 2020)".

## 3. Results and Discussion

*3.1. Method Development*

Different ethanol/water concentrations between the ranges of 40 and 90% ethanol were investigated as greener mobile mixtures for method development in order to create a valid chromatogram for concurrent detection of CZN and PCT by the proposed HPTLC-UV methodology. The development of all the suggested green mobile phase mixtures took place under saturation chamber conditions. The outcomes revealed that the green solvent mixture of ethanol and water (70:30, v v$^{-1}$) produced well-separated and intact chromatographic peaks for CZN at $R_f$ = 0.54 ± 0.02 and for PCT at $R_f$ = 0.67 ± 0.02, as shown in Figure 2. In HPTLC chromatograms of CZN and PCT (Figure 2), the zero point of the horizontal coordinate and the zero point of the vertical coordinate were coincided together, which is very common in such chromatograms. Additionally, As values of 1.06

and 1.04, which are particularly reliable for simultaneous detection of CZN and PCT, were discovered for CZN and PCT. As a result, the ethanol/water (70:30, v v$^{-1}$) ratio was chosen as the final mobile phase composition for the proposed HPTLC-UV method for simultaneous detection of CZN and PCT in commercial tablets and capsules. When the spectral bands for CZN and PCT were recorded in densitometry mode, the maximum response was found at a wavelength of 268 nm. As a consequence, the wavelength at which CZN and PCT were simultaneously detected was 268 nm.

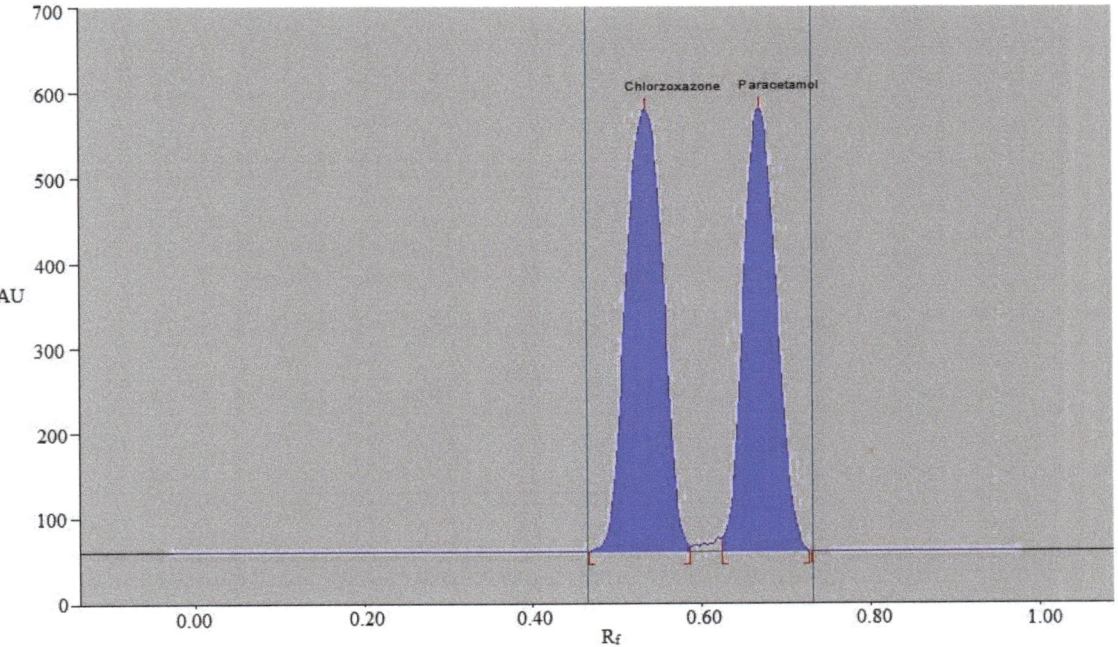

**Figure 2.** Representative high-performance thin-layer chromatography ultraviolet (HPTLC-UV) chromatograms of standard CZN and PCT.

*3.2. Validation Parameters*

To evaluate a number of parameters for simultaneous detection of CZN and PCT, the ICH-Q2-R1 protocol was employed [36]. The outcomes of the HPTLC-UV method's linear regression analysis of the CZN and PCT calibration curves are illustrated in Table 1. CZN and PCT calibration curves were linear in the 40–1600 ng band$^{-1}$ and 30–1600 ng band$^{-1}$ ranges for CZN and PCT, respectively. In the HPTLC method of analysis, the samples are applied in the form of bands. As a result, the most convenient unit for representing concentration is ng band$^{-1}$. Therefore, the concentrations of CZN and PCT are expressed in ng band$^{-1}$ in this study. CZN and PCT's determination coefficients ($R^2$) were predicted to be 0.9990 and 0.9985, respectively. For both medications, the values of $R^2$ were highly significant ($p < 0.05$). These results revealed that for simultaneous detection of CZN and PCT, there was a significant correlation between the concentration and measured response of CZN and PCT.

**Table 1.** Results for linearity evaluation for simultaneous detection of chlorzoxazone (CZN) and paracetamol (PCT) using the greener high-performance thin-layer chromatography ultraviolet (HPTLC-UV) methodology (mean ± SD; n = 6).

| Parameters | CZN | PCT |
|---|---|---|
| Linearity range (ng band$^{-1}$) | 40–1600 | 30–1600 |
| $R^2$ | 0.9990 | 0.9985 |
| Slope ± SD | 19.29 ± 0.56 | 18.73 ± 0.53 |
| Intercept ± SD | 294.32 ± 5.38 | 900.96 ± 9.87 |
| Standard error of slope | 0.22 | 0.21 |
| Standard error of intercept | 2.19 | 4.03 |
| 95% CI of slope | 18.31–20.28 | 17.80–19.67 |
| 95% CI of intercept | 284.86–303.77 | 883.61–918.30 |
| LOD ± SD (ng band$^{-1}$) | 13.86 ± 0.21 | 10.21 ± 0.16 |
| LOQ ± SD (ng band$^{-1}$) | 41.58 ± 0.63 | 30.63 ± 0.48 |

$R^2$: determination coefficient; CI: confidence interval; LOD: limit of detection; LOQ: limit of quantification.

Table 2 documents the details of the system's compatibility for the proposed HPTLC-UV methodology. The $R_f$, As, and N m$^{-1}$ for the proposed HPTLC-UV technique were found to be satisfactory for simultaneous detection of CZN and PCT.

**Table 2.** The values of system suitability parameters for CZN and PCT for the HPTLC-UV methodology (mean ± SD; n = 3).

| Parameters | CZN | PCT |
|---|---|---|
| $R_f$ | 0.54 ± 0.02 | 0.67 ± 0.02 |
| As | 1.06 ± 0.03 | 1.04 ± 0.02 |
| N m$^{-1}$ | 5361 ± 6.12 | 5485 ± 6.28 |

$R_f$: retardation factor; As: asymmetry factor; N m$^{-1}$: number of theoretical plates per meter.

The accuracy of the proposed HPTLC-UV approach was evaluated using the percentage of recovery for both CZN and PCT. Table 3 presents the accuracy evaluation findings for the proposed HPTLC-UV methodology. The percentage recoveries of CZN and PCT at three different QC concentrations were determined to be 98.68–101.42 and 98.69–100.96, respectively, using the proposed HPTLC-UV technique. These findings demonstrated that the proposed HPTLC-UV technique was reliable for simultaneous detection of CZN and PCT.

**Table 3.** Determination of the accuracy of CZN and PCT for the HPTLC-UV methodology (mean ± SD; n = 6).

| Conc. (ng band$^{-1}$) | Conc. Found (ng band$^{-1}$) ± SD | Recovery (%) | CV (%) |
|---|---|---|---|
| | CZN | | |
| 50 | 50.71 ± 0.54 | 101.42 | 1.06 |
| 400 | 404.64 ± 3.12 | 101.16 | 0.77 |
| 1600 | 1578.95 ± 9.94 | 98.68 | 0.62 |
| | PCT | | |
| 50 | 50.48 ± 0.47 | 100.96 | 0.93 |
| 400 | 394.76 ± 1.92 | 98.69 | 0.48 |
| 1600 | 1585.83 ± 7.16 | 99.11 | 0.45 |

CV: coefficient of variance.

For simultaneous detection of CZN and PCT, the precision of the proposed HPTLC-UV approach was assessed as intra- and inter-assay precision, with findings reported as a percentage of the coefficient of variation (CV). The findings of intra- and inter-day precisions for the simultaneous detection of CZN and PCT utilizing the proposed HPTLC-UV approach are illustrated in Table 4. It was discovered that the percent CVs of CZN and

PCT for intra-day variation were 0.49–0.90 and 0.64–0.97 percent, respectively. According to research, the percent CVs of CZN and PCT for intra-day variation are 0.73–0.98 and 0.54–0.96 percent, respectively. These findings showed that the proposed HPTLC-UV technique was precise for simultaneous detection of CZN and PCT.

Table 4. Determination of intra- and inter-day precision of CZN and PCT for the HPTLC-UV methodology (mean ± SD; n = 6).

| Conc. (ng band$^{-1}$) | Intra-Day Precision | | | Inter-Day Precision | | |
|---|---|---|---|---|---|---|
| | Conc. (ng band$^{-1}$) ± SD | Standard Error | CV (%) | Conc. (ng band$^{-1}$) ± SD | Standard Error | CV (%) |
| | | | CZN | | | |
| 50 | 49.39 ± 0.48 | 0.19 | 0.97 | 49.52 ± 0.49 | 0.20 | 0.98 |
| 400 | 392.34 ± 3.08 | 1.25 | 0.78 | 389.88 ± 3.14 | 1.28 | 0.80 |
| 1600 | 1581.41 ± 10.22 | 4.17 | 0.64 | 1610.23 ± 12.21 | 4.98 | 0.75 |
| | | | PCT | | | |
| 50 | 49.61 ± 0.45 | 0.18 | 0.90 | 50.66 ± 0.49 | 0.20 | 0.96 |
| 400 | 406.14 ± 2.14 | 0.87 | 0.52 | 392.87 ± 2.24 | 0.91 | 0.57 |
| 1600 | 1653.54 ± 8.12 | 3.31 | 0.49 | 1581.87 ± 8.55 | 3.49 | 0.54 |

CV: coefficient of variance.

The robustness of the proposed HPTLC-UV approach for concurrent detection of CZN and PCT was assessed by making small, purposeful changes to the greener mobile phase mixtures. Table 5 illustrates the resulting data of robustness analysis utilizing the proposed HPTLC-UV method. It was established that percent CVs for CZN and PCT were 0.94–1.06 and 0.94–0.97 percent, respectively. Additionally, the $R_f$ values of CZN and PCT were found to be 0.53–0.55 and 0.66–0.68, respectively. These findings demonstrated the robustness of the greener HPTLC-UV technique for simultaneous detection of CZN and PCT.

Table 5. Robustness evaluation of CZN and PCT for the HPTLC-UV methodology (mean ± SD; n = 6).

| Conc. (ng band$^{-1}$) | Mobile Phase Composition (Ethanol/Water) | | | Results | | |
|---|---|---|---|---|---|---|
| | Original | Used | | (ng band$^{-1}$) ± SD | % CV | $R_f$ |
| | | | CZN | | | |
| 400 | 70:30 | 72:28 | +2.0 | 386.87 ± 3.68 | 0.95 | 0.53 |
| | | 70:30 | 0.0 | 395.41 ± 3.74 | 0.94 | 0.54 |
| | | 68:32 | −2.0 | 404.64 ± 3.95 | 0.97 | 0.55 |
| | | | PCT | | | |
| 400 | 70:30 | 72:28 | +2.0 | 387.84 ± 3.65 | 0.94 | 0.66 |
| | | 70:30 | 0.0 | 397.63 ± 3.86 | 0.97 | 0.67 |
| | | 68:32 | −2.0 | 405.64 ± 4.30 | 1.06 | 0.68 |

CV: coefficient of variance; $R_f$: retardation factor.

In order to evaluate the sensitivity of the proposed HPTLC-UV methodology for simultaneous detection of CZN and PCT, the "LOD and LOQ" were determined. Table 1 illustrates the computed data of "LOD and LOQ" for CZN and PCT using the proposed HPTLC-UV approach. The "LOD and LOQ" for CZN were calculated to be 13.86 ± 0.21 and 41.58 ± 0.63 ng band$^{-1}$, respectively, using the proposed HPTLC-UV methodology. The "LOD and LOQ" for PCT were calculated to be 10.21 ± 0.16 and 30.63 ± 0.48 ng band$^{-1}$, respectively, using the proposed HPTLC-UV technique. The LOD values for CZN and PCT were much lower than the lowest linearity concentration for both drugs. The linearity range for CZN was 40–1600 ng band$^{-1}$, and its LOD was 13.86 ng band$^{-1}$, which was much lower than the lowest linearity concentration of CZN. Similarly, the linearity range for PCT was 30–1600 ng band$^{-1}$, and its LOD was 10.21 ng band$^{-1}$, which was also much lower than the lowest linearity concentration of PCT. However, the LOQ values for CZN and PCT were slightly higher than the lowest linearity concentration for both drugs. The LOQ is the amount of analyte which can be quantified by the proposed analytical method. The LOQ value is always within the linearity range, which was achieved in the present study [36]. Hence, both LOD and LOQ values for both drugs correlated well with the linearity range. These results revealed that the greener HPTLC-UV method was sensitive enough for simultaneous detection and quantification of CZN and PCT.

The $R_f$ values and overlaid UV-absorption spectra of CZN and PCT in commercially available tablets (Myodol) and capsules (Relaxon) were examined and compared with that of standards CZN and PCT to determine the specificity of the proposed HPTLC-UV method for simultaneous detection of CZN and PCT. The standards CZN and PCT as well as CZN and PCT in commercially available tablets and capsules are shown in Figure 3 along with their overlapping UV-absorption spectra.

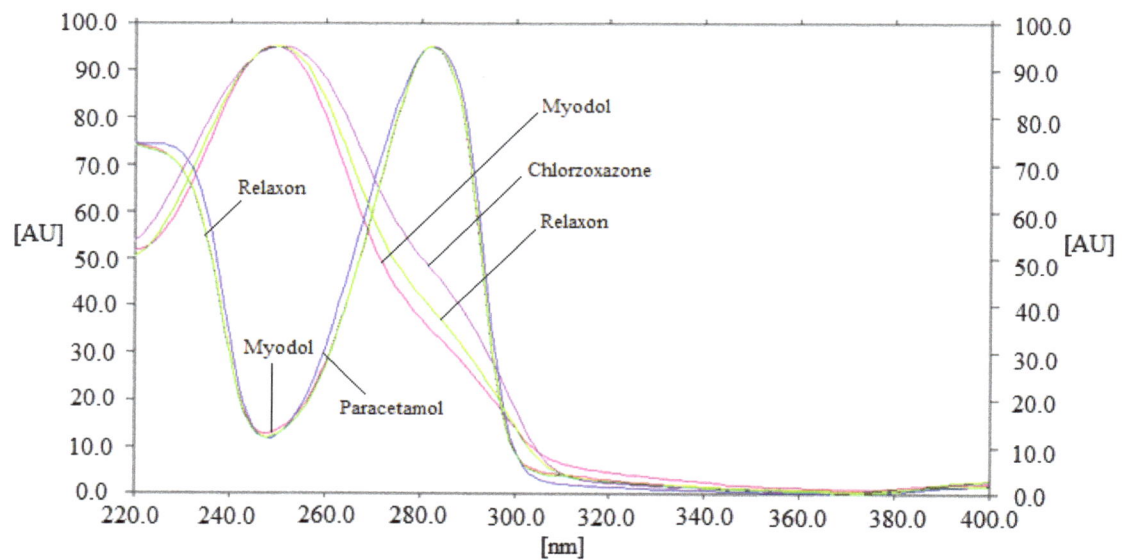

**Figure 3.** Overlaid UV-absorption spectra of standard CZN and PCT and CZN and PCT in commercial Relaxon capsules and commercial Myodol tablets.

The maximum densitometry response of CZN and PCT in standards, commercially available tablets, and commercially available capsules was observed at a wavelength of 268 nm. The identical UV-absorption spectra, $R_f$ values, and wavelengths of CZN and PCT in standards and commercial tablets and capsules demonstrated the specificity of the proposed HPTLC-UV approach for concurrent detection of CZN and PCT.

*3.3. Application of the HPTLC-UV Method in Simutaneous Detection of CZN and PCT in Marketed Tablets and Capsules*

For simultaneous detection of CZN and PCT in their commercially available tablets and capsules, the greener HPTLC-UV method has been proposed as an alternative to routine liquid chromatography methods. The chromatograms of CZN and PCT from marketed tablets and capsules were verified by comparing their TLC bands at $R_f = 0.54 \pm 0.02$ for CZN and $R_f = 0.67 \pm 0.02$ for PCT with that of standards CZN and PCT using the proposed HPTLC-UV method. Figure 4 illustrates the obtained densitograms of CZN and PCT in marketed capsules Relaxon (Figure 4A) and tablets Myodol (Figure 4B), which revealed the densitograms of CZN and PCT similar to those of standard CZN and PCT in marketed capsules and tablets.

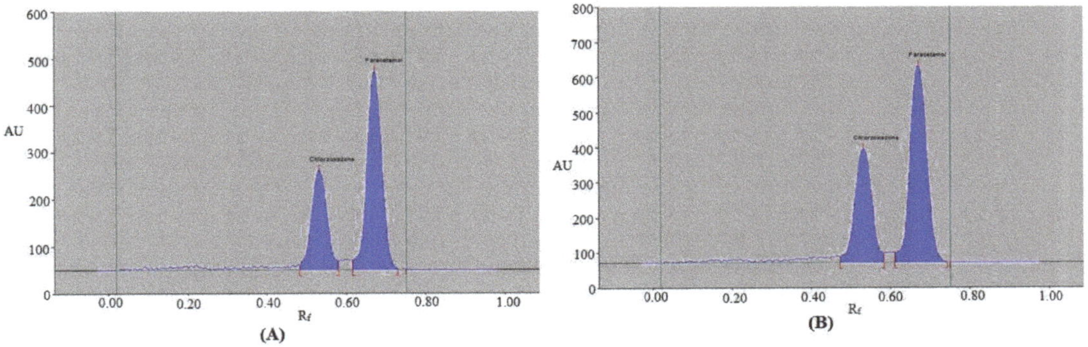

**Figure 4.** HPTLC-UV densitograms of CZN and PCT in (**A**) marketed Relaxon capsules and (**B**) marketed Myodol tablets.

The percent assay of CZN in commercially available capsules (Relaxon) and tablets (Myodol) was found to be 99.01 ± 1.53 and 100.87 ± 1.61 percent, respectively, using the proposed HPTLC-UV approach. The percent assay of PCT in commercially available capsules (Relaxon) and tablets (Myodol) was discovered to be 98.31 ± 1.38 and 101.21 ± 1.67%, respectively, utilizing the proposed HPTLC-UV method. These pharmaceutical assay results demonstrated that the greener HPTLC-UV method was suitable and acceptable for simultaneous detection of CZN and PCT in marketed capsules and tablets.

*3.4. Greenness Evaluation Using the AGREE Approach*

For the estimation of the greenness of analytical techniques, a number of quantitative and qualitative methodologies have been published [31–35]. Only AGREE, however, makes use of all twelve GAC principles for this objective [33]. As a result, the suggested HPTLC-UV approach's greenness was assessed using the AGREE methodology. Figure 5 shows the typical pictogram for the AGREE index of the suggested HPTLC-UV approach. Using all twelve GAC principles, 0.79 was the expected value for the AGREE index. These results revealed that the suggested HPTLC-UV methodology for concurrent detection of CZN and PCT had an outstanding greenness profile.

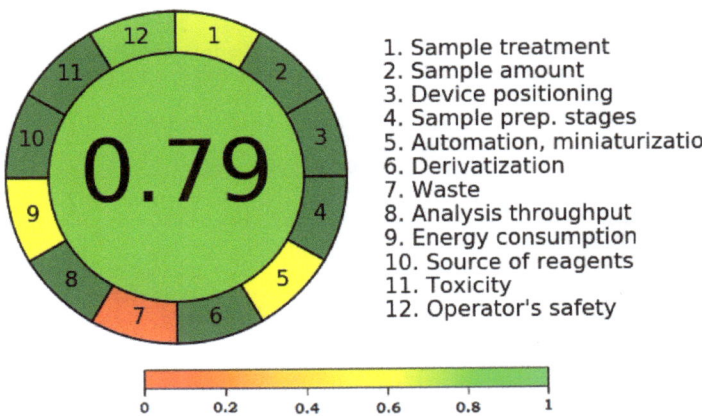

**Figure 5.** "Analytical GREEnness (AGREE)" score for the greener HPTLC-UV method.

## 4. Conclusions

The literature shows a scarcity in greener analytical methods for simultaneous detection of CZN and PCT. Therefore, the goal of the current work was to design and validate a rapid (analysis time: 30 min for 24 samples), sensitive, and greener HPTLC-UV method for simultaneous detection of CZN and PCT in their commercialized tablets and capsules. The proposed HPTLC-UV approach is linear, rapid, affordable, accurate, precise, robust, specific, and sensitive for concurrent detection of CZN and PCT. The amounts of CZN and PCT in marketed tablets and capsules were found to be acceptable using the proposed HPTLC-UV method. The AGREE evaluation revealed an excellent greenness profile of the proposed HPTLC-UV method. These observations and results demonstrated that the proposed HPTLC-UV method can be successfully utilized for simultaneous detection of CZN and PCT in commercially available combined dosage forms, such as tablets and capsules. Due to its successful application in tablets and capsules, the proposed HPTLC-UV method can also be applied to simultaneous detection of CZN and PCT in other dosage forms.

**Author Contributions:** Conceptualization—P.A. and F.S.; methodology—A.I.F., P.A., M.H.A. and T.M.A.; validation—S.A. and M.H.A.; data curation—T.M.A.; funding acquisition—S.A.; project administration—P.A.; supervision—P.A. and F.S.; software—S.A. and F.S.; writing original draft—F.S.; writing—review and editing—A.I.F. and S.A. All authors have read and agreed to the published version of the manuscript.

**Funding:** This research was funded by the Researchers Supporting Project (Number RSP-2021/146), King Saud University, Riyadh, Saudi Arabia. The APC was funded by RSP.

**Institutional Review Board Statement:** Not applicable.

**Informed Consent Statement:** Not applicable.

**Data Availability Statement:** This study did not report any data.

**Acknowledgments:** The authors are thankful to the Researchers Supporting Project (Number RSP-2020/146) at King Saud University, Riyadh, Saudi Arabia for supporting this work.

**Conflicts of Interest:** The authors declare no conflict of interest.

## References

1. Toubar, S.S.; Hegazy, M.A.; Elshaheda, M.S.; Helmy, M.I. Novel pure component contribution, mean centering of ratio spectra and factor based algorithms for simultaneous resolution and quantification of overlapped spectral signals: An application to recently co-formulated tablets of chlorzoxazone, aceclofenac and paracetamol. *Spectrochim. Acta Part A* **2016**, *163*, 89–95.
2. Jimenez, J.A.; Martinez, F. Thermodynamic study of the solubility of acetaminophen in propylene glycol + water cosolvent mixtures. *J. Braz. Chem. Soc.* **2006**, *17*, 125–134. [CrossRef]
3. Shakeel, F.; Alanazi, F.K.; Alsarra, I.A.; Haq, N. Solubilization behavior of paracetamol in Transcutol-water mixtures at $T$ = (298.15 to 333.15) K. *J. Chem. Eng. Data* **2012**, *58*, 3551–3556. [CrossRef]
4. Petsalo, A.; Turpeinen, M.; Pelkonen, O.; Tolonen, A. Analysis of nine drugs and their cytochrome P450-specific probe metabolites from urine by liquid chromatography tandem mass spectrometry utilizing sub 2 µm particle size column. *J. Chromatogr. A* **2008**, *1215*, 107–115. [CrossRef]
5. Joshi, M.; Tyndale, R.F. Regional and cellular distribution of CYP2E1 in monkey brain and its induction by chronic nicotine. *Neuropharmacology* **2006**, *50*, 568–575. [CrossRef]
6. McCrae, J.C.; Morrison, E.E.; MacIntyre, I.M.; Dear, J.W.; Webb, D.J. Long-term adverse effects of paracetamol-a review. *Br. J. Clin. Pharmacol.* **2018**, *84*, 2218–2230. [CrossRef]
7. Richards, B.L.; Whittle, S.L.; Buchbinder, R. Muscle relaxants for pain management in rheumatoid arthritis. *Cochrane Database Syst. Rev.* **2012**, *1*, CD008922. [CrossRef]
8. Chatterjee, P.K.; Jain, C.L.; Sethi, P.D. Simultaneous estimation of chlorzoxazone and acetaminophen in combined dosage forms by an absorbance ratio technique and difference spectrophotometry. *J. Pharm. Biomed. Anal.* **1989**, *7*, 693–698. [CrossRef]
9. Garg, C.; Saraf, W.; Saraf, S. Simultaneous estimation of aceclofenac, paracetamol and chlorzoxazone in tablets. *Indian J. Pharm. Sci.* **2007**, *69*, 692–694.
10. Joshi, R.; Pawar, N.; Sawant, R.; Gaikwad, P. Simultaneous estimation of paracetamol, chlorzoxazone and ibuprofen by validated spectrophotometric methods. *Anal. Chem. Lett.* **2012**, *2*, 118–124. [CrossRef]

11. El-Bagary, R.I.; El-Kady, E.F.; Al-Matari, A.A. Simultaneous spectrophotometric determination of diclofenac sodium, paracetamol, and chlorzoxazone in ternary mixture using chemometric and artificial neural networks techniques. *Asian J. Pharm. Clin. Res.* **2017**, *10*, 225–230. [CrossRef]
12. Mohammed, M.A.; Reid, I.O.A.; Elawni, A. H-point standard additions method for the simultaneous determination of paracetamol and chlorzoxazone in tablets using addition of both analytes and absorbance increments (ΔA). *Int. J. Adv. Pharm. Anal.* **2017**, *7*, 1–5.
13. Ravisankar, S.; Vasudevan, M.; Gandhimathi, M.; Suresh, B. Reversed-phase HPLC method for the estimation of acetoaminophen, ibuprofen and chlorzoxazone in formulations. *Talanta* **1998**, *46*, 1577–1581. [CrossRef]
14. Ali, M.S.; Rafiuddin, S.; Ghori, M.; Kahtri, R.A. Simultaneous determination of paracetamol, chlorzoxazone, and related impurities 4-aminophenol, 4'-chloroacetanilide, and p-chlorophenol in pharmaceutical preparations by high-performance liquid chromatography. *J. AOAC Int.* **2007**, *90*, 82–93. [CrossRef]
15. Shaikh, K.A.; Devkhile, A.B. Simultaneous determination of aceclofenac, paracetamol, and chlorzoxazone by RP-HPLC in pharmaceutical dosage form. *J. Chromatogr. Sci.* **2008**, *46*, 649–652. [CrossRef] [PubMed]
16. Karthikeyan, V.; Vaidhyalingam, V.; Yuvaraj, G.; Nema, R.K. Simultaneous estimation of paracetamol, chlorzoxazone and aceclofenac in pharmaceutical formulation by HPLC method. *Int. J. ChemTech. Res.* **2009**, *1*, 457–460.
17. Badgujar, M.A.; Pingale, S.G.; Mangaonkar, K.V. Simultaneous determination of paracetamol, chlorzoxazone and diclofenac sodium in tablet dosage form by high performance liquid chromatography. *E-J. Chem.* **2011**, *8*, 1206–1211. [CrossRef]
18. More, S.J.; Tandulwadkar, S.S.; Nikam, A.R.; Rathore, A.S.; Sathianarayanan, L.; Mahadik, K.R. Application of HPLC for the simultaneous determination of paracetamol, chlorzoxazone, and nimesulide in pharmaceutical dosage form. *ISRN Chromatogr.* **2012**, *2012*, E252895. [CrossRef]
19. Vidya, P.; Gawande, V.; Chandewar, A.V.; Dewani, A.P. Development and validation of HPLC-PDA method for simultaneous estimation of famotidine, paracetamol, chlorzoxazone and diclofenac potassium in combined solid dosage from. *Der. Pharm. Lett.* **2017**, *10*, 15–27.
20. El-Yazbi, A.F.; Guirguis, K.M.; Bedair, M.M.; Belal, T.S. Validated specific HPLC-DAD method for simultaneous estimation of paracetamol and chlorzoxazone in the presence of five of their degradation products and toxic impurities. *Drug Dev. Ind. Pharm.* **2020**, *46*, 1853–1861. [CrossRef]
21. Prema Kumari, K.B.; Murugan, V.; Ezhilarasan, V.; Joseph, S.R. Simultaneous determination of paracetamol, aceclofenac and chlorzoxazone in pharmaceutical dosage form by UHPLC method. *J. Chromatogr. Sep. Technol.* **2017**, *8*, E1000384.
22. Bari, V.R.; Dhorda, U.J.; Sundaresan, M. A simultaneous packed column supercritical fluid chromatographic method for ibuprofen, chlorzoxazone and acetoaminophen in bulk and dosage forms. *Talanta* **1997**, *45*, 297–302. [CrossRef]
23. Khan, H.; Ali, M.; Ahuja, A.; Ali, J. Validated UPLC/Q-TOF-MS method for simultaneous determination of aceclofenac, paracetamol and chlorzoxazone in human plasma and its application to pharmacokinetic study. *Asian J. Pharm. Anal.* **2017**, *7*, 93. [CrossRef]
24. Mohamed, D.; Hegazy, M.A.; Elshahed, M.S.; Toubar, S.S.; Helmy, M.I. Liquid chromatography-tandem MS/MS method for simultaneous quantification of paracetamol, chlorzoxazone and aceclofenac in human plasma: An application to a pharmacokinetic study. *Biomed. Chromatogr.* **2018**, *32*, E4232. [CrossRef]
25. Sane, R.T.; Gadgil, M. Simultaneous determination of paracetamol, chlorzoxazone, and nimesulide by HPTLC. *J. Planar Chromatogr.* **2002**, *15*, 76–78. [CrossRef]
26. Yadav, S.S.; Jagtap, A.S.; Rao, J.R. Simultaneous determination of paracetamol, lornoxicam and chlorzoxazone in tablets by high performance thin layer chromatography. *Der. Pharm. Lett.* **2012**, *4*, 1798–1802.
27. Abdelaleem, E.A.; Abdelwahab, N.S. Stability-indicating TLC-densitometry method for simultaneous determination of paracetamol and chlorzoxazone and their toxic impurities. *J. Chromatogr. Sci.* **2013**, *51*, 187–191. [CrossRef]
28. Chhalotiya, U.K.; Patel, D.B.; Shah, D.A.; Mehta, F.A.; Bhatt, K.K. Simultaneous estimation of chlorzoxazone, paracetamol, famotidine and diclofenac potassium in their combined dosage form by thin layer chromatography. *J. Pharm. Pharmacol. Sci.* **2017**, *2*, E111. [CrossRef]
29. Salih, M.E.; Aqel, A.; Abdulkhair, B.Y.; Alothman, Z.A.; Abdulaziz, M.A.; Badjah-Hadj-Ahmed, A.Y. Simultaneous determination of paracetamol and chlorzoxazone in their combined pharmaceutical formulations by reversed-phase capillary liquid chromatography using a polymethacrylate monolithic column. *J. Chromatogr. Sci.* **2018**, *56*, 819–827. [CrossRef]
30. Alam, P.; Shakeel, F.; Ali, A.; Alqarni, M.H.; Foudah, A.I.; Aljarba, T.M.; Alkholifi, F.K.; Alshehri, S.; Ghoneim, M.M.; Ali, A. Simultaneous determination of caffeine and paracetamol in commercial formulations using normal-phase and reversed-phase HPTLC methods: A contrast of validation parameters. *Molecules* **2022**, *27*, 405. [CrossRef]
31. Abdelrahman, M.M.; Abdelwahab, N.S.; Hegazy, M.A.; Fares, M.Y.; El-Sayed, G.M. Determination of the abused intravenously administered madness drops (tropicamide) by liquid chromatography in rat plasma; an application to pharmacokinetic study and greenness profile assessment. *Microchem. J.* **2020**, *159*, E105582. [CrossRef]
32. Duan, X.; Liu, X.; Dong, Y.; Yang, J.; Zhang, J.; He, S.; Yang, F.; Wang, Z.; Dong, Y. A green HPLC method for determination of nine sulfonamides in milk and beef, and its greenness assessment with analytical eco-scale and greenness profile. *J. AOAC Int.* **2020**, *103*, 1181–1189. [CrossRef]
33. Pena-Pereira, F.; Wojnowski, W.; Tobiszewski, M. AGREE-Analytical GREEnness metric approach and software. *Anal. Chem.* **2020**, *92*, 10076–10082. [CrossRef]

34. Foudah, A.I.; Shakeel, F.; Alqarni, M.H.; Alam, P. A rapid and sensitive stability-indicating green RP-HPTLC method for the quantitation of flibanserin compared to green NP-HPTLC method: Validation studies and greenness assessment. *Microchem. J.* **2021**, *164*, E105960. [CrossRef]
35. Alam, P.; Salem-Bekhit, M.M.; Al-Joufi, F.A.; Alqarni, M.H.; Shakeel, F. Quantitative analysis of cabozantinib in pharmaceutical dosage forms using green RP-HPTLC and green NP-HPTLC methods: A comparative evaluation. *Sustain. Chem. Pharm.* **2021**, *21*, E100413. [CrossRef]
36. International Conference on Harmonization (ICH). *Q2 (R1): Validation of Analytical Procedures–Text and Methodology*; International Conference on Harmonization (ICH): Geneva, Switzerland, 2005.

Article

# Comparison of Validation Parameters for the Determination of Vitamin D3 in Commercial Pharmaceutical Products Using Traditional and Greener HPTLC Methods

Mohammed H. Alqarni [1], Faiyaz Shakeel [2], Ahmed I. Foudah [1], Tariq M. Aljarba [1], Aftab Alam [1], Sultan Alshehri [2] and Prawez Alam [1,*]

1 Department of Pharmacognosy, College of Pharmacy, Prince Sattam Bin Abdulaziz University, Al-Kharj 11942, Saudi Arabia
2 Department of Pharmaceutics, College of Pharmacy, King Saud University, Riyadh 11451, Saudi Arabia
* Correspondence: p.alam@psau.edu.sa or prawez_pharma@yahoo.com

**Abstract:** Several analytical methods are documented for the estimation of vitamin D3 (VD3) in pharmaceuticals, food supplements, nutritional supplements, and biological samples. However, greener analytical methods for VD3 analysis are scarce in the literature. As a consequence, attempts were made to design and validate a greener "high-performance thin-layer chromatography (HPTLC)" method for VD3 estimation in commercial pharmaceutical products, as compared to the traditional HPTLC method. The greenness indices of both approaches were predicted by utilizing the "Analytical GREENness (AGREE)" method. Both traditional and greener analytical methods were linear for VD3 estimation in the 50–600 ng band$^{-1}$ and 25–1200 ng band$^{-1}$ ranges, respectively. The greener HPTLC strategy outperformed the traditional HPTLC strategy for VD3 estimation in terms of sensitivity, accuracy, precision, and robustness. For VD3 estimation in commercial tablets A–D, the greener analytical strategy was better in terms of VD3 assay over the traditional analytical strategy. The AGREE index of the traditional and greener analytical strategies was estimated to be 0.47 and 0.87, respectively. The AGREE analytical outcomes suggested that the greener analytical strategy had a superior greener profile to the traditional analytical strategy. The greener HPTLC strategy was regarded as superior to the traditional HPTLC methodology based on a variety of validation factors and pharmaceutical assays.

**Keywords:** AGREE; greener HPTLC; traditional HPTLC; validation; vitamin D3

## 1. Introduction

Vitamin D3 (VD3), also known as "cholecalciferol", is a fat-soluble vitamin used in the treatment of rickets [1–3]. It metabolizes to an active metabolite "25-hydroxyvitamin D3 (calcifediol)" which plays an important role in several biochemical processes [4,5]. Most of the population of Saudi Arabia suffers from VD3 deficiency [6,7]. VD3 is present in several pharmaceutical products, food supplements, and plant products. As a result, the determination of VD3 in a variety of products, including pharmaceutical products, is necessary both qualitatively and quantitatively.

An exhaustive literature survey demonstrated several analytical approaches for VD3 analysis in commercial pharmaceutical products, food supplements, and biological fluids. For the determination of VD3 in various food, feed, pharmaceutical, and environmental samples, a spectrophotometry method was reported [8]. Several "high-performance liquid chromatography (HPLC)" methods were reported for VD3 analysis in various food products, nutritional supplements, pharmaceutical products, and edible fungus [9–17]. A number of HPLC approaches were also reported to determine VD3 and its metabolites in human plasma and serum samples [18–20]. Additionally, various "liquid-chromatography mass-spectrometry (LC-MS)" assays were reported for the determination of VD3 and its metabolites in foodstuffs, plasma, and serum samples [12,21,22]. An ultra-high-performance liquid chromatography method was used for the detection of VD3 in dietary supplements [23]. A fast supercritical fluid chromatography (SFC) method was also reported for the quantitative determination of VD3 and its related impurities [24]. A few SFC-mass spectrometry (SFC-MS) methods were proposed for the determination of VD3 and its metabolites in human milk and plasma samples [25,26]. An electrochemical strategy was also utilized for VD3 estimation in dosage forms [27]. A single "high-performance thin-layer chromatography (HPTLC)" method was reported for VD3 analysis in fish oil [28]. A single greener HPLC approach was reported for the determination of VD3 in thermodynamic solubility samples [29]. The range of analytical approaches for VD3 analysis was found in published literature. Some green analytical methods, such as SFC, SFC-MS, and HPLC methods, were utilized for the quantification of VD3 in a variety of sample matrices [24–26,29]. However, the greenness indices of the literature pharmaceutical assays were not determined. Furthermore, no VD3 detection has been carried out using the greener HPTLC approach. The literature has employed a variety of qualitative and quantitative methods to evaluate the analytical assays' greenness profiles [30–34]. Although, only the "Analytical GREENness (AGREE)" methodology utilizes all twelve green analytical chemistry (GAC) principles for the determination of the greenness profile [32]. Accordingly, the "AGREE approach" was utilized for the evaluation of the greenness profile of the present analytical assays [32].

Based on these assumptions, the objective of the current research was to create and verify a greener reverse-phase HPTLC strategy for VD3 detection in pharmaceutical products in comparison to the traditional normal-phase HPTLC strategy. The traditional solvent combinations were utilized as the mobile phase in the traditional analytical strategy. However, the greener analytical strategy used green solvent combinations as the mobile phase. Traditional and greener analytical strategies for VD3 detection have proven effective using "The International Council for Harmonization (ICH)" Q2-R1 recommendations [35].

## 2. Materials and Methods

### 2.1. Materials

VD3 sample (purity > 98%) was procured from "Sigma Aldrich (St. Louis, MO, USA)". The HPLC-grade solvents such as chloroform (CHL), diethyl ether (Et2O), ethanol (E2OH), and methanol (MeOH) were procured from "E-Merck (Darmstadt, Germany)". The HPLC-grade water was obtained from the Milli-Q unit. The commercial tablets of VD3 (A–D) (each tablet containing 5000 IU or 125 µg VD3) were procured from the local pharmacy shop in Riyadh, Saudi Arabia. All other materials and reagents used were of analytical grades.

### 2.2. Instrumentation and Analytical Conditions

The "HPTLC CAMAG TLC system (CAMAG, Muttenz, Switzerland)" was utilized for the VD3 analysis in commercial tablets A–D. The samples were prepared and spotted as 6 mm bands utilizing a "CAMAG Automatic TLC Sampler 4 (ATS4) Sample Applicator (CAMAG, Geneva, Switzerland)". The "CAMAG microliter Syringe (Hamilton, Bonaduz, Switzerland)" was attached with the sample applicator. The application rate for VD3 detection was set at 150 nL s$^{-1}$ and remained constant. The TLC plates were developed in a "CAMAG automated developing chamber 2 (ADC2) (CAMAG, Muttenz, Switzerland)" in linear ascending mode at an 80 mm distance. The preparation chamber was saturated with the appropriate mobile phase vapors for 30 min at 22 °C. VD3 was identified at a wavelength of 272 nm. Scan speed was set at 20 mm s$^{-1}$, and the slit size was adjusted to $4 \times 0.45$ mm$^2$. For each experiment, three or six replicates were used. The software used was "WinCAT's (version 1.4.3.6336, CAMAG, Muttenz, Switzerland)".

Both the traditional normal-phase HPTLC strategy and the greener reverse-phase HPTLC strategy utilized the same analytical conditions and instruments. The main distinctions between traditional and greener analytical strategies were the TLC plates and mobile phase mixtures. The TLC plates used in the traditional HPTLC strategy were "glass plates (plate size: $10 \times 20$ cm$^2$) pre-coated with normal-phase silica gel (particle size: 5 µm) 60F254S plates (E-Merck, Darmstadt, Germany)" while the TLC plates used in the greener HPTLC strategy were "RP-60F254S plates (E-Merck, Darmstadt, Germany)". The traditional mobile phase in the traditional analytical strategy was CHL-Et2O (90-10, v v$^{-1}$), whereas the greener mobile phase in the greener analytical strategy was E2OH-water (70-30, v v$^{-1}$). Due to the use of RP-TLC plates and green solvent mixtures in the greener analytical method, it is considered a reverse-phase HPTLC method.

*2.3. Calibration Curves and Quality Control (QC) Sample for VD3*

The necessary quantity of VD3 was dispensed into the specified volume of the mobile phase to create the VD3 stock solution, which had a final concentration of 100 µg mL$^{-1}$. The traditional HPTLC strategy was used to obtain VD3 concentrations in the 50–600 ng band$^{-1}$ range, whilst the greener analytical strategy—which entailed adjusting the amount of VD3 stock solution—was used to obtain concentrations in the 25–1200 ng band$^{-1}$ range. For the traditional and greener analytical strategies, 200 µL of each concentration of VD3 were spotted onto normal-phase and reverse-phase TLC plates, respectively. Both methods were used to record the VD3 concentration spot area. Plotting VD3 concentrations versus observed spot area over six replicates (n = 6) resulted in the creation of VD3 calibration curves. For the evaluation of many validation parameters, three separate QC samples were produced fresh.

*2.4. Sample Processing for the Estimation of VD3 in Marketed Tablets A–D*

The average weight of ten marketed tablets of each brand (A–D) (each containing 125 µg of VD3) was noted. The VD3-containing tablets were crushed and finely pulverized using a glass pestle and mortar. MeOH was utilized to extract the weight of powder containing 250 µg of VD3. Each brand (A–D) separately had 50 mL of MeOH redispersed into it after the MeOH had been evaporated at 40 °C [36]. The collected sample served as a test sample for both methods to figure out the quantity of VD3 in the marketed tablets.

*2.5. Validation Parameters*

Traditional and greener analytical methods for VD3 estimation were validated for different parameters following the ICH-Q2-R1 guidelines [35]. By graphing VD3 concentrations versus measured spot area, VD3 linearity was discovered. In the 50–600 ng band$^{-1}$ range (n = 6), the linearity of the traditional analytical strategy for VD3 was determined. For the greener analytical strategy, VD3 linearity was determined in the 25–1200 ng band$^{-1}$ range (n = 6).

The determination of the retardation factor ($R_f$), asymmetry factor (As), and theoretical plate number per meter (N m$^{-1}$) were utilized to assess the system suitability parameters for traditional and greener analytical methods for VD3 analysis. The "$R_f$, As, and N m$^{-1}$" values for both processes were determined using their published equations [34].

Utilizing the percent recovery method, the accuracy of traditional and greener analytical strategies for the analysis of VD3 was assessed. To assess the accuracy of the traditional analytical strategy, VD3 was measured at three QC concentrations of standard VD3 solution: low QC (LQC; 100 ng band$^{-1}$), middle QC (MQC; 300 ng band$^{-1}$), and high QC (HQC; 600 ng band$^{-1}$). To assess the accuracy of the greener analytical strategy, VD3 was also measured at three QC concentrations of standard VD3 solution: LQC (50 ng band$^{-1}$), MQC (400 ng band$^{-1}$), and HQC (1200 ng band$^{-1}$). For both analytical strategies at each QC level, the percent recovery for VD3 was computed (n = 6).

The intra/inter-assay precision of traditional and greener analytical strategies was compared for VD3. The estimation of freshly produced VD3 samples at LQC, MQC, and HQC on the same day for both analytical strategies (n = 6) was used to determine the intra-assay precision for VD3. The assessment of freshly produced VD3 samples at LQC, MQC, and HQC for three consecutive days for both strategies (n = 6) allowed for the determination of the VD3 inter-assay precision (n = 6).

By purposefully changing the mobile phase compositions, the VD3 robustness was assessed for both analytical strategies. For the traditional analytical strategy, the traditional mobile phase CHL-Et2O (90-10, v v$^{-1}$) for VD3 was changed to CHL-Et2O (92-8, v v$^{-1}$), and CHL-Et2O (88-12, v v$^{-1}$), and the variations in measured response and $R_f$ values were recorded (n = 6). Additionally, the changes in measured response and $R_f$ values were recorded (n = 6) when the greener mobile phase E2OH-water (70-30, v v$^{-1}$) for VD3 was changed to E2OH-water (72:28, v v$^{-1}$) and E2OH-water (68-32, v v$^{-1}$) for the greener analytical strategy.

Using a "standard deviation" methodology, the sensitivity of the traditional and greener analytical methods for VD3 was evaluated in terms of "limit of detection (LOD) and limit of quantification (LOQ)". The VD3 "LOD and LOQ" values were obtained using their reported formulae for both analytical procedures (n = 6) [35].

To assess the specificity of the traditional and greener analytical strategies for VD3 estimation, the $R_f$ values and UV absorption spectra of VD3 in the marketed formulations A–D were compared to a VD3 standard.

*2.6. Application of Traditional and Greener Analytical Strategies in the Estimation of VD3 in Marketed Tablets A–D*

For the traditional analytical procedure, the processed samples of commercial tablets A–D were applied to normal-phase TLC plates and reversed-phase TLC plates for the greener analytical procedure. For both analytical procedures, the chromatographic responses were recorded using the same experimental procedures utilized for the determination of standard VD3 (n = 3). For analytical procedures, the percent assay of VD3 in commercial tablets A–D was obtained using a VD3 calibration curve.

*2.7. Greenness Evaluation*

The greenness profile for the traditional and greener analytical strategies for VD3 estimation was assessed using the AGREE methodology [32]. The AGREE index (0.0–1.0) for the traditional and greener analytical strategies was determined using "AGREE: The Analytical Greenness Calculator (version 0.5, Gdansk University of Technology, Gdansk, Poland, 2020)".

*2.8. Statistical Analysis*

Several validation parameters of the traditional and greener analytical methods were determined and compared utilizing the Student's *t*-test, which was determined using MS Excel 2013 program. A value of $p < 0.05$ was taken as a significant value.

## 3. Results and Discussion

*3.1. Method Development*

In order to develop a suitable band for VD3 estimation by the traditional analytical procedure, different concentrations of CHL and Et2O, including CHL-Et2O (40-60, v v$^{-1}$), CHL-Et2O (50-50, v v$^{-1}$), CHL-Et2O (60-40, v v$^{-1}$), CHL-Et2O (70-30, v v$^{-1}$), CHL-Et2O (80-20, v v$^{-1}$), and CHL-Et2O (90-10, v v$^{-1}$) were evaluated as the traditional mobile phase mixtures. The chamber saturation conditions were applied to develop all mobile phase compositions. A typical TLC plate for the standard and commercial formulations is presented in Figure 1.

It was discovered that the traditional mobile phases, including CHL-Et2O (40-60, v v$^{-1}$), CHL-Et2O (50-50, v v$^{-1}$), CHL-Et2O (60-40, v v$^{-1}$), CHL-Et2O (70-30, v v$^{-1}$), and CHL-Et2O (80-20, v v$^{-1}$), provided unfavorable VD3 chromatographic peaks with higher As values (As >1.15). It was discovered that the traditional mobile phase CHL-Et2O (90-10, v v$^{-1}$) provided a well-resolved and intact VD3 chromatographic peak at $R_f$ = 0.34 ± 0.01 (Figure 2A) when tested. VD3 was also found to have an As values of 0.97, which is acceptable. As a consequence, the CHL-Et2O (90-10, v v$^{-1}$) was optimized as the final traditional mobile phase for the traditional analytical method of VD3 measurement.

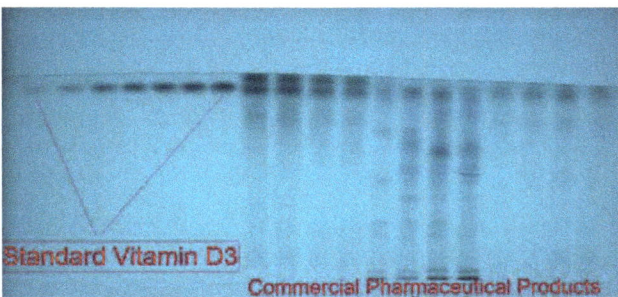

**Figure 1.** A typical thin-layer chromatography (TLC) plate of standard vitamin D3 (VD3) and commercial formulations developed using ethanol-water (70:30 v v$^{-1}$) as the greener mobile phase for the greener high-performance TLC (HPTLC) method.

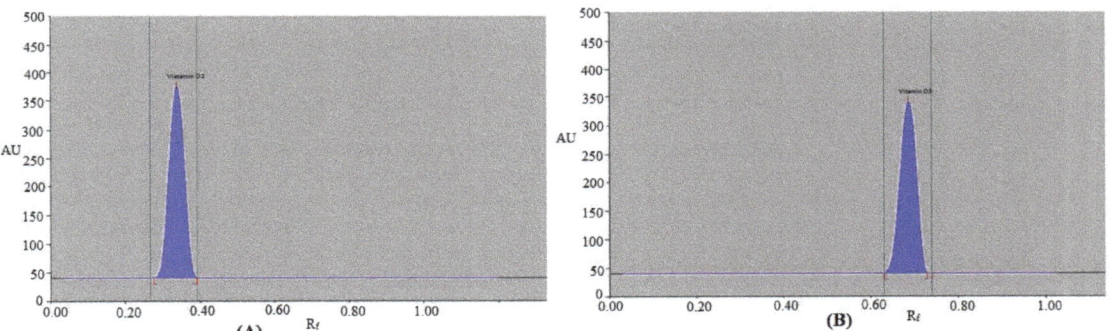

**Figure 2.** Representative chromatograms of standard VD3 recorded using (**A**) traditional normal-phase HPTLC and (**B**) greener reversed-phase HPTLC methods.

In order to develop a suitable band for VD3 estimation using the greener analytical method, different concentrations of E2OH and water, such as E2OH-water (40-60, v v$^{-1}$), E2OH-water (50-50, v v$^{-1}$), E2OH-water (60-40, v v$^{-1}$), E2OH-water (70-30, v v$^{-1}$), E2OH-water (80-20, v v$^{-1}$), and E2OH-water (90-10, v v$^{-1}$), were evaluated as the greener mobile phase mixtures. It was discovered that the greener mobile phase mixtures, including E2OH-water (40-60, v v$^{-1}$), E2OH-water (50-50, v v$^{-1}$), E2OH-water (60-40, v v$^{-1}$), E2OH-water (80-20, v v$^{-1}$), and E2OH-water (90-10, v v$^{-1}$), provided unfavorable VD3 chromatographic peaks with higher As values (As >1.20). It was discovered that the greener mobile phase E2OH-water (70-30, v v$^{-1}$) provided a well-resolved and intact VD3 chromatographic peak at $R_f = 0.69 \pm 0.02$ (Figure 2B) when tested. VD3 was also found to have an As values of 1.04, which is acceptable. As a consequence, the E2OH-water (70-30, v v$^{-1}$) was optimized as the final greener mobile phase for the greener analytical method of VD3 measurement. When the spectral bands for VD3 were recorded in densitometry mode, the greatest TLC response for VD3 was discovered at a wavelength of 272 nm. Thus, the complete VD3 study was performed at 272 nm.

*3.2. Validation Parameters*

The ICH-Q2-R1 recommendations were used to obtain a number of parameters for VD3 measurement [35]. The outcomes of the linear regression analysis of the VD3 calibration curves for both analytical methods are shown in Table 1. The VD3 calibration curve for the traditional analytical strategy was linear in the 50–600 ng band$^{-1}$ range. The VD3 calibration curve was linear in the 25–1200 ng band$^{-1}$ range for the greener analytical strategy. The determination coefficient ($R^2$) and regression coefficient (R) for VD3 were estimated to be 0.9919 and 0.9959, respectively, for the traditional analytical assay. The $R^2$ and R values for VD3 were predicted to be 0.9955 and 0.9977, respectively, for the greener analytical assay. The findings showed a significant correlation between the measured area and VD3 levels. All these outcomes demonstrated the reliability of both analytical approaches for VD3 estimation. On the other hand, the greener analytical assay was linear over a wider range than the traditional analytical assay.

**Table 1.** Results for the linearity of vitamin D3 (VD3) for the traditional normal-phase high-performance thin-layer chromatography (HPTLC) and greener reversed-phase HPTLC methods (mean ± SD; n = 6).

| Parameters | Traditional HPTLC | Greener HPTLC |
| --- | --- | --- |
| Linearity range (ng band$^{-1}$) | 50–600 | 25–1200 |
| Regression equation | y = 16.975x + 922.55 | y = 18.446x + 1213.2 |
| $R^2$ | 0.9919 | 0.9955 |
| R | 0.9959 | 0.9977 |
| Traditional error of slope | 0.40 | 0.41 |
| Traditional error of intercept | 11.81 | 4.63 |
| 95% confidence interval of slope | 15.25–18.69 | 16.65–20.23 |
| 95% confidence interval of intercept | 871.70–973.39 | 1193.25–1233.14 |
| LOD ± SD (ng band$^{-1}$) | 17.54 ± 0.24 | 8.47 ± 0.12 |
| LOQ ± SD (ng band$^{-1}$) | 52.62 ± 0.72 | 25.41 ± 0.36 |

$R^2$: determination coefficient; R: regression coefficient; LOD: limit of detection; LOQ: limit of quantification.

Table 2 presents the system suitability parameters for both the traditional and greener analytical assays. For VD3 estimation, the $R_f$, As, and N m$^{-1}$ values for the traditional analytical assay were obtained as 0.34, 0.97, and 4875, respectively, which were acceptable. For the greener analytical assay, the $R_f$, As, and N m$^{-1}$ results for VD3 estimation were 0.69, 1.04, and 4798, respectively, which were also acceptable values.

**Table 2.** System suitability parameters of traditional and greener HPTLC methods for VD3 estimation (mean ± SD; n = 3).

| Parameters | Traditional HPTLC | Greener HPTLC |
| --- | --- | --- |
| $R_f$ | 0.34 ± 0.01 | 0.69 ± 0.02 |
| As | 0.97 ± 0.01 | 1.04 ± 0.02 |
| N m$^{-1}$ | 4875 ± 4.19 | 4798 ± 4.12 |

$R_f$: retardation factor; As: asymmetry factor; N m$^{-1}$: number of theoretical plates per meter.

The accuracy of both analytical methods for VD3 estimation was measured in terms of percent recovery. Table 3 illustrates the accuracy outcomes for both analytical methods. The % recoveries of VD3 at three QC concentrations were uncovered as 94.83–103.52% using the traditional analytical assay. The VD3 % recoveries at three QC concentrations were uncovered as 98.74–100.85% for the greener analytical assay. Both assays were expected to be accurate for VD3 estimation based on these outcomes. However, the % recoveries of VD3 using the greener analytical assay were significant compared to the traditional analytical assay ($p < 0.05$). As a result, for VD3 estimation, the greener analytical assay was demonstrated to be more accurate than the traditional analytical assay.

The intra/inter-assay precision of both analytical assays was studied, and the data for VD3 estimation were expressed as the percent of the relative standard deviation (% RSD). For both analytical assays of VD3 estimation, Table 4 illustrates the outcomes of the intra/inter-day precisions. The % RSD of VD3 for intra-day precision was uncovered as 2.69–3.13% for the traditional analytical assay. The % RSD of VD3 for inter-day precision was uncovered as 2.99–3.14% for the traditional analytical assay. For the greener analytical assay, the % RSD of VD3 for intra-day precision was uncovered as 0.61–0.77%. For the greener analytical assay, the % RSD of VD3 for inter-day precision was uncovered as 0.61–0.86%. These outcomes revealed that both assays for VD3 estimation were precise. The precisions of VD3 using the greener analytical assay were significant compared to the traditional analytical assay ($p < 0.05$). Therefore, the greener analytical assay showed to be more precise than the traditional analytical assay for VD3 estimation.

Table 3. Accuracy analysis of VD3 for traditional and greener HPTLC methods (mean ± SD; n = 6).

| Conc. (ng band$^{-1}$) | Conc. Found (ng band$^{-1}$) ± SD | Recovery (%) | RSD (%) |
|---|---|---|---|
| Traditional HPTLC | | | |
| 100 | 103.24 ± 3.23 | 103.24 | 3.12 |
| 300 | 284.51 ± 8.65 | 94.83 | 3.04 |
| 600 | 621.14 ± 16.97 | 103.52 | 2.73 |
| Greener HPTLC | | | |
| 50 | 49.91 ± 0.38 | 99.82 | 0.76 |
| 400 | 394.98 ± 2.97 | 98.74 | 0.75 |
| 1200 | 1210.23 ± 7.61 | 100.85 | 0.62 |

Table 4. Evaluation of VD3 intra/inter-day precision for traditional and greener HPTLC methods (mean ± SD; n = 6).

| Conc. (ng band$^{-1}$) | Intra-Day Precision | | | Inter-Day Precision | | |
|---|---|---|---|---|---|---|
| | Conc. Found (ng band$^{-1}$) ± SD | Standard Error | RSD (%) | Conc. Found (ng band$^{-1}$) ± SD | Standard Error | RSD (%) |
| Traditional HPTLC | | | | | | |
| 100 | 94.87 ± 2.97 | 1.21 | 3.13 | 93.61 ± 2.94 | 1.20 | 3.14 |
| 300 | 316.54 ± 9.12 | 3.72 | 2.88 | 318.21 ± 9.68 | 3.95 | 3.04 |
| 600 | 581.45 ± 15.67 | 6.39 | 2.69 | 618.31 ± 18.54 | 7.57 | 2.99 |
| Greener HPTLC | | | | | | |
| 50 | 50.23 ± 0.39 | 0.15 | 0.77 | 50.64 ± 0.44 | 0.17 | 0.86 |
| 400 | 405.61 ± 3.01 | 1.22 | 0.74 | 393.65 ± 3.10 | 1.26 | 0.78 |
| 1200 | 1194.51 ± 7.35 | 3.00 | 0.61 | 1206.32 ± 7.41 | 3.02 | 0.61 |

The robustness of both analytical methodologies for VD3 estimation was determined by intentionally altering mobile phase components. Table 5 illustrates the outcomes of the robustness analysis for both analytical strategies. The VD3 % RSD for the traditional analytical strategy was uncovered as 3.63–3.71%. The VD3 $R_f$ values were predicted to be 0.33–0.36 for the traditional analytical strategy. For the greener analytical strategy, the % RSD for VD3 was uncovered as 0.67–0.71%. The VD3 $R_f$ values were uncovered as 0.68–0.70 for the greener analytical strategy. These outcomes demonstrated that both analytical strategies for VD3 estimation were robust. When compared to the traditional analytical strategy, the greener analytical strategy significantly reduced the %RSD of VD3 ($p < 0.05$). Accordingly, the greener analytical strategy fared better than the traditional analytical strategy when it came to VD3 estimation.

Table 5. Measurement of VD3 robustness for traditional and greener HPTLC methods (mean ± SD; n = 6).

| Conc. (ng band$^{-1}$) | Mobile Phase Mixture (Chloroform-Diethyl Ether) | | | Results | | |
|---|---|---|---|---|---|---|
| | Original | Used | | Conc. (ng band$^{-1}$) ± SD | RSD (%) | $R_f$ |
| Traditional HPTLC | | | | | | |
| 300 | 90:10 | 92:8 | +2.0 | 288.71 ± 10.12 | 3.50 | 0.33 |
| | | 90:10 | 0.0 | 294.61 ± 10.95 | 3.71 | 0.34 |
| | | 88:12 | −2.0 | 308.41 ± 11.21 | 3.63 | 0.36 |
| Greener HPTLC | | | | | | |
| Mobile phase mixture (ethanol-water) | | | | | | |
| 400 | 70:30 | 72:28 | +2.0 | 389.51 ± 2.64 | 0.67 | 0.68 |
| | | 70:30 | 0.0 | 394.25 ± 2.75 | 0.69 | 0.69 |
| | | 68:32 | −2.0 | 403.67 ± 2.89 | 0.71 | 0.70 |

To assess the sensitivity of both VD3 estimation assays, the "LOD and LOQ" were utilized. Table 1 illustrates the outcomes of the "LOD and LOQ" calculations for VD3 utilizing both analytical strategies. According to the traditional HPTLC strategy, VD3s "LOD and LOQ" were found to be 17.54 ± 0.24 and 52.62 ± 0.72 ng band$^{-1}$, respectively. The "LOD and LOQ" of VD3 using the greener HPTLC strategy were determined to be 8.47 ± 0.12 and 25.41 ± 0.36 ng band$^{-1}$, respectively. These outcomes demonstrated that both analytical strategies were sensitive to VD3 estimation. When compared to the traditional analytical strategy, the "LOD and LOQ" values of VD3 when employing the greener analytical strategy were significant ($p < 0.05$). Accordingly, the greener analytical strategy was demonstrated to be more sensitive than the traditional analytical strategy for VD3 estimation.

The specificity of the proposed analytical strategies of VD3 estimation was evaluated by comparing the $R_f$ values and UV absorption spectra of VD3 in commercial tablets A–D with that of standard VD3. Figure 3 shows the overlaid UV absorption spectra of standard VD3 and VD3 in commercial tablets A–D. The peak response of standard VD3 and commercial tablets A–D was measured at 272 nm. By recoding the similar UV absorption spectra, $R_f$ values, and wavelengths of VD3 in standard and commercial formulations A–D, we demonstrated the specificity of both analytical strategies for VD3 determination.

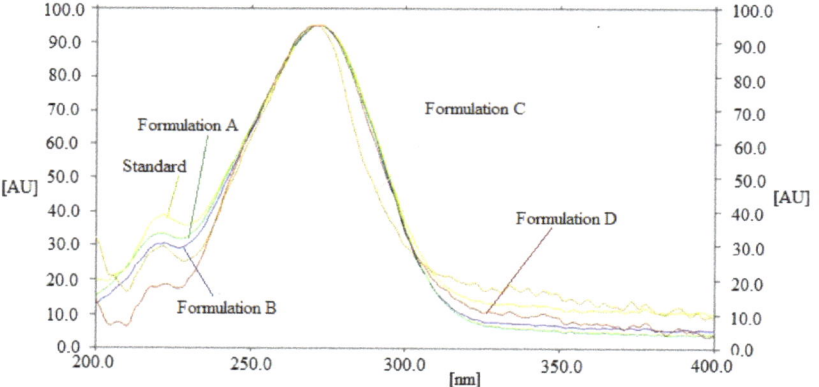

**Figure 3.** Superimposed UV absorption spectra of standard VD3 and various commercial formulations.

*3.3. Application of Traditional and Greener HPTLC Strategies in the Estimation of VD3 in Marketed Tablets A–D*

For the estimation of VD3 in commercial tablets A–D, both analytical strategies were applied as alternative approaches to routine liquid chromatography methods. The chromatograms of VD3 from commercial formulations A–D were identified by comparing the TLC spot at $R_f = 0.34 \pm 0.01$ for VD3 with the standard VD3 utilizing the traditional analytical strategy. The chromatographic peaks of VD3 in commercial tablets A–D were similar to that of standard VD3 when using the traditional analytical assay. The chromatograms of VD3 from commercial formulations A–D were also identified by comparing the TLC spot at $R_f = 0.69 \pm 0.02$ for VD3 with the standard VD3 utilizing the greener analytical strategy. The chromatographic peaks of VD3 in the commercial tablets A–D were also similar to that of standard VD3 when using the greener analytical strategy. Furthermore, no additional peaks of excipients were found in the commercial tablets using both analytical strategies, indicating no interaction between VD3 and tablet excipients. The calibration curve of VD3 was used to determine its content using traditional and greener analytical strategies, and the results are illustrated in Table 6. Using the traditional analytical assay, the % assay of VD3 in the commercial tablets A–D ranged from 87.64–95.36%. Using the greener analytical strategy, the % assay of VD3 in commercial tablets A–D ranged from 98.19–101.12%.

Table 6. Application of traditional and greener HPTLC strategies in the estimation of VD3 in commercial tablets A–D.

| Samples | Label Claim (µg) | Content Found (µg) ± SD | Assay (%) |
|---|---|---|---|
| | | Traditional HPTLC | |
| Formulation A | 125 | 119.21 ± 2.12 | 95.36 |
| Formulation B | 125 | 116.41 ± 2.06 | 93.12 |
| Formulation C | 125 | 111.51 ± 1.97 | 89.20 |
| Formulation D | 125 | 109.56 ± 1.88 | 87.64 |
| | | Greener HPTLC | |
| Formulation A | 125 | 126.41 ± 2.18 | 101.12 |
| Formulation D | 125 | 125.98 ± 2.14 | 100.78 |
| Formulation C | 125 | 124.14 ± 2.13 | 99.31 |
| Formulation D | 125 | 122.74 ± 2.15 | 98.19 |

Using the greener analytical strategy rather than the traditional analytical strategy, it was discovered that all commercial tablets had more VD3. This observation may have been made feasible by the employment of different solvent mixtures in the traditional and greener analytical methods. Overall, the greener analytical strategy was considered superior to the traditional analytical strategy for VD3 pharmaceutical assay.

*3.4. Greenness Assessment*

For the evaluation of analytical techniques' greenness, various qualitative and quantitative methodologies are presented [30–34]. However, only AGREE makes use of all twelve GAC components for evaluating greenness [32]. Accordingly, the greener profiles of both analytical strategies were evaluated using the AGREE method. Figure 4 illustrates a representative pictogram of the AGREE index of the traditional and greener analytical strategies. For traditional and greener analytical strategies, the AGREE index was predicted to be 0.47 and 0.87, respectively. These results showed that the greener analytical strategy for VD3 analysis had a superior greenness profile to the traditional analytical strategy.

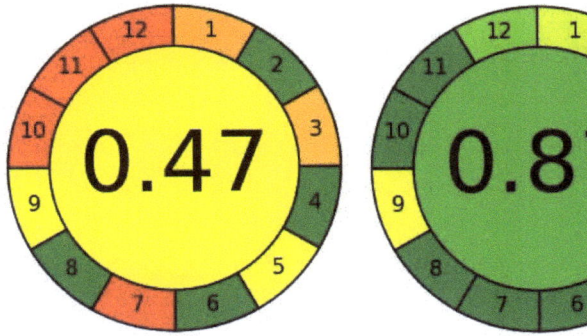

Figure 4. The pictograms of the AGREE indices for the traditional and the greener HPTLC strategies recorded utilizing the "AGREE: The Analytical Greenness Calculator".

## 4. Conclusions

There is a scarcity of greener analytical assays for VD3 estimation in the literature. In contrast to the traditional HPTLC methodology, this research aimed to design and validate a sensitive and greener HPTLC assay for VD3 estimation in marketed tablets. The amount of VD3 in all commercial tablets was discovered to be much higher in terms of % assay when comparing the greener analytical method to the traditional analytical strategy. According to the AGREE outcomes, the greener analytical strategy had a higher level of greenness than the traditional analytical strategy. Based on a number of validation criteria and the results of pharmaceutical assays, the greener HPTLC approach was declared to be superior to the traditional HPTLC approach for VD3 estimation in commercial tablets. In conclusion, these outcomes suggest that the greener HPTLC assay can be used for the estimation of VD3 in commercially available products. Overall, the greener HPTLC strategy is more accurate, precise, robust, and sensitive than the traditional HPTLC strategy for the determination of VD3 in commercial formulations.

**Author Contributions:** Conceptualization, M.H.A. and P.A.; methodology, A.A., P.A. and T.M.A.; software, A.I.F. and F.S.; validation, S.A. and M.H.A.; formal analysis, A.I.F. and T.M.A.; investigation, P.A. and A.A.; resources, S.A.; data curation, A.I.F. and F.S.; writing—original draft preparation, F.S.; writing—review and editing, M.H.A., P.A. and S.A.; visualization, S.A.; supervision, P.A.; project administration, P.A.; funding acquisition, S.A. All authors have read and agreed to the published version of the manuscript.

**Funding:** This research was funded by the Researchers Supporting Project (number RSP-2021/146) at King Saud University, Riyadh, Saudi Arabia and The APC was funded by RSP.

**Institutional Review Board Statement:** Not applicable.

**Informed Consent Statement:** Not applicable.

**Data Availability Statement:** Not applicable.

**Acknowledgments:** The authors are thankful to the Researchers Supporting Project (number RSP-2021/146) at King Saud University, Riyadh, Saudi Arabia for supporting this research.

**Conflicts of Interest:** The authors declare no conflict of interest.

## References

1. Christakos, S.; Dhawan, P.; Benn, B.; Porta, A.; Hediger, M.; Oh, G.T.; Jeung, E.B.; Zhong, Y.; Ajibade, D.; Dhawan, K.; et al. Vitamin D-molecular mechanism of action. *Ann. N. Y. Acad. Sci.* **2007**, *1116*, 340–348. [CrossRef]
2. Pike, J.W. Vitamin-D3 receptors-structure and function in transcription. *Annu. Rev. Nutr.* **1991**, *11*, 189–216. [CrossRef]
3. Holick, M.F. Resurrection of vitamin D deficiency and rickets. *J. Clin. Investig.* **2006**, *116*, 2062–2072. [CrossRef] [PubMed]
4. Deluca, H.D. Metabolism and molecular mechanism of action of vitamin-D. *Biochem. Soc. Trans.* **1982**, *10*, 147–158. [CrossRef]
5. Jurutka, P.W.; Bartik, L.; Whitfield, G.K.; Mathern, D.R.; Barthel, T.K.; Gurevich, M.; Hsieh, J.C.; Kaczmarska, M.; Haussler, C.A.; Haussler, M.R. Vitamin D receptor: Key roles in bone mineral pathophysiology, molecular mechanism of action, and novel nutritional ligands. *J. Bone Miner. Res.* **2007**, *22*, V2–V10. [CrossRef] [PubMed]
6. Alsuwdia, A.O.; Frag, Y.M.; Sayyari, A.A.; Mousa, D.H.; Alhijaili, F.F.; Al-Harbi, A.S.; Housawi, A.A.; Mittal, B.V.; Singh, A.K. Prevalence of vitamin D deficiency in Saudi adults. *Saudi Med. J.* **2013**, *34*, 814–818.
7. AlBuhairan, F.S.; Tamim, H.; Al-Dubayee, M.; AlDhukair, S.; Al-Shehri, S.; Tamimi, W.; El-Bcheraoui, C.; Magzoub, M.E.; de Vries, N.; Al-Alwan, I. Time for an adolescent health surveillance system in Saudi Arabia: Findings from "Jeeluna". *J. Adolesc. Health* **2015**, *57*, 263–269. [CrossRef] [PubMed]
8. Rahman, A.; Rahman, M.M.; Hossain, M.S.; Jahan, M.S.; Akter, N.J.; Bari, M.L. A simple and alternative UV spectrometric method for the estimation of vitamin D3. *Microb. Bioact.* **2019**, *2*, 98–105.
9. Johnsson, H.; Halen, B.; Hessel, H.; Nyman, A.; Thorzell, K. Determination of vitamin D3 in margarines, oils and other supplemented food products using HPLC. *Int. J. Vitam. Nutr. Res.* **1989**, *59*, 262–268.
10. Sarioglu, K.; Celebi, S.S.; Mutlu, M. A rapid method for determination of vitamins D2 and D3 in pharmaceutical preparations by HPLC. *J. Liq. Chromatogr. Relat. Technol.* **2001**, *24*, 973–982. [CrossRef]
11. Jakobsen, J.; Clausen, I.; Leth, T.; Ovesen, L. A new method for the determination of vitamin D3 and 25-hydroxyvitamin D3 in meat. *J. Food Compos. Anal.* **2004**, *17*, 777–787. [CrossRef]
12. Bilodeau, L.; Dufresne, G.; Deeks, J.; Clement, G.; Bertrand, J.; Turcotte, S.; Rbichaud, A.; Beraldin, F.; Fouquet, A. Determination of vitamin D3 and 25-hydroxyvitamin D3 in foodstuffs by HPLC UV-DAD and LC–MS/MS. *J. Food Compos. Anal.* **2011**, *24*, 441–448. [CrossRef]
13. Kumar, S.; Chawla, D.; Tripathi, K. An improved and sensitive method for vitamin D3 estimation by RP-HPLC. *Pharm. Anal. Acta* **2015**, *6*, E1000410.
14. Temova, Z.; Roskar, R. Stability-indicating HPLC–UV method for vitamin D3 determination in solutions, nutritional supplements and pharmaceuticals. *J. Chromatogr. Sci.* **2016**, *54*, 1180–1186. [CrossRef]
15. Farag, A.M.; Rizk, M.S.; El-Bassel, H.A.; Youssif, M.H. Determination of vitamin D3 content in high, low and zero fat food using high performance liquid chromatography. *Med. J. Cairo Univ.* **2018**, *86*, 3911–3918.

16. Huang, B.-F.; Pan, X.-D.; Zhang, J.-S.; Xu, J.-J.; Cai, Z.-X. Determination of vitamins D2 and D3 in edible fungus by reversed-phase two-dimensional liquid chromatography. *J. Food Qual.* **2020**, *2020*, E8869279. [CrossRef]
17. Rashidi, L.; Nodeh, H.R.; Shahabuddin, S. Determination of vitamin D in the fortified sunflower oil: Comparison of two developed methods. *Food Anal. Methods* **2022**, *15*, 330–337. [CrossRef]
18. Brunetto, M.R.; Obando, M.A.; Gallignani, M.; Alarcon, O.M.; Nieto, E.; Salinas, R.; Burguera, J.L.; Burguera, M. HPLC determination of vitamin D3 and its metabolite in human plasma with on-line sample cleanup. *Talanta* **2004**, *64*. 1364–1370. [CrossRef]
19. Keyfi, F.; Nahid, S.; Mokhtariye, A.; Nayerabadi, S.; Alaei, A.; Varasteh, A.-R. Evaluation of 25-OH vitamin D by high performance liquid chromatography: Validation and comparison with electrochemiluminescence. *J. Anal. Sci. Technol.* **2018**, *9*, E25. [CrossRef]
20. Babat, N.; Turkmen, Y. Determination of serum vitamin D3 level by high performance liquid chromatography (HPLC) in patients with coronary artery ectasia. *Cardiol. Cardiovasc. Med.* **2020**, *4*, 97–104. [CrossRef]
21. Mirza, T.; Qadeer, K.; Ahmad, I. Clinical analysis of vitamin D and metabolites. *J. Baqai Med. Univ.* **2009**, *12*, 25–28.
22. Shah, I.; James, R.; Barker, J.; Petroczi, A.; Naughton, D.P. Misleading measures in vitamin D analysis: A novel LC-MS/MS assay to account for epimers and isobars. *Nutr. J.* **2011**, *10*, E46. [CrossRef]
23. Becze, A.; Fuss, V.L.B.; Scurtu, D.A.; Tomoaia-Cotisel, M.; Mocanu, A.; Cadar, O. Simultaneous determination of vitamins D3 (calcitriol, cholecalciferol) and K2 (menaquinone-4 and menaquinone-7) in dietary supplements by UHPLC. *Molecules* **2021**, *26*, 6982. [CrossRef]
24. Andri, B.; Lebrun, P.; Dispas, A.; Klinkenberg, R.; Streel, B.; Ziemons, E.; Marini, R.D.; Hubert, P. Optimization and validation of a fast supercritical fluid chromatography method for the quantitative determination of vitamin D3 and its related impurities. *J. Chromatogr. A* **2017**, *1491*, 171–181. [CrossRef]
25. Oberson, J.M.; Benet, S.; Redeuil, K.; Campos-Gimenez, E. Quantitative analysis of vitamin D and its main metabolites in human milk by supercritical fluid chromatography coupled to tandem mass spectrometry. *Anal. Bioanal. Chem.* **2020**, *412*, 365–375. [CrossRef]
26. Socas-Rodriguez, B.; Pilarova, V.; Sandahl, M.; Holm, C.; Turner, C. Simultaneous determination of vitamin D and its hydroxylated and esterified metabolites by ultrahigh-performance supercritical fluid chromatography–tandem mass spectrometry. *Anal. Chem.* **2022**, *94*, 3065–3073. [CrossRef]
27. Durovic, A.; Stojanovic, Z.; Kravic, S.; Kos, J.; Richtera, L. Electrochemical determination of vitamin D3 in pharmaceutical products by using boron doped diamond electrode. *Electroanalysis* **2020**, *32*, 741–748. [CrossRef]
28. Demchenko, D.V.; Pozharitskaya, O.N.; Shikov, A.N.; Makarov, V.G. Validated HPTLC method for quantification of vitamin D in fish oil. *J. Planar Chromatogr.* **2011**, *24*, 487–490. [CrossRef]
29. Almarri, F.; Haq, N.; Alanazi, F.K.; Mohsin, K.; Alsarra, I.A.; Aleanizy, F.S.; Shakeel, F. An environmentally benign HPLC-UV method for thermodynamic solubility measurement of vitamin D3 in various (Transcutol + water) mixtures. *J. Mol. Liq.* **2017**, *242*, 798–806. [CrossRef]
30. Abdelrahman, M.M.; Abdelwahab, N.S.; Hegazy, M.A.; Fares, M.Y.; El-Sayed, G.M. Determination of the abused intravenously administered madness drops (tropicamide) by liquid chromatography in rat plasma; an application to pharmacokinetic study and greenness profile assessment. *Microchem. J.* **2020**, *159*, E105582. [CrossRef]
31. Duan, X.; Liu, X.; Dong, Y.; Yang, J.; Zhang, J.; He, S.; Yang, F.; Wang, Z.; Dong, Y. A green HPLC method for determination of nine sulfonamides in milk and beef, and its greenness assessment with analytical eco-scale and greenness profile. *J. AOAC Int.* **2020**, *103*, 1181–1189. [CrossRef]
32. Pena-Pereira, F.; Wojnowski, W.; Tobiszewski, M. AGREE-Analytical GREEnness metric approach and software. *Anal. Chem.* **2020**, *92*, 10076–10082. [CrossRef]
33. Alam, P.; Salem-Bekhit, M.M.; Al-Joufi, F.A.; Alqarni, M.H.; Shakeel, F. Quantitative analysis of cabozantinib in pharmaceutical dosage forms using green RP-HPTLC and green NP-HPTLC methods: A comparative evaluation. *Sustain. Chem. Pharm.* **2021**, *21*, E100413. [CrossRef]
34. Foudah, A.I.; Shakeel, F.; Alqarni, M.H.; Alam, P. A rapid and sensitive stability-indicating green RP-HPTLC method for the quantitation of flibanserin compared to green NP-HPTLC method: Validation studies and greenness assessment. *Microchem. J.* **2021**, *164*, E105960. [CrossRef]
35. International Conference on Harmonization (ICH). *Q2 (R1): Validation of Analytical Procedures–Text and Methodology*; ICH: Geneva, Switzerland, 2005.
36. Alam, P.; Shakeel, F.; Ali, A.; Alqarni, M.H.; Foudah, A.I.; Aljarba, T.M.; Alkholifi, F.K.; Alshehri, S.; Ghoneim, M.M.; Ali, A. Simultaneous determination of caffeine and paracetamol in commercial formulations using greener normal-phase and reversed-phase HPTLC methods: A contrast of validation parameters. *Molecules* **2022**, *27*, 405. [CrossRef]

Article

# Evaluation of Pulsed Electric Field-Assisted Extraction on the Microstructure and Recovery of Nutrients and Bioactive Compounds from Mushroom (*Agaricus bisporus*)

Mara Calleja-Gómez [1], Juan Manuel Castagnini [1,*], Ester Carbó [2], Emilia Ferrer [1], Houda Berrada [1] and Francisco J. Barba [1,*]

[1] Department of Preventive Medicine and Public Health, Food Science, Toxicology and Forensic Medicine, Faculty of Pharmacy, Universitat de València, Avda. Vicent Andrés Estellés s/n, Burjassot, 46100 València, Spain

[2] Department of Environmental Quality and Soils, Centro de Investigaciones Sobre Desertificación-CIDE (CSIC-Universitat de Valencia-GV), 46113 Valencia, Spain

* Correspondence: juan.castagnini@uv.es (J.M.C.); francisco.barba@uv.es (F.J.B.); Tel.: +34-963-544-972 (F.J.B.)

**Abstract:** Pulsed electric field (PEF) is a sustainable innovative technology that allows for the recovery of nutrients and bioactive compounds from vegetable matrices. *A. bisporus* was chosen for its nutritional value and the effect of PEF pretreatment was evaluated using different conditions of electric field (2–3 kV/cm), specific energy (50–200 kJ/kg) and extraction time (0–6 h) to obtain the best conditions for nutrient and bioactive compound extraction. Spectrophotometric methods were used to evaluate the different compounds, along with an analysis of mineral content by inductively coupled plasma mass spectrometry (ICP-MS) and the surface was evaluated using scanning electron microscopy (SEM). In addition, the results were compared with those obtained by conventional extraction (under constant shaking without PEF pretreatment). After evaluating the extractions, the best extraction conditions were 2.5 kV/cm, 50 kJ/kg and 6 h which showed that PEF extraction increased the recovery of total phenolic compounds in 96.86%, carbohydrates in 105.28%, proteins in 11.29%, and minerals such as P, Mg, Fe and Se. These results indicate that PEF pretreatment is a promising sustainable technology to improve the extraction of compounds and minerals from mushrooms showing microporation on the surface, positioning them as a source of compounds of great nutritional interest.

**Keywords:** pulsed electric field; bioactive compounds; optimization; mushrooms; *Agaricus bisporus*

## 1. Introduction

World mushroom production has increased by more than 30 times since 1978 (from 4.2 million kg in 1978 to 34 billion kg in 2013) currently reaching a 4.3% production increase every year [1,2], with five genera (*Lentinula, Pleurotus, Auricularia, Agaricus* and *Flammulina*) comprising 85% of the world supply. Among these five genera, *Lentinula edodes*, *Pleurotus* spp. and *Agaricus bisporus* mushrooms represent 22%, 19% and 15% of the supply, respectively. At European and Spanish levels, the production of *A. bisporus* stands out, representing 90% of the total cultivable mushrooms [3,4].

Furthermore, a worldwide mushroom consumption increase of almost five times has been observed over the last decades (from 1 kg/person in 1997 to 4.7 kg/person in 2013) [1], observing the same trend at the European and Spanish levels, with an increase in domestic consumption of approximately 15% in recent decades with *A. bisporus* being the most consumed mushroom [5]. Considering the growing interest in *A. bisporus*, the number of studies evaluating its nutritional and bioactive properties has increased considerably. Besides being used as food, mushrooms are of great interest as a source of compounds that can be used in the formulation of supplements and/or food additives [6].

Mushrooms are rich in macronutrients (proteins and polysaccharides), micronutrients (vitamins and minerals) and bioactive compounds (e.g., polyphenols), and have a low lipid content, which makes them a food of high dietary value [7]. In fact, *A. bisporus* and *L. edodes* mushrooms are colloquially known as "vegetable meat" due to their high protein content and nutritional value [8]. This is particularly interesting for specific population groups such as people following a vegetarian/vegan diet as possible alternatives to meat products. Regarding polysaccharides, *A. bisporus* has a high content of β-glucans, which have been associated with immune-regulating, hypoglycemic and anticoagulant properties [9]. Moreover, mushrooms have a high content of natural antioxidants with the ability to reduce the damage caused by oxidative stress, which is one of the main causes of cellular aging [10]. It is also remarkable for its high micronutrient content, being a good source of potassium, phosphorus, magnesium and selenium, as well as vitamin A, B (thiamine, niacin and folic acid), C and D vitamins [11]. This high micronutrient content combined with bioactive compounds shows biological properties (antioxidant, antimicrobial and antitumor) [12,13].

Considering the nutritional value of mushrooms, there is a growing interest in the extraction of these compounds. Traditionally, conventional methods have been used for this purpose, which is not highly ecological and efficient since they require high temperatures, long extraction times and organic solvents, being in many cases toxic. Therefore, in recent years, pulsed electric field-assisted extraction (PEF) technology has been used for the recovery of compounds from plant matrices [14–16]. The use of PEF allows for the sustainable and economical obtaining of compounds by using water as a solvent, thus reducing the use of organic solvents that are much more polluting. In addition, it is a technology that reduces the temperature and time required for the extraction of the different compounds, thus reducing the degradation of thermolabile components, making it of special interest at the industrial level when seeking greater process efficiency [9].

This technology is based on the application of electrical pulses between two electrodes inside the treatment chamber, allowing the formation of micropores in eukaryotic cell membranes and increasing cell permeability, which allows for the selective extraction of intracellular compounds. The efficiency of the PEF extraction process to permeabilize the membrane changes depending on the strength of the electric field applied, the specific energy, treatment time, temperature and the properties of the material used, such as pH, conductivity and the characteristics of the matrix cells to be extracted [17].

The electroporation produced by this extraction method can be observed on the surface of the sample using techniques such as scanning electron microscopy (SEM) [18]. In this way, it is possible to compare the conventional extraction or untreated sample versus alternative methods in relation to the impact on the surface of the food.

Therefore, the present study aims to evaluate the recovery of high added-value compounds from *Agaricus bisporus* using an optimization strategy based on response surface methodology (RSM) and to study the influence of PEF on the mushroom surface using scanning electron microscopy (SEM).

## 2. Materials and Methods

### 2.1. Chemicals and Reagents

Sodium carbonate ($Na_2CO_3$) was purchased from VWR (Saint-Prix, France). AAPH (2,2′-Azobis (2-methylpropanimidamide) dihydrochloride), ABTS (2,2′-Azinobis (3-ethylbenzothiazoline-6-sulfonic acid)), Trolox (6-hydroxy-2,5,7,8-tetramethylchroman-2-carboxylic acid), potassium persulfate ($K_2S_2O_8$), Folin–Ciocalteu reagent, gallic acid ($C_7H_6O_5$), fluorescein ($C_{20}H_{12}O_5$), mineral standards (Ca, P, Mg, Fe, Zn and Se) and internal standards of Sc and Ge were purchased from Sigma–Aldrich (Steinheim, Baden-Württemberg, Germany). Disodium phosphate ($Na_2HPO_4$) and potassium dihydrogen phosphate ($KH_2PO_4$) were purchased from VWR International Eurolab S.L. (Barcelona, Spain).

Sulfuric acid ($H_2SO_4$) and phenol ($C_6H_6O$) were purchased from Thermo Fisher Scientific (Waltham, MA, USA). Ethanol (99%) was purchased from Baker (Deventer, The

Netherlands). Distilled water was obtained from Milli-Q SP Reagent Water System (Millipore Corporation, Bedford, MA, USA).

## 2.2. Sample Preparation

White button mushroom (*A. bisporus*) samples were obtained from a local supermarket (Valencia, Spain) and used the day after purchase. All mushroom samples were stored in plastic containers in a refrigerator at 4 °C for 24 h. Subsequently, they were cut into 3 × 5 × 3 mm slices manually with a kitchen knife to obtain 20 g of fresh sample for each of the replicates in the PEF and conventional extraction and for the lyophilization necessary to compare the samples by SEM [8]. The initial *A. bisporus* moisture content (g water/100 g sample) was 0.88 ± 0.01.

## 2.3. Extraction Conditions

A PEF-Cellcrack III (German Institute of Food Technologies (DIL)) equipment (ELEA, Quakenbrück, Germany) located at the Faculty of Pharmacy of the University of Valencia (València, Spain) was used for the extraction. Specifically, for each extraction, 20 g of fresh sample previously cut into slices were taken and placed in contact with 200 mL of water (100 mL distilled water and 100 mL tap water) until the conductivity was approximately 700–800 µS/cm, using an extraction chamber with a capacity of 900 mL and a distance between electrodes of 10 cm. In addition, pulse duration was 100 ms using a 2.00 Hz frequency. The samples were pre-treated according to the optimal conditions obtained after performing RSM with 17 samples (Table 1); electric field strength values varied between 2–3 kV/cm, specific energy values between 50–200 kJ/kg and total extraction time ranged between 0–6 h.

**Table 1.** Pulsed electric field (PEF)-assisted extraction experimental conditions.

| Sample | Weight (g) | Field strength (kV/cm) | Specific Energy (kJ/kg) | Time (h) |
|---|---|---|---|---|
| 1 | 220 | 3.00 | 50 | 6 |
| 2 | 220 | 2.00 | 125 | 3 |
| 3 | 220 | 3.00 | 200 | 0 |
| 4 | 220 | 3.00 | 50 | 0 |
| 5 | 220 | 2.00 | 50 | 6 |
| 6 | 220 | 2.50 | 125 | 3 |
| 7 | 220 | 2.00 | 200 | 6 |
| 8 | 220 | 2.50 | 125 | 0 |
| 9 | 220 | 2.00 | 200 | 0 |
| 10 | 220 | 3.00 | 125 | 3 |
| 11 | 220 | 2.50 | 200 | 3 |
| 12 | 220 | 2.50 | 125 | 6 |
| 13 | 220 | 2.50 | 51 | 3 |
| 14 | 220 | 2.50 | 125 | 3 |
| 15 | 220 | 2.00 | 50 | 0 |
| 16 | 220 | 3.00 | 200 | 6 |
| 17 | 220 | 2.50 | 125 | 3 |

The temperature and conductivity of each sample was measured before and after PEF treatment with the ProfiLine Cond 3310 conductometer (WTW, Xylem Analytics, Weilheim in Oberbayern, Germany). According to previous studies, the minimum electric field to produce cellular changes is 1 kV/cm, because of that the application of 2–3 kV/cm is enough to produce electroporation [19,20].

Extracts with extraction time 0 h were filtered and centrifuged (4000 rpm, 15 min) to remove solid residues and stored frozen at −20 °C until their use in chemical analysis; those extracts with time higher than 0 h were kept after PEF pre-treatment in agitation using a magnetic stirrer for a certain period depending on the number of samples. They were then filtered and centrifuged under the same conditions as the samples at time 0 h.

Conventional extracts were subsequently obtained and stored under the same conditions without the PEF pre-treatment.

Response surface methodology (RSM) was used as a method to optimize the extraction conditions. This methodology includes a variety of techniques used to study the relationship between factors or independent variables with one or more responses or dependent variables, to optimize them [21]. In this case, the study was carried out to determine how the variation of factors related to the PEF technology (electric field strength, specific energy, and extraction time) affects the concentration of different mushroom compounds in the extracts obtained by PEF (proteins, carbohydrates, and antioxidant compounds). A central composite design with 17 experiments and 3 central points (3 samples with the same PEF extraction conditions) was applied. The inclusion of central points allows for estimating the experimental error and avoiding the generation of a model that leads to incorrect conclusions.

After applying SRM, 20 g of fresh *A. bisporus* sample was treated under optimal PEF extraction conditions to validate the result. The results were compared with a control obtained under the same conditions except for PEF pre-treatment.

### 2.4. Chemical Analyses

2.4.1. Total Protein Content

The bicinchoninic acid (BCA) assay was used to determine the protein content of the extracts. The working solution was prepared according to the Pierce BCA kit Protein Assay (Thermo Fisher Scientific, Waltham, MA, USA). Bovine serum albumin (0–2000 mg/L) was used as standard. To prepare the analysis, 10 µL of sample/standard was added to the microplate combined with 200 µL of the working solution of BCA, subsequently mixed and incubated at 37 °C for 30 min. Finally, the absorbance of the samples was measured at 562 nm. The results are expressed in mg of bovine serum albumin/g dry matter (mg BSA/g DM).

2.4.2. Total Antioxidant Capacity

ORAC determination was performed according to Cao et al. [22]. This assay measures the oxidative degradation of a fluorescent molecule, fluorescein, after the addition of AAPH, measuring the antioxidant capacity of the sample compared to the Trolox standard. Phosphate buffer pH 7.0–7.4 was used for the blank and 1 mM Trolox as the standard. Fluorescein (50 µL) was added to the microplate along with 50 µL of Trolox/blank/sample and incubated for 10 min at 37 °C. Subsequently, 25 µL of AAPH was added and wavelengths 480 nm excitation and of 520 nm emission were set using Wallac 1420 VICTOR 2 plate reader (Perkin–Elmer, Jügesheim, Germany) and the measurements were collected every minute for 60 min. Finally, the blank was subtracted from the results obtained; one ORAC unit indicates that the antioxidant capacity of the sample is equivalent to 1 µM Trolox. The results were expressed in µmol Trolox equivalents/g dry matter (µmol TE/g DM).

The TEAC assay is used to observe the capacity of the extracts to neutralize the ABTS+ radical. To perform the assay, 25 mL of ABTS (7 mM) and 440 µL $K_2S_2O_8$ (140 mM) were mixed, the solution was incubated in the dark at 20 °C for 16 h to obtain the working solution with the ABTS+ radical. Subsequently, this working solution was mixed with 96% ethanol to reach an absorbance at 734 nm of 0.70 ± 0.02. For the measurement, 2 mL of ethanol solution was taken and 100 µL of the sample was added, the initial and final absorbance was recorded at 734 nm. A Trolox standard curve was used as a reference at different concentrations (0–250 µM). The results were expressed in µmol Trolox equivalents/g dry matter (µmol TE/g DM).

2.4.3. Total Phenolic Compounds (TPC)

The Folin–Ciocalteu method was used for the total phenolic compounds (TPC) determination, according to the method proposed by Singleton et al. [23], which is based on the capacity of phenols to react against oxidizing compounds. Thus, the Folin–Ciocalteu reagent, which contains molybdate and sodium tungstate, can react with the phenolic compounds found in the sample forming phosphomolybdic and phosphotungstic complexes. As these compounds are found in a basic medium, they are reduced forming a blue-colored compound that is proportional to the phenolic concentration. Folin–Ciocalteu reagent at 50% $v/v$ was used together with $Na_2CO_3$ solution and gallic acid standards. In each tube, 100 µL of standard/sample, 3 mL of $Na_2CO_3$ and finally 100 µL of Folin–Ciocalteu were added. The samples were incubated for 60 min in the dark and measured at 750 nm using a Perkin–Elmer UV/V is Lambda 2 spectrophotometer (Perkin-Elmer, Jügesheim, Germany). Results were expressed as mg gallic acid equivalents/g dry matter (mg GAE/g DM).

### 2.4.4. Total Carbohydrate Content

Total carbohydrate content was determined by the phenol-sulfuric method described by Dubois et al. [24], which allows knowing the concentration of total sugars by acid catalysis of these by adding sulfuric acid, obtaining furfural and hydroxymethyl-furfural that condensed with phenols give rise to yellow-orange products proportional to the total concentration of carbohydrates. For their determination, D-glucose solutions (10–100 mg/L) was used as standard. One milliliter of sample was taken with 1/10 dilution or the D-glucose standard solutions along with 0.5 mL of 5% phenol solution and 2.5 mL of sulfuric acid. After mixing the reagents, they were incubated for 30 min at 25 °C. Finally, absorbance was measured at a wavelength of 490 nm. The results were expressed in mg glucose/g dry matter (mg glu/g DM).

### 2.4.5. Mineral Content

To determine the Ca, Mg, Fe, Zn, P and Se content of liquid extracts from PEF and conventional extraction, 1 mL of each extract was taken and digested with 1 mL of 69% nitric acid ($HNO_3$) along with 250 µL of hydrogen peroxide ($H_2O_2$) in a 180 °C microwave oven. Subsequently, it was brought to a volume of 5 mL with ultrapure water (UW), 100 µL were taken and brought again to a final volume of 10 mL with UW.

On the other hand, to determine the content of the above-mentioned minerals in fresh samples of *A. bisporus*, 10 mg were weighed and digested with 1 mL of 69% $HNO_3$ followed by 250 µL of $H_2O_2$ in a microwave oven at 180 °C. Afterward, it was brought to 10 mL with UW to take an aliquot of 100 µL and add 9 mL of UW. The multi-elemental determination was performed by inductively coupled plasma mass spectrometry (ICP-MS) using a 20 µg/g Sc and Ge solution as an internal standard. The results are expressed in mg/100 g for Ca, Mg, Fe, Zn and P, and µg/100 g for Se.

### 2.5. Scanning Electron Microscopy (SEM)

To observe the surface of *A. bisporus*, small fragments belonging to the pileus were used both from the sample subjected only to freeze-drying (control sample) and from the samples subjected to freeze-drying and PEF/conventional treatment. For sample preparation, a carbon film was taken on which the sample is placed, and the fragments were treated for 2 min to produce metallization of the sample with a thin layer of Au and Pd. Subsequently, the sample was placed under the microscope and the difference between the control and the treated sample was observed on the surface, searching for the electroporation produced by the PEF pre-treatment in *A. bisporus*.

### 2.6. Statistical Analysis

The data were analyzed using an analysis of variance (ANOVA), where the parameters of the PEF pre-treatment (electric field and specific energy) with extraction time were the factors and the values of TEAC, ORAC, TPC, carbohydrates and total proteins were the variables. A $p$-value < 0.05 was considered a significant difference. All statistical analyses concerning MSR were performed with Statgraphics Centurion XVII software (Statpoint Technologies, Inc., Warrenton, VA, USA), while ANOVA analysis of data obtained from analysis of extracts under optimal PEF extraction conditions was performed with GraphPad

Prism 8 software (GraphPad Software, San Diego, CA, USA). Each analysis was performed in triplicate assuming a significance level of 5%. Standard deviations are represented in the figures using error bars.

## 3. Results

*3.1. Effect of Extraction Time, Electric Field Strength and Specific Energy on the Selective Extraction of Nutrients and Bioactive Compounds*

PEF-assisted water extraction of *A. bisporus* samples was optimized according to the response surface methodology (RSM) with three central points to maximize the values obtained for the following factors: TPC (mg GAE/g DM), TEAC (µmol TE/g DM), ORAC (µmol TE/g DM), total carbohydrate content (mg glu/g DM) and total protein content (mg BSA/g DM).

### 3.1.1. Macronutrients

The range of values for protein and total carbohydrate was from $7.79 \pm 0.83$ to $140.72 \pm 15.14$ mg BSA/g DM and $0.86 \pm 0.07$ to $39.51 \pm 0.59$ mg glu/g DM, respectively.

Figure 1 shows the influence of electric field strength, specific energy and extraction time on carbohydrate recovery. It can be observed that the extraction time had a significant effect ($p < 0.05$) on the recovery of carbohydrate with a linear rise as time increased. Moreover, an increase in carbohydrate content was observed as the electric field and specific energy increased ($p < 0.05$), reaching an optimum point at 2.5 kV/cm and 125 kJ/kg. Once these values were reached, a decrease in the carbohydrate values was obtained as both factors continued to increase. According to the obtained results, the maximum carbohydrate extraction ($39.51 \pm 0.59$ mg glu/g DM) in *A. bisporus* was observed after applying 2.5 kV/cm, 125 kJ/kg and 6 h of extraction.

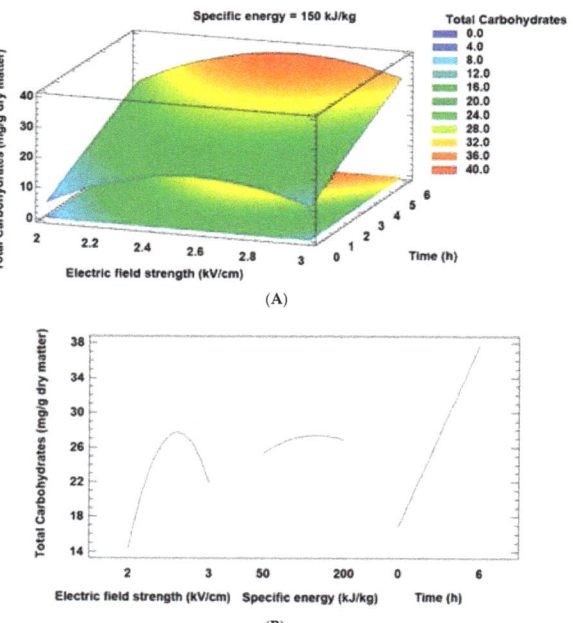

**Figure 1.** Influence of the different extraction conditions (**A**) and main effects chart (**B**) for electric field strength, specific energy and time on carbohydrate recovery yield (mg glucose/g dry matter). The least relevant factor (highest *p*-value) has been set at its optimal value.

The protein content was also influenced by the factors mentioned above (Figure 2). Similarly to carbohydrates, time had a significant impact ($p < 0.05$) on protein extraction, obtaining a higher recovery of protein with the elapse of treatment time. However, an opposite trend was observed for specific energy, finding a decrease in protein content as the specific energy increased. On the other hand, an increase in protein recovery was observed after increasing the electric field ($p > 0.05$) from 2 to 2.5 kV/cm, reaching a plateau after these conditions, then decreasing the recovery.

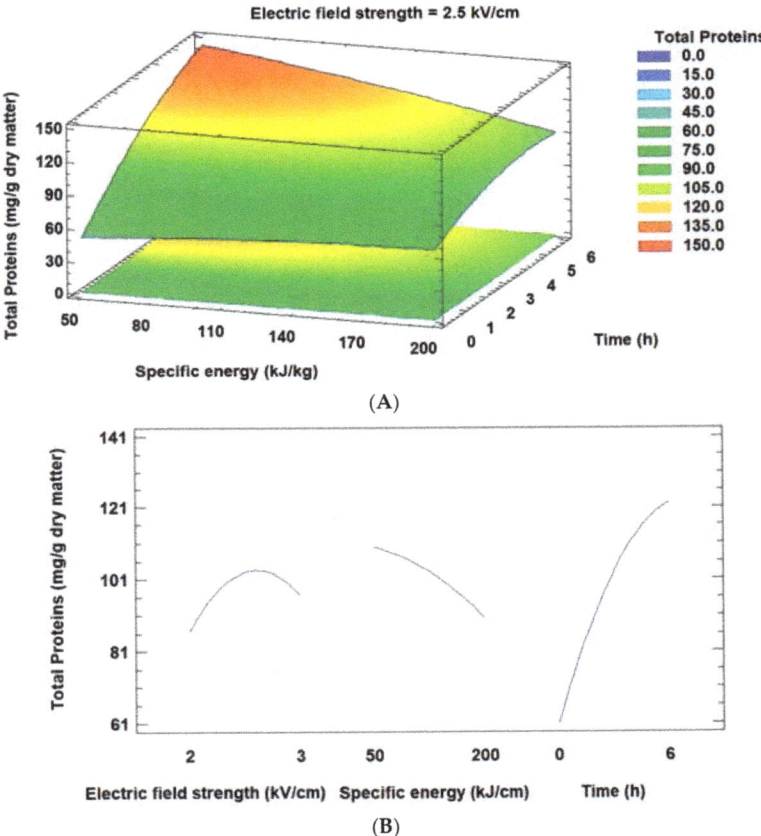

**Figure 2.** Influence of the different extraction conditions (**A**) and main effects chart (**B**) for electric field strength, specific energy and time on protein recovery yield (mg bovine serum albumin/g dry matter). The least relevant factor (highest *p*-value) has been set at its optimal value.

After evaluating the optimal extraction conditions, the maximum protein recovery (140.72 ± 15.14 mg BSA/g DM) was obtained after applying a PEF pre-treatment of 2 kV/cm, 50 kJ/kg and 6 h of extraction.

To evaluate the influence of PEF pre-treatment to reduce the time required for carbohydrates and protein extraction, other authors compared PEF-assisted extraction with conventional methods. In this sense, these authors indicated possible limitations when extracting carbohydrates in water must be considered since, although carbohydrates are mostly polar and highly soluble, mushrooms present certain insoluble polysaccharides that would need longer extraction times, higher temperatures or alkaline solvents [25]. In addition, chitinous compounds and high molecular weight polysaccharides, which are more viscous, also require longer extraction times or higher temperatures [26,27], which would promote the degradation of other compounds and protein denaturation. Finally, PEF technology could generate hydroxyl radicals that degrade polysaccharides such as chitosan. Moreover, Dellarosa et al. reported that the variation in water holding capacity of the mushroom observed could be explained by a lower molecular weight due to degradation of the chitosan after PEF pre-treatment [28].

3.1.2. Total Antioxidant Capacity and Total Phenolic Compounds

The TPC, TEAC and ORAC values range from $1.49 \pm 0.16$ to $25.20 \pm 1.81$ mg GAE/g DM, $9.22 \pm 0.45$ to $65.83 \pm 1.14$ µmol TE/g DM and $14.53 \pm 0.58$ to $145.68 \pm 17.80$ µmol TE/g DM, respectively.

The ANOVA analysis performed showed a significant effect of extraction time on TPC recovery, as well as TEAC and ORAC values, this parameter also having the greatest influence ($p < 0.05$) on the extraction of antioxidant compounds.

The factor having the strongest influence on the recovery was the extraction time, after performing the ANOVA analysis, observing significant ($p < 0.05$) changes in ORAC and TEAC values. However, the specific energy of 50 kJ/kg is the value coincident as optimal for the results of the three antioxidant capacity analyses, while the time presented a slight variability according to the specific analyses.

Figure 3 shows the values resulting from the three tests, obtaining the maximum recovery by setting a value of 50 kJ/kg, which is the lowest value of the chosen range. Regarding time, a similar behavior to that of carbohydrates was found for ORAC values, where the maximum obtained corresponds to 6 h; however, for TPC and TEAC values, the maximum values were not obtained after 6 h, but presented a maximum recovery at 5 h, reaching a plateau after this time and then decreasing.

The electric field strength was the lowest significant factor in all studies; however, it was set at 2.5 kV/cm, obtaining the highest values for TEAC, ORAC, carbohydrates and total proteins. However, the behavior in the extraction of polyphenols was different, presenting its maximum at 3 kV/cm. This behavior agrees with that observed by several authors. For instance, in the study conducted by Darra et al. [29] on grape pomace, it was observed that increasing the electric field from 400 to 800 V/cm improved the extraction of polyphenols, although this fact was also observed on *A. bisporus*, where an intense increase in recovery was observed when increasing the electric field strength [30]; however, regarding the antioxidant components evaluated by TEAC and ORAC, the behavior was similar to that observed for macronutrients, where an increase in the electric field strength promotes their degradation.

In addition, a decrease in the compounds was observed when increasing the specific energy supplied (related to the number of pulses), which could indicate a degradation caused by the conditions as they are very sensitive compounds, such as vitamin C measured through the TEAC assay, where the slight increase in temperature caused due to the increase in the number of pulses would be responsible for this loss, as it has been observed in several studies [31,32].

Therefore, the maximum recovery of antioxidant compounds would be obtained by applying 50 kJ/kg of specific energy maintaining a total extraction time of 5.6 h with a theoretical value of 67.94 µmol TE/g DM for TEAC, 161.41 µmol TE/g DM for ORAC and 22.16 mg GAE/g DM for TPC.

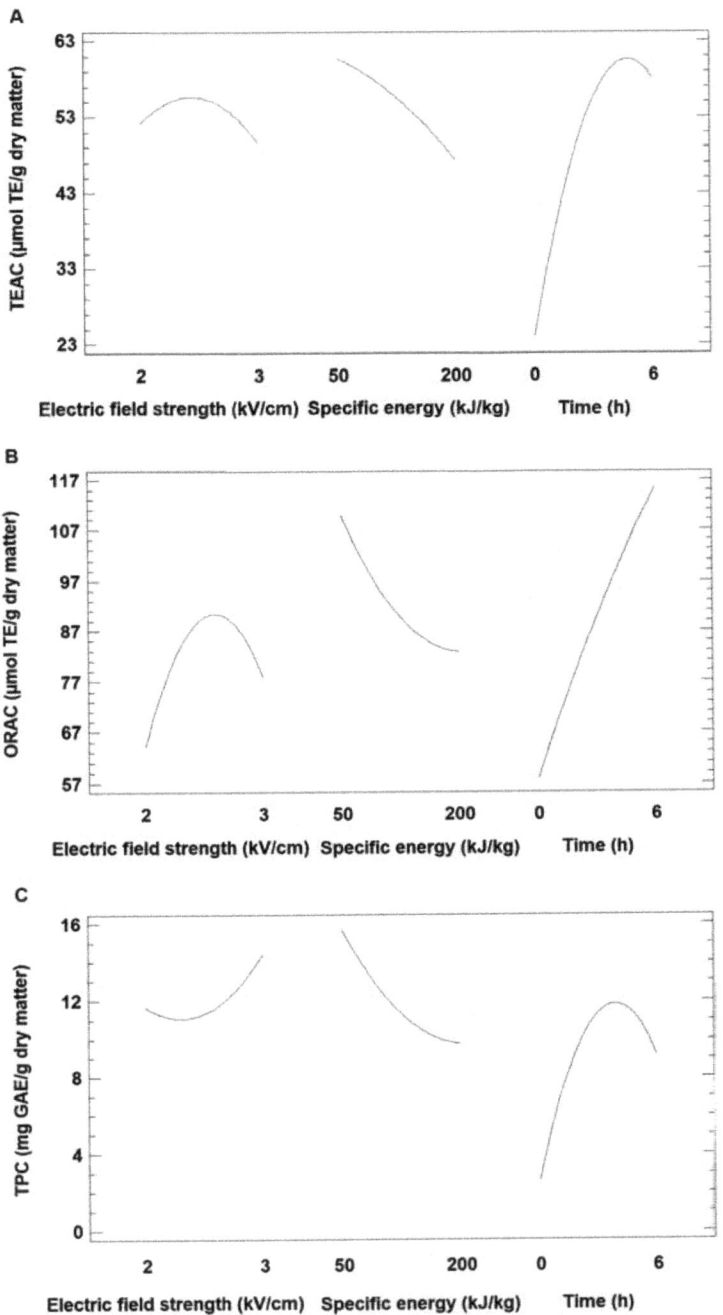

**Figure 3.** Main effect chart of different extraction conditions on (**A**) Trolox equivalent antioxidant capacity (TEAC), (**B**) oxygen radical antioxidant capacity recovery yield (μmol Trolox equivalent/g dry matter), and (**C**) total phenolic compounds (TPC) recovery yield (mg gallic acid equivalent/g dry matter).

3.1.3. Optimization

The simultaneous optimization of all the responses was carried out by the desirability function, in such a way that the extraction of all the compounds was maximized. The optimum conditions obtained were 50 kJ/kg for specific energy, 2.5 kV/cm for electric field strength and 6 h of total extraction time. As can be seen in Figure 4, the desirability obtained at the optimum conditions was 0.88. This result is due to the variability in the behavior of each of the compounds in relation to the factors; for example, a decrease in the extraction of antioxidant compounds measured by TEAC and TPC was found when an extraction time of 6 h was set, while the content of carbohydrates, proteins and antioxidant compounds measured by ORAC was maximized, presenting the opposite behavior.

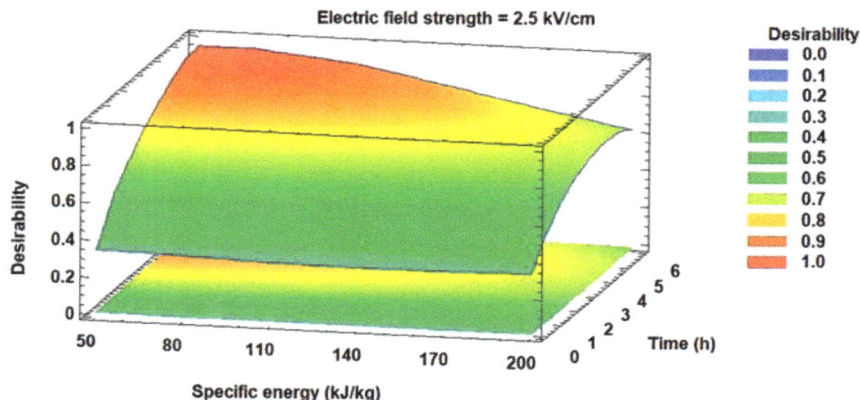

**Figure 4.** Influence of different extraction conditions on the recovery yield of antioxidant compounds, total carbohydrates and total proteins. The least relevant factor (highest $p$-value) has been set at its optimum value.

Therefore, it would be necessary to further investigate whether a longer extraction time could increase the recovery of these macronutrients and antioxidant compounds. However, that prolonged extraction could end up degrading certain compounds such as proteins, where a curvature in the behavior with respect to time is observed (Figure 2). On the other hand, it would not be desirable to increase the specific energy and the electric field strength since this would result in a loss of the compounds studied.

Finally, it should be considered that the knowledge of PEF technology is more extensive for pasteurization and food preservation, but it is limited in the effect on the recovery of bioactive compounds, whose specific mechanism is still partially unknown [31].

### 3.2. Recovery of Nutrients and Bioactive Compounds in PEF-Assisted Extraction at Optimal Conditions

3.2.1. Macronutrient Content

The carbohydrate content obtained, shown in Figure 5, ranged from 8.14 ± 0.74 to 16.71 ± 0.72 mg glu/g DM, belonging to conventional and PEF-assisted extraction using the optimal conditions previously described in the preceding sections, respectively. Therefore, the application of PEF technology on *A. bisporus* samples increases carbohydrate recovery by 105.28% compared to conventional extraction, showing statistical difference ($p < 0.05$) between methodologies.

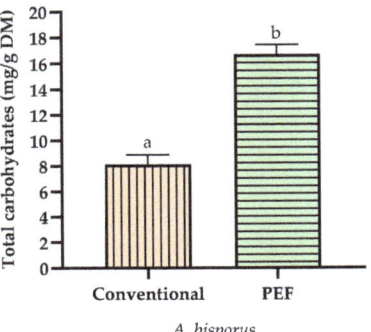

**Figure 5.** Total carbohydrate content (mg glucose (glu)/g dry matter (DM)) in conventional and pulsed electric field-assisted extraction (PEF) of *A. bisporus*. Different lowercase letters in the same parameter indicate statistical differences related to the extraction methodology.

Increased carbohydrate recovery was observed by other authors, who obtained higher polysaccharide and protein content after exposing *A. bisporus* to PEF pretreatment compared to conventional extraction at 95 °C for 1 h [30]. Parniakov et al. [9] compared the efficiency and stability of extraction from *A. bisporus* with different methodologies, which showed that the combination of PEF methodology along with the pressure application exhibited the highest polysaccharide content compared to conventional aqueous extraction at high temperatures.

Likewise, the protein recovery is noteworthy, with 107.02 ± 10.13 mg BSA/g DM for conventional extraction and 119.11 ± 11.05 mg BSA/g DM for PEF pretreatment shown in Figure 6, observing an increase of 11.29% compared to conventional extraction. Therefore, the application of PEF technology for the recovery of the macronutrients studied is a promising extraction method, enabling an increase in macronutrient recovery without the application of high temperature.

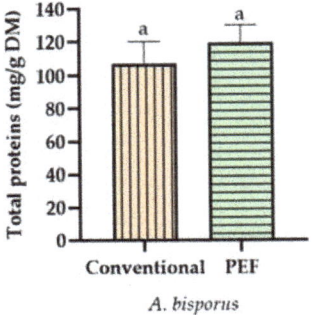

**Figure 6.** Total protein content (mg bovine serum albumin (BSA)/g dry matter (DM)) in conventional and pulsed electric field-assisted extraction (PEF) of *A. bisporus*. Different lowercase letters in the same parameter indicate statistical differences related to the extraction methodology.

Both protein and carbohydrate recovery have been evaluated in several studies after the application of PEF pretreatment compared to conventional aqueous extraction, which led to a loss of protein quality due to coagulation when high temperatures were applied, and other methods such as pressure application showed similar results [9,30]. In addition, the application of PEF is especially interesting in protein extraction since mushrooms, such as *A. bisporus*, are considered good sources of protein, not only in terms of protein quantity but also in terms of amino acid composition compared to animal protein. Kakon et al. [33]

reported that mushroom proteins contain nine essential amino acids, making *A. bisporus* a food suitable as a substitute for meat protein. However, the variation observed in the protein content when the growing substrate of *A. bisporus* changes should be noted, with ranges of protein content indicated from 11.01% by Sadiq et al. [34] to 29.14% by Ahlawat et al. [35].

3.2.2. Total Antioxidant Capacity and Total Phenolic Compounds

The global recovery of antioxidant compounds is shown in Figure 7, with values of 14.63 ± 2.76 and 28.80 ± 2.86 mg GAE/g DM for TPC, 53.15 ± 1.2 and 57.37 ± 5.40 µmol TE/g DM for TEAC and 133.48 ± 7.61 and 202.20 ± 21.19 µmol TE/g DM for ORAC comparing conventional and PEF-assisted extraction, respectively. Similarly to the protein and carbohydrate content, PEF-assisted extraction showed a higher recovery of these compounds measured through three assays, increasing all of them and doubling the value obtained for TPC with the application of PEF technology. Furthermore, significant differences ($p > 0.05$) were observed for TPC and ORAC values concerning the extraction methodology, with a 96.86% and 51.48% increase, respectively.

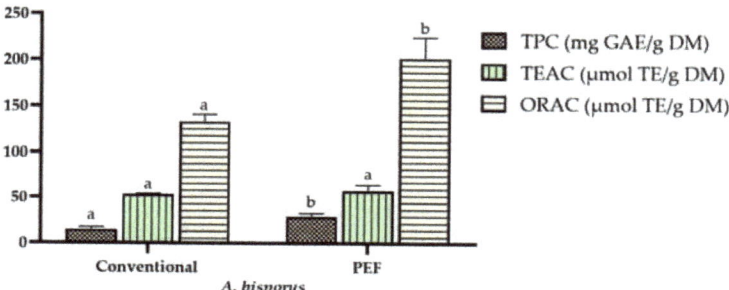

**Figure 7.** Total phenolic compounds (TPC) (mg gallic acid equivalents (GAE)/g dry matter (DM)), Trolox equivalent antioxidant capacity (TEAC) and oxygen radical antioxidant capacity (ORAC) (µmol Trolox equivalents (TE)/g dry matter (DM)) values in conventional and pulsed electric field-assisted extraction of *A. bisporus*. Different lowercase letters in the same parameter indicate statistical differences related to the extraction methodology.

In general, the results obtained were in agreement with those indicated by other authors, showing that the application of moderate PEF intensity increases the production of bioactive compounds with antioxidant capacity in *A. bisporus*, establishing a high positive correlation between antioxidant capacity and phenolic compounds [30,36]. However, the great variability of specific results should be considered due to the affectation of the food by cultivation and storage conditions, and those related to PEF technology, since although the overall result was an increase in the recovery of these compounds, it depends on the conditions applied and the sample processing [31].

According to the values obtained, the antioxidant capacity of *A. bisporus* mainly attributed to its phenolic compounds has shown anti-inflammatory capacity reducing bleeding and damage to the intestinal mucosa of mice with colitis, also attenuating myeloperoxidase activity and overproduction of TNF-α as a consequence of the disease when this mushroom is introduced in their diet [37].

3.2.3. Mineral Content

The mineral content per 100 g and the percentage of the dietary reference intake (DRI) established by the Spanish Agency for Food Safety and Nutrition (AESAN) [38] covered by a portion are shown in Table 2 after the analysis of the fresh samples. It was observed that *A. bisporus* is a source of P and Se, covering 31.32% and 50.07% of the INR, respectively, especially the Se content with 24.30 µg/100 g, covering half of the INR in a portion. In contrast to the remaining minerals, the Ca content of *A. bisporus* does not exceed 1% of the DRI, so it is not considered a source of this mineral, showing contents of 3.02 mg/100 g.

These results agreed with the results obtained by other authors, where it was observed that the major minerals in mushrooms are K and P with contents higher than or comparable to those of most vegetables [8,39]. The low Ca content of mushrooms has also been reported by several authors [11,40], especially in *A. bisporus* and *P. ostreatus* along with Fe, so the results are consistent with these observations. However, the great variability in the mineral content of mushrooms attributable to different soil conditions, cultivation and location produced mixed results such as those observed in the study published by Mattila et al. [11].

The mineral contents obtained suggest that *A. bisporus* could contribute to the decrease in population deficiencies shown in the nutritional assessment study published by AESAN based on data from the National Survey of Dietary Intake (ENIDE) [41]. Both P and Se are obtained mainly from animal sources (72% and 67%, respectively), so the introduction of *A. bisporus* in the vegan population would increase the intake of these minerals, avoiding population deficiencies. On the other hand, although the contribution is lower than in the minerals indicated above, the consumption of *A. bisporus* can contribute to the daily intake of Fe, especially in fertile women and children, and of Zn, whose main source is fish, with a worldwide population deficiency of 30%.

**Table 2.** Mineral content (mg or µg/100 g) of *A. bisporus* and dietary reference intake percentage of each mineral covering a portion (150 g) [38,42].

|  | A. bisporus | | | | | |
|---|---|---|---|---|---|---|
|  | Ca (mg) | Fe (mg) | Zn (mg) | Se (µg) | P (mg) | Mg (mg) |
| In 100 g | 3.02 | 0.86 | 0.99 | 24.30 | 146.10 | 14.10 |
| % DRI [1] | 0.48 | 14.18 | 13.50 | 52.07 | 31.32 | 6.04 |

[1] Dietary reference intake.

In addition, the mineral recovery percentages analyzed with each extraction methodology compared to the fresh solid matrix differed according to the mineral, giving mixed results. As shown in Figure 8, the recovery of Mg, P, Se and Fe was higher in PEF extraction showing significant differences ($p < 0.05$), with a PEF recovery of 20.64% to 61.73% belonging to Fe and Se, respectively. The PEF recovery of the previously mentioned minerals does not decrease below 20% while the conventional extraction does not exceed 24.69%, obtained for Se. On the other hand, conventional extraction has shown a higher recovery in Ca recovery with 2.25% compared to 1.30% obtained by PEF showing significant differences ($p < 0.05$), however, low recoveries were shown for both extractions.

Therefore, PEF technology is useful in the extraction of P, Mg, Fe and Se compared to conventional technology; in addition, the recovery of Se and P, abundant minerals in the fresh matrix, was noteworthy. Nevertheless, it is not appropriate for Ca recovery due to the low recovered value added to the limited quantity of this mineral in the fresh matrix. Finally, although *A. bisporus* contains a considerable amount of Zn, both methodologies were not effective for its recovery, showing non-significant differences between them ($p > 0.05$) as can be seen in Figure 8.

3.2.4. Evaluation of the Extraction Methodology Effect on the Mushroom Surface by SEM

Figure 9 shows the effects of a conventional extraction (20 ± 4 °C, 6 h) and a PEF pre-treated extraction (2.5 kV/cm, 50 kJ/kg, 6 h) on the microstructure of freeze-dried samples of *A. bisporus*. The results were compared with those obtained for an untreated sample (freeze-dried sample of fresh mushroom). Clear differences in microstructure were obtained when comparing the pretreated sample with the control sample. In this regard, the structure of the untreated samples difficult the diffusion of the compounds to the exterior by presenting fibers intertwined with each other, as can be observed in the control sample of *A. bisporus*.

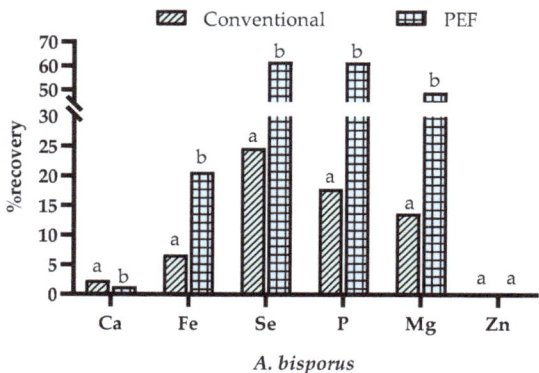

**Figure 8.** The percentage recovery of each mineral in conventional and pulsed electric field-assisted (PEF) extracts from fresh samples of *A. bisporus*. Different lowercase letters in the same mineral suggest significant differences in relation to the methodology applied.

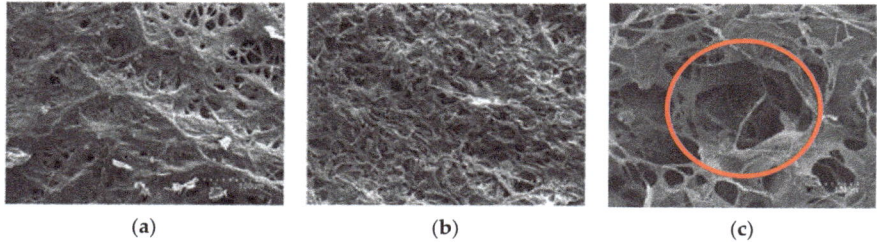

**Figure 9.** Microstructure (presented as scanning electron microscopy images; ×300 magnification) of freeze-dried *A. bisporus* after different processes: (**a**) untreated sample (control), (**b**) conventional aqueous extraction and (**c**) pulsed electric field (PEF) pretreated sample.

On the other hand, in the samples pretreated with PEF, the presence of pores or cavities on the surface was observed, leading to a structural change that allows an improvement in the diffusion of the compounds and selective extraction of these at the cell level, since the disintegration of the cell membrane is caused [43], compared to a more disorganized structure resulting from conventional extraction. This observation was consistent with the results obtained in the study by Li et al. [44], in which the same results were presented by treating *L. edodes* samples using PEF and observing the effect of microporation on their surface. In addition, comparing the PEF methodology with other extraction techniques such as ultrasound application, and despite the fact that in the application of ultrasound the energy supplied to the sample is higher, it was observed that PEF induces a greater degree of cell disruption that is reflected in changes in the microstructure caused by a rupture of the cell membrane [28].

The presence of micropores on the surface of samples pretreated with PEF was associated with the increased recovery of macronutrients and antioxidant compounds in the extracts pretreated with PEF. In fact, electroporation-assisted processes were proposed as a method to improve the extraction of beneficial compounds from plant tissues, including mushrooms [9,45], which is an interesting process for the extraction of thermosensitive components. Therefore, there is a correlation between the content of nutrients and compounds with antioxidant capacity obtained in PEF and conventional extracts with the microstructure of the solid after treatment.

## 4. Conclusions

From the results obtained, it is possible to conclude that the optimal conditions for the extraction of nutrients and bioactive antioxidant compounds from *A. bisporus*, using response surface methodology based on all the variables analyzed, were an electric field strength of 2.5 kV/cm, 50 kJ/kg of specific energy and 6 h of time. Moreover, the influence of the studied parameters (electric field strength, specific energy, and extraction time) differed according to the target compound analyzed, showing different behaviors in relation to the parameters depending on the compound. Likewise, it was observed that an increase in the extraction time increased the recovery of carbohydrates, proteins, and antioxidant compounds, the last mentioned with a maximum of 5 h in general, while the increase in the electric field strength showed a positive effect on the recovery of all compounds with a maximum at 2.5 kV/cm, a field from which a decrease in the recovery was observed, except for total phenolic compounds. The specific energy showed mixed results, the increase in the energy supplied caused a decrease in the recovery of proteins and antioxidant compounds, and an increase in the recovery of carbohydrates with a maximum of 150 kJ/kg. On the other hand, SEM results showed that PEF pretreatment changes the microstructure of the mushrooms causing surface electroporation which allows an increase in the recovery of compounds observable in the extracts. The application of PEF technology under optimal conditions to mushrooms increases the extraction of carbohydrates, proteins, antioxidant compounds and minerals such as P, Mg, Fe and Se compared to conventional methodology. In addition, *A. bisporus* is an optimal matrix for the high content of bioactive compounds and micronutrients, as a source of P and Se, and containing considerable amounts of Fe, Zn and Mg, making them foods of great interest in the diet, especially for people with a vegetarian/vegan diet, as they largely supply nutrients abundant in animal products.

**Author Contributions:** Conceptualization, F.J.B.; methodology, M.C.-G.; formal analysis, M.C.-G., and J.M.C.; software, M.C.-G. and J.M.C.; resources, E.C., E.F., H.B. and F.J.B.; writing—original draft preparation, M.C.-G., J.M.C. and F.J.B.; writing—review and editing, J.M.C., E.C., E.F., H.B. and F.J.B.; supervision, J.M.C., E.C., E.F., H.B. and F.J.B.; funding acquisition, E.C. and F.J.B. All authors have read and agreed to the published version of the manuscript.

**Funding:** This research was supported by the Spanish Ministry of Science and Innovation project (PID2021-123628OB-C42—Eco-innovative extraction of nutrients and bioactive compounds from agri-food co-products for the design of healthier foods. Study of biological activities) funded by MCIN/AEI/10.13039/501100011033/. Moreover, authors would like to acknowledge the University of Valencia through the project OTR2021-21736INVES, supported by the University of Vigo.

**Institutional Review Board Statement:** Not applicable.

**Informed Consent Statement:** Not applicable.

**Data Availability Statement:** Not applicable.

**Acknowledgments:** Juan Manuel Castagnini is the beneficiary of the grant (ZA21-028) for the requalification of the Spanish university system from the Ministry of Universities of the Government of Spain, modality "Maria Zambrano", financed by the European Union, NextGeneration EU through the project "Extraction of bioactive compounds from food matrices using innovative and sustainable technologies (EXTRABIO)". Moreover, Mara Calleja Gómez would like to acknowledge the Ministry of Education for providing her a collaboration grant. Francisco J. Barba is member of the CYTED network "P320RT0186—Aprovechamiento sostenible de recursos biomásicos vegetales iberoamericanos en cosmética (BIOLATES)". The authors would also like to acknowledge Generalitat Valenciana for financial support (IDIFEDER/2018/046—Procesos innovadores de extracción y conservación: pulsos eléctricos y fluidos supercríticos) through the European Union ERDF funds (European Regional Development Fund).

**Conflicts of Interest:** The authors declare no conflict of interest.

## References

1. Royse, D.J.; Baars, J.; Tan, Q. Current overview of mushroom production in the world. In *Edible and Medicinal Mushrooms*; John Wiley & Sons, Ltd.: Chichester, UK, 2017; pp. 5–13. [CrossRef]
2. Vaishnavi, M.; Durga Prasad, M.; Sharma, A.; Tiwari, J.; Singh, S.; Sharma, S. Production of edible mushrooms to meet the food security: A review. *J. Posit. Sch. Psychol.* **2022**, *6*, 4316–4325.
3. Eurostat. Crop Production in European Union Standard Humidity. Available online: https://ec.europa.eu/eurostat/databrowser/view/APRO_CPSH1__custom_2950901/default/table?lang=en (accessed on 20 June 2022).
4. Ministerio de Agricultura, Pesca y Alimentación. *Superficies y Producciones Anuales de Cultivos*. Available online: https://www.mapa.gob.es/es/estadistica/temas/estadisticas-agrarias/agricultura/superficies-producciones-anuales-cultivos/ (accessed on 8 May 2022).
5. Ministerio de Agricultura, Pesca y Alimentación. *Informe del consumo de alimentación en España*. Available online: https://www.mapa.gob.es/es/alimentacion/temas/consumo-tendencias/informe-anual-consumo-2020-v2-nov2021-baja-res_tcm30-562704.pdf (accessed on 8 May 2022).
6. Leuci, R.; Brunetti, L.; Poliseno, V.; Laghezza, A.; Loiodice, F.; Tortorella, P.; Piemontese, L. Natural compounds for the prevention and treatment of cardiovascular and neurodegenerative diseases. *Foods* **2020**, *10*, 29. [CrossRef] [PubMed]
7. Kumar, K.; Mehra, R.; Guiné, R.P.F.; Lima, M.J.; Kumar, N.; Kaushik, R.; Ahmed, N.; Yadav, A.N.; Kumar, H. Edible mushrooms: A comprehensive review on bioactive compounds with health benefits and processing aspects. *Foods* **2021**, *10*, 2996. [CrossRef] [PubMed]
8. Torija, M.E. Principios Inmediatos y Elementos Minerales en Hongos Comestibles. Ph.D. Thesis, Universidad Complutense de Madrid, Madrid, Spain, 2015.
9. Parniakov, O.; Lebovka, N.I.; van Hecke, E.; Vorobiev, E. Pulsed electric field assisted pressure extraction and solvent extraction from mushroom (*Agaricus bisporus*). *Food Bioprocess Technol.* **2014**, *7*, 174–183. [CrossRef]
10. Gazaryan, I.G.; Ratan, R.R. Oxidative damage in neurodegeneration and injury. In *The Curated Reference Collection in Neuroscience and Biobehavioral Psychology*; Elsevier: Amsterdam, The Netherlands, 2017; pp. 327–336. [CrossRef]
11. Mattila, P.; Könkö, K.; Eurola, M.; Pihlava, J.M.; Astola, J.; Vahteristo, L.; Hietaniemi, V.; Kumpulainen, J.; Valtonen, M.; Piironen, V. Contents of vitamins, mineral elements, and some phenolic compounds in cultivated mushrooms. *J. Agric. Food Chem* **2001**, *49*, 2343–2348. [CrossRef]
12. Roselló-Soto, E.; Parniakov, O.; Deng, Q.; Patras, A.; Koubaa, M.; Grimi, N.; Boussetta, N.; Tiwari, B.K.; Vorobiev, E.; Lebovka, N.; et al. Application of non-conventional extraction methods: Toward a sustainable and green production of valuable compounds from mushrooms. *Food Eng. Rev.* **2016**, *8*, 214–234. [CrossRef]
13. Khatun, S. Research on mushroom as a potential source of nutraceuticals: A review on Indian perspective. *Am. J. Exp. Agric.* **2012**, *2*, 47–73. [CrossRef]
14. Athanasiadis, V.; Pappas, V.M.; Palaiogiannis, D.; Chatzimitakos, T.; Bozinou, E.; Makris, D.P.; Lalas, S.I. Pulsed electric field-based extraction of total polyphenols from *Sideritis raiseri* using hydroethanolic mixtures. *Oxygen* **2022**, *2*, 91–98. [CrossRef]
15. Barros, M.; Redondo, L.; Rego, D.; Serra, C.; Miloudi, K. Extraction of essential oils from plants by hydrodistillation with pulsed electric fields (PEF) pre-treatment. *Appl. Sci.* **2022**, *12*, 8107. [CrossRef]
16. Bocker, R.; Silva, E.K. Pulsed electric field-assisted extraction of natural food pigments and colorings from plant matrices. *Food Chem.* **2022**, *15*, 100398. [CrossRef]
17. Martí-Quijal, F.J.; Ramon-Mascarell, F.; Pallarés, N.; Ferrer, E.; Berrada, H.; Phimolsiripol, Y.; Barba, F.J. Extraction of antioxidant compounds and pigments from Spirulina (*Arthrospira platensis*) assisted by pulsed electric fields and the binary mixture of organic solvents and water. *Appl. Sci.* **2021**, *11*, 7629. [CrossRef]
18. Vernon-Parry, K.D. Scanning electron microscopy: An introduction. *III-Vs Rev.* **2000**, *13*, 40–44. [CrossRef]
19. De Vito, F.; Ferrari, G.; Lebovka, N.I.; Shynkaryk, N.V.; Vorobiev, E. Pulse duration and efficiency of soft cellular tissue disintegration by pulsed electric fields. *Food Bioprocess Technol.* **2008**, *1*, 307–313. [CrossRef]
20. Luengo, E.; Raso, J. Pulsed electric field-assisted extraction of pigments from *Chlorella vulgaris*. In *Handbook of Electroporation*; Springer: Berlin/Heidelberg, Germany, 2017; pp. 2939–2954. ISBN 9783319328867.
21. Ovalles-Rodríguez, G.A. Uso de la metodología de superficies de respuesta en la optimización de procesos. *Eco. Mat.* **2011**, *2*, 16–20. [CrossRef]
22. Cao, G.; Alessio, H.M.; Cutler, R.G. Oxygen-radical absorbance capacity assay for antioxidants. *Free Radic. Biol. Med.* **1993**, *14*, 303–311. [CrossRef]
23. Singleton, V.L.; Orthofer, R.; Lamuela-Raventos, R.M. Analysis of total phenols and other oxidation substrates and antioxidants by means of Folin–Ciocalteu reagent. In *Methods in Enzymology*; Elsevier: Amsterdam, The Netherlands, 1999; pp. 152–178. [CrossRef]
24. Dubois, M.; Gilles, K.A.; Hamilton, J.K.; Rebers, P.A.; Smith, F. Colorimetric method for determination of sugars and related substances. *Anal. Chem.* **1956**, *28*, 350–356. [CrossRef]
25. Smiderle, F.R.; Morales, D.; Gil-Ramírez, A.; de Jesus, L.I.; Gilbert-López, B.; Iacomini, M.; Soler-Rivas, C. Evaluation of microwave-assisted and pressurized liquid extractions to obtain β-d-glucans from mushrooms. *Carbohydr. Polym.* **2017**, *156*, 165–174. [CrossRef]

26. Akram, K.; Shahbaz, H.M.; Kim, G.R.; Farooq, U.; Kwon, J.H. Improved extraction and quality characterization of water-soluble polysaccharide from gamma-irradiated *Lentinus edodes*. *J. Food Sci.* **2017**, *82*, 296–303. [CrossRef]
27. Morales, D.; Smiderle, F.R.; Villalva, M.; Abreu, H.; Rico, C.; Santoyo, S.; Iacomini, M.; Soler-Rivas, C. Testing the effect of combining innovative extraction technologies on the biological activities of obtained β-glucan-enriched fractions from *Lentinula edodes*. *J. Funct. Foods* **2019**, *60*, 103446. [CrossRef]
28. Dellarosa, N.; Frontuto, D.; Laghi, L.; Dalla Rosa, M.; Lyng, J.G. The impact of pulsed electric fields and ultrasound on water distribution and loss in mushrooms stalks. *Food Chem.* **2017**, *236*, 94–100. [CrossRef]
29. El Darra, N.; Grimi, N.; Vorobiev, E.; Louka, N.; Maroun, R. Extraction of polyphenols from red grape pomace assisted by pulsed ohmic heating. *Food Bioprocess Technol.* **2013**, *6*, 1281–1289. [CrossRef]
30. Xue, D.; Farid, M.M. Pulsed electric field extraction of valuable compounds from white button mushroom (*Agaricus bisporus*). *Innov. Food Sci. Emerg. Technol.* **2015**, *29*, 178–186. [CrossRef]
31. Vivanco, D.; Ardiles, P.; Castillo, D.; Puente, L. Emerging technology: Pulsed electric fields (PEF) for food treatment and its effect on antioxidant content. *Rev. Chil. Nutr.* **2021**, *48*, 609–619. [CrossRef]
32. Agostini, L.R.; Morón Jiménez, M.J.; Ramón, A.N.; Ayala Gómez, A. Determinación de la capacidad antioxidante de flavonoides en frutas y verduras frescas y tratadas térmicamente. *Arch. Latinoam. Nutr.* **2004**, *54*, 89–92.
33. Kakon, A.J.; Bazlul, M.; Choudhury, K.; Saha, S. Mushroom is an ideal food supplement. *J. Dhaka Natl. Med. Coll. Hosp.* **2012**, *18*, 58–62. [CrossRef]
34. Saiqa, S.; Haq, N.B.; Muhammad, A.H.; Muhammad, A.A.; Rehman, A. Studies on chemical composition and nutritive evaluation of wild edible mushrooms. *Iran J. Chem. Chem. Eng.* **2008**, *27*, 151–154. [CrossRef]
35. Ahlawat, O.P.; Manikandan, K.; Singh, M. Proximate composition of different mushroom varieties and effect of UV light exposure on vitamin D content in *Agaricus bisporus* and *Volvariella volvacea*. *Mushroom Res.* **2016**, *25*, 1–8.
36. Mendiola, M. Caracterización de Compuestos Bioactivos y Efecto de la Aplicación de Pulsos Eléctricos de Moderada Intensidad de Campo en Setas Cultivadas en La Rioja. Ph.D. Thesis, Universitat de Lleida, Lérida, Spain, 2017.
37. Moro, C. Obtención de Extractos Metanólicos Ricos en Compuestos Fenólicos a Partir de Hongos Comestibles. Valoración, In Vitro, de la Actividad Antioxidante y Antiinflamatoria de los Extractos. Ph.D. Thesis, Universidad de Valladolid, Valladolid, Spain, 2015.
38. Agencia Española de Seguridad Alimentaria y Nutrición. Informe del Comité Científico de la Agencia Española de Seguridad Alimentaria y Nutrición (AESAN) Sobre Ingestas Nutricionales de Referencia para la Población Española. Available online: https://www.aesan.gob.es/AECOSAN/docs/documentos/seguridad_alimentaria/evaluacion_riesgos/informes_comite/INR.pdf (accessed on 8 May 2022).
39. Kalač, P. A review of chemical composition and nutritional value of wild-growing and cultivated mushrooms. *J. Sci. Food Agric.* **2013**, *93*, 209–218. [CrossRef]
40. Bano, Z.; Rajarathnam, S. *Pleurotus* mushrooms. Part II. Chemical composition, nutritional value, post-harvest physiology, preservation, and role as human food. *Crit. Rev. Food Sci. Nutr.* **1988**, *27*, 87–158. [CrossRef]
41. Agencia Española de Seguridad Alimentaria y Nutrición. Evaluación Nutricional de la Dieta Española II SOBRE datos de la Encuesta Nacional de Ingesta Dietética (ENIDE). Available online: http://www.laboratoriolcn.com/wp-content/uploads/2019/11/Valoracion_nutricional_ENIDE_micronutrientes.pdf (accessed on 24 May 2022).
42. Fundación española de la Nutrición. Alimentos y Bebidas: Verduras y Hortalizas. Available online: https://www.fen.org.es/vida-saludable/alimentos-bebidas (accessed on 8 May 2022).
43. Saulis, G. Electroporation of cell membranes: The fundamental effects of pulsed electric fields in food processing. *Food Eng. Rev.* **2010**, *2*, 52–73. [CrossRef]
44. Li, X.; Li, J.; Wang, R.; Rahaman, A.; Zeng, X.A.; Brennan, C.S. Combined effects of pulsed electric field and ultrasound pretreatments on mass transfer and quality of mushrooms. *LWT* **2021**, *150*, 112008. [CrossRef]
45. Vorobiev, E.; Lebovka, N. Pulsed-electric-fields-induced effects in plant tissues: Fundamental aspects and perspectives of applications. In *Electrotechnologies for Extraction from Food Plants and Biomaterials*; Springer: New York, NY, USA, 2009; pp. 39–81. ISBN 9780387793733.

Article

# Analytical Determination of Allergenic Fragrances in Indoor Air

Catia Balducci [1], Marina Cerasa [1], Pasquale Avino [2,*], Paolo Ceci [3], Alessandro Bacaloni [4] and Martina Garofalo [1,4]

[1] Institute of Atmospheric Pollution Research (CNR-IIA), National Research Council of Italy, Via Salaria Km 29.3, Monterotondo, P.O. Box 10, 00015 Rome, Italy; balducci@iia.cnr.it (C.B.); marina.cerasa@iia.cnr.it (M.C.); garofalo.1651813@studenti.uniroma1.it (M.G.)

[2] Department of Agricultural, Environmental and Food Sciences (DiAAA), University of Molise, Via De Sanctis, 86100 Campobasso, Italy

[3] Institute of Atmospheric Pollution Research, Division of Rome, c/o Ministry of Environment, Land and the Sea, Via Cristoforo Colombo 44, 00147 Rome, Italy; paolo.ceci@iia.cnr.it

[4] Department of Chemistry, Sapienza University of Rome, Piazzale Aldo Moro 5, 00185 Rome, Italy; alessandro.bacaloni@uniroma1.it

* Correspondence: avino@unimol.it; Tel.: +39-0874-404634

**Abstract:** Among all the emerging contaminants, fragrances are gaining more relevance for their proven allergenic and, in some cases, endocrine-disrupting properties. To date, little information exists on their concentration in the air. This study aims to fill this gap by developing a method for the determination of semivolatile fragrances in the indoor gaseous phase with sampling protocols usually adopted for the collection of atmospheric particulate matter (sampling time 24 h, flow rate 10 L min$^{-1}$) and instrumental analysis by gas chromatography coupled with mass spectrometry. The method was developed on 66 analytes and tested at three concentration levels: 20 compounds showed analytical recoveries $\geq 72\%$ with percentage standard deviations always better than 20%. For most compounds, negligible sampling breakthroughs were observed. The method was then applied to real samples collected in a coffee bar and in a private house. Considering the fragrances for which the method has shown good effectiveness, the highest concentrations were observed for carvone in the coffee bar (349 ng m$^{-3}$) and camphor in the house (157 ng m$^{-3}$). As concerns certain or suspected endocrine disruptors, lilyal and galaxolide were detected at both sites, α-isomethylionone was the second most concentrated compound in the house (63.2 ng m$^{-3}$), musk xylene and musk ketone were present at lower concentration ($\approx$ 1 or 2 ng m$^{-3}$).

**Keywords:** indoor air quality; fragrances; indoor pollution; endocrine disruptors; analytical method; GC-MS; musks fragrances; emerging contaminants; α-isomethylionone

**Citation:** Balducci, C.; Cerasa, M.; Avino, P.; Ceci, P.; Bacaloni, A.; Garofalo, M. Analytical Determination of Allergenic Fragrances in Indoor Air. *Separations* **2022**, *9*, 99. https://doi.org/10.3390/separations9040099

Academic Editor: Victoria Samanidou

Received: 16 February 2022
Accepted: 12 April 2022
Published: 13 April 2022

**Publisher's Note:** MDPI stays neutral with regard to jurisdictional claims in published maps and institutional affiliations.

**Copyright:** © 2022 by the authors. Licensee MDPI, Basel, Switzerland. This article is an open access article distributed under the terms and conditions of the Creative Commons Attribution (CC BY) license (https://creativecommons.org/licenses/by/4.0/).

## 1. Introduction

In recent years, interest in indoor air quality has been increasing considerably [1,2]. It is ascertained that in the developed countries, people (and, in particular, sensitive subjects such as children and the elderly) spend up to 90% of their time indoors [3,4] where generally inadequate ventilation, high temperatures, humidity, together with slow degradation processes can increase the concentration of pollutants compared to outdoor ones [5]. As a result, inhalation exposure peaks indoors due to residence times and higher concentrations of contaminants [6]. Among all pollutant classes, fragrances have been identified as one of the main causes of indoor pollution [7,8]. Recent studies report that more than 2600 fragrances are contained in everyday products and especially in the most developed countries, the demand and consumption of perfumed products such as incense or candles, personal care, or cleaning products have increased. In the last period, toxicological and clinical studies on the effects of this class have increased together with their diffusion, but for many substances, there is little information on the ability to interact with biological tissues (bioactivity), and on their behavior once dispersed in the environment [9–13]. The main health effects related to fragrances are linked to their allergenic properties [12], to which dermatological effects (e.g., irritation, dermatitis, eczema, psoriasis) and respiratory problems are also added [7,8]. Above all, some fragrances are recognized as endocrine disruptors [14,15], among which lylial, α-isomethylionone, benzyl salicylate, and nitro-musks are listed.

Despite all the evidence reported, there are a few studies on fragrances in the air, especially as regards indoor environments where their concentrations are presumably higher [16]. Of all fragrances, light terpenes, such as limonene and pinenes, are the most investigated due to their allergenic properties or to the ability to react with ozone to generate secondary particles [4,17,18].

Less information exists on higher molecular weight fragrances in the indoor gas phase. As there are no pre-established protocols for the analysis of these substances, among the few studies carried out, the procedures adopted are different. Furthermore, there are no studies that evaluate the volatility and degradation of these substances or that consider their distribution between the gas and the particulate phase. Lamas et al. [19,20] reported two studies on higher molecular weight fragrances in the indoor gas phase by adopting two different analytical procedures obtaining satisfactory results. Both involved sample collection by active sampling, followed by glass funnel microfiltration and ultrasound-assisted extraction in one case, and solid-phase microextraction in the other. The latter procedure was also used by Regueiro et al. [21] for the analysis of nine synthetic musks. Ramirez et al. measured musks by using Tenax cartridges and thermal desorption in GC-MS (gas chromatography coupled with mass spectrometry) [22]. Balci et al. [23], evaluated the ability of synthetic musks to diffuse indoors, through an experiment in controlled conditions, by using Amberlite XAD-2 resin sandwiched between polyurethane foam followed by extraction with an ultrasonic bath. Also, concerning the indoor concentrations of fragrances in the particulate phase, so far little information is available [24,25].

This study aimed to provide information in this field through the development of a method for assessing the indoor concentration of semivolatile fragrances at sampling conditions applied for the capture of atmospheric particulate matter (PM) in interiors. At this stage, the results concerning fragrances in the gaseous phase are reported. These results are preparatory to subsequent studies in which the investigation will be extended to the simultaneous collection of fragrances and other classes in both aerial and particulate phases. The feasibility of the method has been assessed for 66 fragrances, although the focus was not on the lighter ones such as limonene, for which extensive studies already exist. After verifying the efficiency of the method, it was applied to real samples collected in indoor environments, in which a semi-quantitative or quantitative determination of the compounds was carried out, depending on the goodness of accuracy and precision of the method itself.

## 2. Materials and Methods

*2.1. Materials*

All standards, marketed by Supelco, Sigma-Aldrich, Merck, Germany, were of purity grade ≥98%. Concerning fragrances, they were Fragrance allergen mix A1 (24 components, product no. 89131), Fragrance allergen mix A2 (40 components, no. 16558), Musk ketone solution (no. 46377), and Musk xylene solution (no. 46383). All tests for the development of the analytical method in the laboratory were performed using a solution of all the standards previously mentioned (Fragrance Standard solution = FS solution). The deuterated Internal Standards (IS solution at 5 ng $\mu L^{-1}$) used as references for quantitative analysis were naphthalene-d8 (catalog product no. 176044), diethyl phthalate-d4 (no. 492221), and benzo[a]anthracene-d12 (no. 456306). To assess the recovery of the IS solution, a mixed solution of 2-metylnaphtalene-d10 (Product no. 454249) and Pyrene-d10 (no. 490695) was used (Syringe Standard solution = SS solution at 10 ng $\mu L^{-1}$). The adsorbent used for air sampling was Supelco Amberlite XAD-2 (20–60 mesh) (Restek, Bellafonte, PA, USA). The solvents were of super purity grade (produced by Romyl Ltd., Cambridge, UK) and were purchased from Deltek srl (Naples, Italy).

## 2.2. Gas Chromatography–Mass Spectrometry

The GC–MS analyses were performed using a Trace 1300 GC Ultra coupled with the mass spectrometer ISQ 7000 Series (ThermoFisher Scientific, Rodano, Italy). The temperatures of the transfer line and ion source were set at 280 and 250 °C, respectively. Separation was carried out on Rxi-5Sil MS (Restek, Bellafonte, PA, USA) capillary column (30 m × 0.25 mm i.d., 0.25 µm film thickness). Helium was the carrier gas, employed at a constant column flow of 1.0 mL min$^{-1}$ while the injection was performed in splitless mode at 280 °C. The GC oven temperature was programmed from 55 °C (held 1.30 min.) to 90 °C (at 40 °C min$^{-1}$, held 3 min), to 260 °C at 5 °C min$^{-1}$. A final ramp to 300 °C at 20 °C min$^{-1}$ (held 5 min) was set to ensure the elution of all solutes from the column.

A full scan acquisition, in the $m/z$ range 40–320, was used for compound identification through the mass spectrum observation. Four minutes of solvent delay were applied to the acquisition to prevent filament damage. Three ions for each compound were chosen to define the selected ion monitoring acquisition. One ion was chosen for the quantitative determination, the other two were used to confirm the correct determination. For a number of selected compounds, the internal standards and syringe standards, the retention times together with quantification and confirm ions are listed in Table 1. Tables S1 and S2 report these parameters for all the analytes. Each chromatogram was processed using the Thermo Scientific™ Xcalibur™ software.

**Table 1.** Molecular weight (MW), CAS number, retention time, and $m/z$ ratios were used for the fragrance determination.

| Compound | MW | CAS | Retention Time (min) | Quantitative ($m/z$) | Confirmation ($m/z$) |
|---|---|---|---|---|---|
| Salicylaldehyde | 122.12 | 90-02-8 | 6.28 | 122 | 65, 121 |
| Camphor | 152.23 | 464-49-1 | 8.88 | 81 | 95, 108 |
| Folione | 154.21 | 111-12-6 | 10.42 | 123 | 95, 111 |
| Neral | 152.23 | 5392-40-5 | 11.63 | 119 | 69, 84 |
| Carvone | 150.22 | 2244-16-8 | 11.74 | 108 | 54, 82 |
| Geranial | 152.23 | 141-27-5 | 12.51 | 152 | 84, 83 |
| DMBCA | 192.25 | 151-05-1 | 14.06 | 132 | 91, 117 |
| Geranyl acetate | 196.29 | 105-87-1 | 16.00 | 136 | 68, 93 |
| β-Damascenone | 190.28 | 23696-85-7 | 16.10 | 175 | 69, 190 |
| δ-Damascone | 192.30 | 57378-68-4 | 16.38 | 123 | 69, 192 |
| β-Damascone | 192.30 | 23726-91-2 | 17.02 | 177 | 123, 192 |
| Coumarin | 229.16 | 91-64-5 | 17.66 | 146 | 89, 118 |
| α-Isomethylionone | 206.32 | 127-51-5 | 18.98 | 150 | 135, 206 |
| Eugenyl acetate | 206.24 | 93-28-7 | 20.26 | 164 | 131, 149 |
| 3-Propylidenephthalide | 174.2 | 17369-59-4 | 21.50 | 159 | 104, 174 |
| α-Amylcinnamaldehyde | 202.29 | 78605-96-6 | 23.41 | 129 | 201, 202 |
| ISO E® γ | 234.38 | 68155-67-9 | 23.65 | 191 | 109, 121 |
| Musk xylene | 297.26 | 81-15-2 | 28.33 | 282 | 127, 297 |
| Musk ketone | 294.30 | 81-14-1 | 30.77 | 279 | 280, 294 |
| Benzyl cinnamate | 238.28 | 103-41-3 | 32.76 | 131 | 192, 193 |
| Naphthalene-d8 | 136.22 | 1146-65-2 | 9.91 | 108 | 136 |
| 2-Methylnaphthalene-d10 | 152.26 | 7297-45-2 | 13.11 | 122 | 152 |
| Diethyl phthalate-d4 | 226.26 | 93952-12-6 | 22.04 | 181 | 153 |
| Pyrene-d10 | 212.31 | 7297-45-2 | 33.06 | 106 | 212 |
| Benzo[a]anthracene-d12 | 240.36 | 1718-53-2 | 30.08 | 120 | 240 |

Given the high number of target compounds, two different SIMs were used (SIM_A and SIM-B), to minimize the effects of interference between nearby peaks. The first SIM was dedicated to the most polar compounds, less suitable for the type of stationary phase used for chromatographic separation and to the particularly volatile ones. These substances generally showed less affinity with the method developed and greater variability in results.

The MS detector response curves were drawn by plotting the peak area ratios between the fragrances and the respective deuterated internal standard. Five concentration levels of target fragrances ranging from 0.02 to 1.6 ng $\mu L^{-1}$ were processed with three replicates, while the concentration of the internal standards was kept constant at 0.2 ng $\mu L^{-1}$.

The instrumental limit-of-detection ($LOD_{inst}$) was set equal to the analyte concentration corresponding to three times the signal-to-noise ratio (S/N). The instrumental LOD was determined by evaluating the S/N for the analytes starting from the lowest point of the calibration curve (0.02 ng $\mu L^{-1}$) and proceeding by dilutions. The instrumental limit of quantification (LOQ) was set equal to analyte concentration corresponding to 10 times the signal-to-noise ratio.

*2.3. Sampling System and Analytical Procedure*

The sampling was performed with a low-flow air sampler (Silent Sequential Air Sampler FAI Instruments, Palombara, Italy) using a constant airflow rate of 10 L $min^{-1}$ with a sampling cycle set at 24 h, for a total volume of about 14.4 $m^3$.

To collect the molecules present in the gaseous phase, 3 g of XAD-2 contained in a 2 cm diameter glass cartridge was used.

The target compounds collected on the adsorbent cartridge were extracted by solvent elution. This extraction technique allowed the minimization of handling and processing of the samples, improving the quality of the blanks and reducing the risk of analyte losses. The samples were eluted with 30 mL of acetone after adding the IS solution used for quantitative analysis (100 ng). The eluate was then reduced to 500 µL (in a graduated tube) under a gentle stream of nitrogen. Finally, 1 µL of the sample was injected into the GC-MS after spiking with 100 ng of the SS solution.

*2.4. Indoor Sampling*

Real samplings were carried out to test the method in two indoor environments in Rome (Italy), namely, a private house and a coffee bar. The house was a flat of 65 $m^2$ on the third floor of a building. Two people lived there. The sampler was placed in the corridor where the air was influenced by the activities carried out in all the rooms. The coffee bar had an area of about 70 $m^2$. The place was quite frequented and inside there were four tables occupied most of the time. Samplings started at 9:00 a.m. and lasted 24 h. They were carried out in September 2021 with an average temperature of about 23 °C. The windows in the house and the doors in the coffee bar were open ensuring good ventilation in the sampling locations. In addition, during the hottest hours of the day, an air conditioning system worked in the coffee bar.

## 3. Results and Discussion

*3.1. Method Setting and Effectiveness*

The setting of the method involved several phases: the optimization of the instrumental method and the choice of the best adsorbent for the collection of fragrances in the gaseous phase. This is both in terms of adaptability to the operating conditions required by the method adopted and in terms of capture efficiency evaluated through the breakthrough phenomenon, the optimization, and validation of the analytical procedure for the analyte determination.

All the tests for the method development and the method validation were carried out in triplicate.

3.1.1. GC Analysis

The full-scan chromatogram of the target substances is shown in Figure 1. Injector and column temperature settings were optimized for compounds with higher volatility, therefore the most volatile substances (retention time 4.2–6.3 min), such as terpenes, showed a low resolution and tailed peaks.

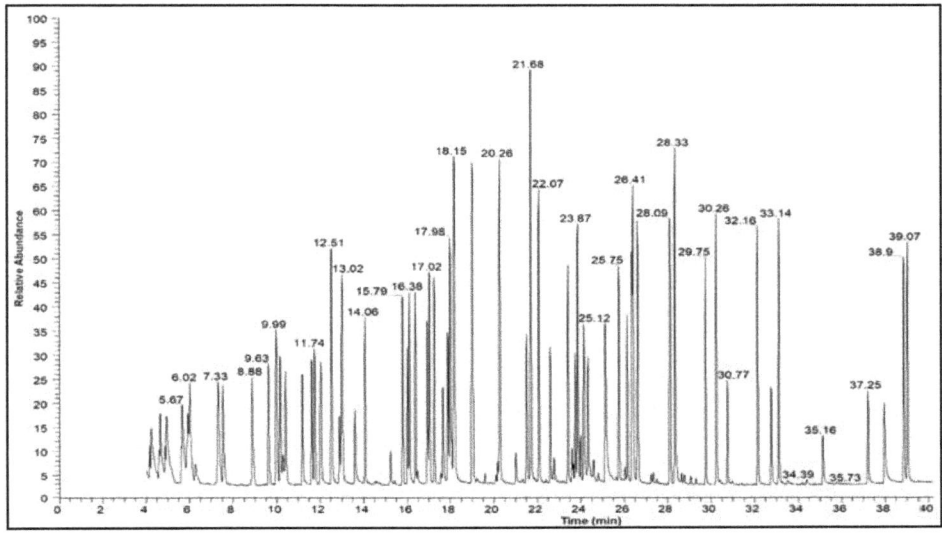

**Figure 1.** GC-MS full-scan chromatogram of the FS solution.

The curve correlation coefficients (R) were in the range 0.9999–0.9943 with the lower associated with eugenol and santalol.

The $LOD_{inst}$ values were between 0.001 ng $\mu L^{-1}$ and 0.02 ng $\mu L^{-1}$. For example, the S/N ratio was equal to 3:1 at a concentration of 0.001 ng $\mu L^{-1}$ for eugenyl acetate and benzyl benzoate and equal to 0.02 ng $\mu L^{-1}$ for ebanol. Table 2 reports the calibration curves, the respective values of R, and the $LOD_{inst}$ for a number of selected fragrances.

**Table 2.** Calibration curves, curve correlation coefficient R, and instrumental LOD of selected fragrances.

| Compound | Calibration Curve | R | $LOD_{inst}$ |
|---|---|---|---|
| Salicylaldehyde | Y = −0.0153816 + 1.12523 X | 0.9959 | 0.001 |
| Camphor | Y = −0.00116375 + 1.10579 X | 0.9997 | 0.001 |
| Folione | Y = −0.0066146 + 0.477204 X | 0.9997 | 0.004 |
| Neral | Y = −0.00723682 + 0.355397 X | 0.9998 | 0.006 |
| Carvone | Y = −0.0138041 + 0.775145 X | 0.9997 | 0.003 |
| Geranial | Y = −0.00914425 + 0.353934 X | 0.9994 | 0.007 |
| DMBCA | Y = −0.0171582 + 1.34708 X | 0.9998 | 0.001 |
| Geranyl acetate | Y = −0.00736214 + 0.312877 X | 0.9997 | 0.012 |
| β-Damascenone | Y = −0.0567008 + 2.59694 X | 0.9995 | 0.001 |
| δ-Damascone | Y = −0.0204197 + 0.950358 X | 0.9993 | 0.004 |
| β-Damascone | Y = −0.0483275 + 2.02308 X | 0.9993 | 0.002 |
| Coumarin | Y = −0.027222 + 1.94787 X | 0.9997 | 0.002 |
| α-Isomethylionone | Y = −0.0390646 + 1.8334 X | 0.9994 | 0.002 |
| Eugenyl acetate | Y = −0.0778768 + 3.53241 X | 0.9996 | 0.001 |
| 3-Propylidenephthalide | Y = −0.0430587 + 2.29359 X | 0.9996 | 0.001 |
| α-Amylcinnamaldehyde | Y = −0.0496146 + 1.74195 X | 0.9992 | 0.002 |
| ISO E® γ | Y = −0.00185945 + 0.122102 X | 0.9985 | 0.026 |
| Musk xylene | Y = −0.0189596 + 0.656847 X | 0.9988 | 0.005 |
| Musk ketone | Y = −0.0329199 + 0.791557 X | 0.9967 | 0.005 |
| Benzyl cinnamate | Y = −0.0813413 + 1.95804 X | 0.9954 | 0.002 |

### 3.1.2. Sampling and Extraction Optimization

Three adsorbent materials were tested, Amberlite XAD-2, Amberlite XAD-4, and Florisil. Elution tests showed that XAD-4 was the adsorbent most susceptible to the presence of impurities, even after the cleaning step. Furthermore, it was the most sensitive to degradation when treated with a polar solvent (acetone). As for the Florisil, it was discarded both for the strong resistance shown to the airflow during sampling and for the inhomogeneity and difficulty in packing the cartridges.

The most satisfactory results in terms of cartridge packing and blanks were obtained with XAD-2, which was chosen as the definitive adsorbent.

During the setup phase of the elution procedure, an attempt was made to perform a selective clean-up of the sample by eluting it with solvents of different polarities. For this purpose, elution tests were performed by spiking a known amount of FS solution on packaged cartridges. In one experiment n-hexane and acetone were used in sequence, in another trimethylpentane was followed by DCM. For this, 5.0 mL of the non-polar solvents were used, and subsequently, progressive additions of the more polar solvents were carried out. The eluates were collected separately to evaluate the partitioning of the analytes (10 mL each time). Table S3 reports the results obtained for selected fragrances. For both n-hexane and trimethylpentane, about 10% of the analytes were already eluted by the addition of little milliliters, consequently, it was decided to proceed using a single solvent and collect the elute in a single fraction. The recoveries showed by dichloromethane were, in all cases, lower than those obtained, with acetone. The differences were in the range of 10–60% (for geranyl acetate and musk xylene respectively), therefore acetone was chosen as the eluting solvent. These allow the minimization of the time of the evaporation phase and therefore of the losses of the more volatile target compounds. The total amount of solvent necessary to recover the fragrances was determined by spiking the XAD-2 cartridge with 1000 ng of the target compounds. Based on the results already obtained during the clean-up tests, 20, 5, 5, and 5 mL subsequent additions of acetone were made, and the different fractions were collected separately. After the addition of the IS solution, the samples were evaporated at 500 µL and the SS solution was added before GC-MS analysis. The recovery of analytes in each fraction was evaluated and the final volume of acetone for elution was set at 30 mL.

### 3.1.3. Blank Evaluation

The blank of the method was evaluated by analyzing pre-cleaned XAD-2 (Soxhlet extraction with toluene for 24 h) according to the procedure developed. In order to monitor possible interference, the blank check was repeated at time intervals during the study. In the used operative condition, benzyl alcohol showed the worst blank corresponding to values higher than 100 ng m$^{-3}$. Variable blanks were observed for camphor and ebanol with values in the ranges of 0–26 ng m$^{-3}$ and 0–54 ng m$^{-3}$ respectively. For all the other compounds, the blanks were always lower than 4 ng m$^{-3}$.

### 3.1.4. Effectiveness of the Method

Little information exists about the levels of fragrances in the air, but the results reported are characterized by wide variability [19,22].

For this reason, the evaluation of the method efficiency was carried out by spiking the sampling cartridges with amounts of analytes equal to 100 ng, 3000 ng, and 7500 ng and 100 ng of internal standards (each level in triplicate). After carrying out all the analytical steps (addition of the IS solution, elution, evaporation, addition of the SS solution, and instrumental analysis), recoveries (R%) were evaluated as the ratio between the amount of substance detected through the analysis and that effectively added. Table S4 reports, for all the fragrances investigated, the average values of the R% for each level with the respective standard deviation (SD). They can be considered representative of the precision and accuracy of the method.

As expected, terpenes recoveries were less than 55% on all the three levels, indeed both extraction and instrumental method used are not suitable for these classes of substances [26,27]. Concerning alcohols, good results were not achieved principally due to the chromatographic column which was not suitable for this class of polar compounds

Except for very volatile fragrances or those belonging to the alcohol class, the standard deviations associated with percentage recoveries on individual levels were generally good, even when the tests were carried out on different days. On the contrary, the percentage recoveries on the three spike levels were different for a lot of compounds, and in general, grew with the amount of standard added. This could be due to a higher retention capability of XAD-2 versus analytes at low concentrations.

Since there are no guidelines or standardized methods for the analysis of fragrances, there are no acceptance criteria to refer to in order to evaluate the validity of the developed analytical method.

A range of R% and a maximum SD was established as a criterion for considering the method valid for each compound. Only compounds whose average R% is between 65% and 115% and whose differences between minimum and maximum R% evaluated for all tests did not differ by more than 30% can be quantified. Of the 66 analytes, 20 met these criteria and are reported in Table 3.

**Table 3.** Percentage Recovery (%R) and Standard Deviation (SD) of the method at three levels of native compounds.

| Compound | Level 1 (100 ng) | Level 2 (3000 ng) | Level 3 (7500 ng) |
|---|---|---|---|
| | %R (SD) | %R (SD) | %R (SD) |
| Salicylaldehyde | 77.1 (8) | 92.7 (7) | 83.4 (14) |
| Camphor | 82.7 (7) | 77.9 (1) | 86.9 (8) |
| Folione | 72.8 (4) | 88.7 (3) | 97.7 (0) |
| Neral | 71.4 (7) | 86.9 (4) | 92.2 (1) |
| Carvone | 64.6 (9) | 75.7 (3) | 93.7 (3) |
| Geranial | 66.7 (6) | 85.1 (7) | 91.9 (4) |
| DMBCA | 74.8 (2) | 82 (6) | 92.5 (5) |
| Geranyl acetate | 66.9 (1) | 70.4 (5) | 84.3 (3) |
| ß-Damascenone | 69.3 (1) | 66.2 (5) | 86.2 (6) |
| δ-Damascone | 70.4 (0) | 68.7 (4) | 86.3 (6) |
| ß-Damascone | 72.4 (1) | 73.0 (5) | 91.3 (6) |
| Coumarin | 67.1 (5) | 70.7 (5) | 86.4 (4) |
| a-Isomethylionone | 79.6 (1) | 70.5 (4) | 89.8 (6) |
| Eugenyl acetate | 75.4 (1) | 81.2 (4) | 88.9 (5) |
| 3-Propylidenephthalide | 69.1 (0) | 77.9 (2) | 83.3 (3) |
| a-Amylcinnamaldehyde | 70.9 (1) | 85.1 (3) | 91.4 (4) |
| ISO E® γ | 78.3 (3) | 83.1 (4) | 92.2 (3) |
| Musk xylene | 90 (5) | 90.5 (11) | 92.4 (6) |
| Musk ketone | 65.4 (1) | 76.9 (5) | 73.6 (3) |
| Benzyl cinnamate | 79.3 (5) | 88.1 (5) | 115 (8) |

Finally, for each of the 20 selected fragrances, the LOD of the sample ($LOD_{sample}$) was calculated considering the final extract volume (500 µL), the sample size ($\approx$14.4 m$^3$ of air), and the minus percentage recovery of the respective internal standard (%$IS_{rec}$) found in the method validation phase (evaluated respect to the syringe standard) [23,29]. The used formula is reported below.

$$LOD_{sample} = [(LOD_{inst} \times Final\ extraction\ volume)/Sample\ Volume] \times \%IS_{rec}$$

The *LOQ* of the samples was calculated as the *LOD* of the sample and it was in the range of 0.07 ng m$^{-3}$ to 0.48 ng m$^{-3}$.

### 3.1.5. Breakthrough Evaluation

In order to ensure the representativeness of the samples, experiments were carried out to detect whether analyte leaks from the adsorbent cartridge occurred under the established sampling conditions. It was therefore verified that fragrances did not undergo breakthrough phenomena [30]. For this purpose, a sampling train was set up consisting of a zero air filter of ACF (Activated Carbon Fiber) [31,32], followed by a first cartridge containing XAD-2 marked with native compounds, and downstream a second XAD-2 cartridge. With the aim of maximizing the evidence of leaks, the first cartridge was labeled with a high amount of native compounds (7500 ng of the FS solution). To ensure that the ACF filter blocked all incoming analytes and worked as a zero filter, fluorene-d10 was spiked on it and checked for its absence downstream of the sampling train (the two XAD-2 cartridges). The experiment was performed in duplicate. After sampling, each adsorbent of the sampling train was processed and analyzed separately. The results showed no trace of fluorene-d10 in either of the two XAD-2 cartridges and the absence of breakthrough for most of the compounds. Among compounds considered suitable for quantitative analysis, the highest leaks were associated with salicylaldehyde and camphor and corresponded to 13.5% and 5.9%, respectively. Except for musk ketone, whose losses were equal to 1.7%, values always lower than 1% were observed for the other fragrances. The results, for all the investigated analytes, are reported in Table S5.

### 3.2. Concentrations of Fragrances in the Real Samples

Once validated, the developed method was tested to collect real samples in two different indoor environments, that is, a house and a coffee bar. Figures 2 and 3 show the SIM-B chromatograms of samples collected at the sites. In each case, the total ion chromatogram (TIC) is reported together with selected ion chromatograms for the visualization of specific target compounds. Table 4 reports the aerial concentrations of the 20 fragrances suitable for quantitative determination. Almost all fragrances were detected with the only exception of folione and ISO E® γ. The highest concentrations were observed for carvone in the coffee bar (349 ng m$^{-3}$) and camphor in the house (157 ng m$^{-3}$). Carvone is a terpenoid and it is the main component of the essential oil of various species of mint. It is extensively used in perfumery, as a perfuming agent in cosmetics, and as a flavor in various foods [33]. Its presence in the coffee bar could depend on the presence of mint-flavored food products such as chewing gum, which are included in the products sold at this store. The second most concentrated compound in the bar was neral.

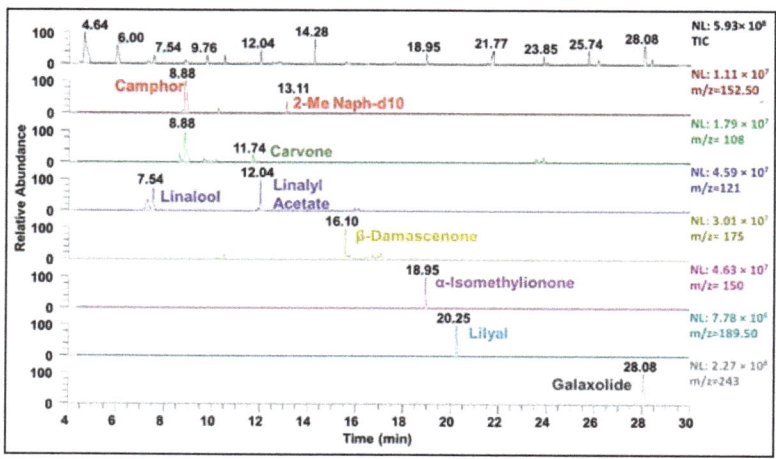

**Figure 2.** SIM-B chromatogram of the sample collected in the house.

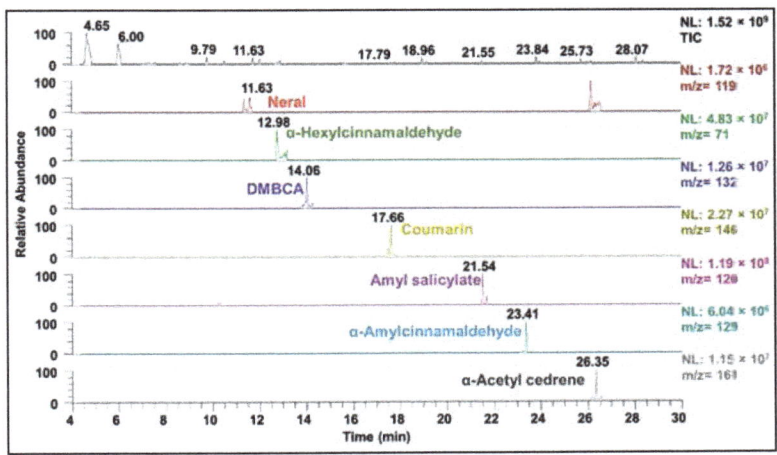

**Figure 3.** SIM-B chromatogram of the sample collected in the coffee bar.

**Table 4.** Concentration (ng m$^{-3}$) of fragrances collected in the house and in the coffee bar.

| Compound | House | Coffee Bar | Compound | House | Coffee Bar |
| --- | --- | --- | --- | --- | --- |
| | (ng m$^{-3}$) | (ng m$^{-3}$) | | (ng m$^{-3}$) | (ng m$^{-3}$) |
| Salicylaldehyde | 54.4 ± 4.3 | 32 ± 2.6 | ß-Damascone | ND [1] | 0.14 ± 0.01 |
| Camphor | 157 ± 11.0 | 75.7 ± 5.3 | Coumarin | 29.6 ± 1.5 | 39.1 ± 1.9 |
| Folione | ND | ND | α-Isomethylionone | 63.2 ± 0.6 | 74.1 ± 0.7 |
| Neral | 46.8 ± 1.8 | 162 ± 6.5 | Eugenyl acetate | 0.45 ± 0.01 | 0.86 ± 0.01 |
| Carvone | 39.5 ± 3.2 | 349 ± 6.5 | 3-Propylidenephthalide | 0.68 ± 0.01 | 1.04 ± 0.01 |
| Geranial | 39.8 ± 2.4 | 19.4 ± 1.2 | α-Amylcinnamaldehyde | 8.33 ± 0.08 | 19.1 ± 0.2 |
| DMBCA | 17.2 ± 0.3 | 17.1 ± 0.3 | ISO E® γ | ND | ND |
| Geranyl acetate | 18.1 ± 0.2 | 17.2 ± 0.2 | Benzyl cinnamate | 0.97 ± 0.04 | 0.96 ± 0.04 |
| ß-Damascenone | 7.85 ± 0.01 | ND | Musk xylene | 2.38 ± 0.02 | 1.32 ± 0.01 |
| δ-Damascone | ND | 1.03 ± 0.07 | Musk ketone | 1.25 ± 0.06 | 1.54 ± 0.07 |

[1] ND not detected.

Neral is an unsaturated aldehyde and together with the geometrical isomer geranial constitutes the citral, which is one of the most widely used flavoring compounds in foods and beverages due to its intense lemon aroma/flavor [34]. Camphor is used in incense and insecticides [35,36]. In pharmacology, it is largely used in topical preparations due to its mild local anesthetizing effect and the production of a circumscribed sensation of heat [37]. In the house, its presence could be connected to the use of mosquito repellent, in fact, also the presence of N, N-diethyl-meta-toluamide (DEET, an insect repellent too) was determined. Concerning endocrine disruptors, α-isomethylionone, was detected in both sites and was the second most concentrated substance in the house (63.2 ng m$^{-3}$), musk xylene and musk ketone were present at lower concentrations (≈1 or 2 ng m$^{-3}$). Finally, a semi-quantitative determination was made for fragrances whose method accuracy and precision were not good enough to allow a quantitative analysis. The measured concentration values are shown in Table S6, but the uncertainty associated with the measurements must be considered (see Table S4). In the bar, the menthol had a concentration certainly higher than 1000 ng m$^{-3}$. Among confirmed or suspected endocrine disruptors, lylial and galaxolide 1 and 2 were detected in both environments at concentrations of approximately 10 ng m$^{-3}$. Concentration values in the order of 100–200 ng m$^{-3}$ were instead observed for benzyl salicylate.

The indoor concentrations of musks observed in this study are comparable with those reported in the literature. According to Ramirez et al. [22], in a study conducted in a chemical laboratory, an office, a medical center, a pharmacy, a hairdresser's shop, and, a flower shop, musk xylene was detected in the range of 2.9–766 ng m$^{-3}$ (lower concentration flower shop, maximum concentration hairdresser), musk ketone was in the range of 1.9–68.5 ng m$^{-3}$ (pharmacy and hairdresser respectively), and galaxolide was in the range of 47.1–1256 ng m$^{-3}$ (min. pharmacy and max hairdresser). In Spain, Reguerio et al. [21] detected aerial galaxolide at 57 ng m$^{-3}$ in a rest facility, while musks were always lower than LOD in all the ambient investigated (LOD ≈ 0.1 ng m$^{-3}$).

According to our knowledge, some fragrances have been measured in indoor environments for the first time in this study. Some others have been investigated in Spain by Lamas et al. [19,20] under normal daily conditions or in environments treated with aerosols diffusion units and different common cleaning products. In the normal conditions recorded in different home places (washroom, laundry room, corridor, living room, kitchen, bedroom, and storage room), the concentrations were comparable with those measured in the present study. Citral was often lower than the LOD (1.9 ng m$^{-3}$), it was detected in only two cases (living room and kitchen) and it was at levels higher than 440 ng m$^{-3}$. α-isomethylionone, lylial, and benzyl salicylate were in the range of 15–765 ng m$^{-3}$, 116–1090 ng m$^{-3}$, and 14–18 ng m$^{-3}$, respectively. Only α-isomethylionone was detected in all samples and its highest levels were reached in the kitchen, while the highest concentrations of lilyal were measured in a washroom and that of benzyl salicylate in a storage room. Concerning menthol and camphor, their presence has been detected but not measured by Cobo-Golpe and coworkers through GG-high resolution mass spectrometry (GC-HRMS) in portable dehumidifiers condensed water [38].

## 4. Conclusions

A method for the determination of gas-phase fragrances indoors was developed. The applied sampling system is constituted by a XAD-2 cartridge sampling 24 h, at a flow rate of 10 L min$^{-1}$. 66 fragrances were tested and for 20 analytes, accuracy and precision tested in triplicate on three levels of concentration were considered good enough to allow quantitative determination (% absolute recovery in the range 72% ± 9%–92% ± 13%). For much of the other compounds, the effectiveness of the method was good enough to gain information on the order of concentration in the monitored sites. The applicability of the technique in real samples was tested in a house and a coffee bar. Almost all fragrances were detected, and results showed that the method is sensitive enough to reveal fragrance concentrations at levels of ng m$^{-3}$ in real samples. This method can be used in the future for the indoor simultaneous collection of gaseous and particulate fragrances together with other classes.

**Supplementary Materials:** The following supporting information can be downloaded at: https://www.mdpi.com/article/10.3390/separations9040099/s1, Table S1. Compounds detected through SIM-A; Table S2. Compounds detected through SIM-B; Table S3. Percent recovery of the standards during the tests for the extraction procedure setting. Table S4. Percentage Recovery (%R) and Standard Deviation (SD) of the method at three levels of native compounds; Table S5. % of ng lost during the sampling for Breakthrough evaluation; Table S6. Results of the semi-quantitative determination of fragrances in the house and in the coffee bar.

**Author Contributions:** Conceptualization, C.B. and M.G.; methodology, P.A. and A.B.; validation, M.G. and M.C.; resources, P.C.; writing—original draft preparation, C.B., M.G.; writing—review and editing, P.A., M.C., A.B. and P.C. All authors have read and agreed to the published version of the manuscript.

**Funding:** This research received no external funding.

**Institutional Review Board Statement:** Not applicable.

**Data Availability Statement:** All data are available from the corresponding author upon request.

**Acknowledgments:** We acknowledge Mattia Perilli for technical support.

**Conflicts of Interest:** The authors declare no conflict of interest.

## References

1. Settimo, G.; Manigrasso, M.; Avino, P. Indoor air quality: A focus on the european legislation and state-of-the-art research in Italy. *Atmosphere* **2020**, *11*, 370. [CrossRef]
2. Maroni, M.; Axelrad, R.; Bacaloni, A. NATO's efforts to set indoor air quality guidelines and standards. *Am. Ind Hyg. Assoc. J.* **1995**, *56*, 499–508. [CrossRef]
3. Takaoka, M.; Norbäck, D. The Indoor Environment in Schools, Kindergartens and Day Care Centres. In *Indoor Environmental Quality and Health Risk toward Healthier Environment for All*; Springer: Berlin/Heidelberg, Germany, 2020; ISBN 9789813291812.
4. Baloch, R.M.; Maesano, C.N.; Christoffersen, J.; Banerjee, S.; Gabriel, M.; Csobod, É.; de Oliveira Fernandes, E.; Annesi-Maesano, I.; Szuppinger, P.; Prokai, R.; et al. Indoor air pollution, physical and comfort parameters related to schoolchildren's health: Data from the European SINPHONIE study. *Sci. Total Environ.* **2020**, *739*, 139870. [CrossRef] [PubMed]
5. Nehr, S.; Hösen, E.; Tanabe, S. Ichi Emerging developments in the standardized chemical characterization of indoor air quality. *Environ. Int.* **2017**, *98*, 233–237. [CrossRef] [PubMed]
6. González-Martín, J.; Kraakman, N.J.R.; Pérez, C.; Lebrero, R.; Muñoz, R. A state–of–the-art review on indoor air pollution and strategies for indoor air pollution control. *Chemosphere* **2021**, *262*, 128376. [CrossRef] [PubMed]
7. Wolkoff, P.; Nielsen, G.D. Effects by inhalation of abundant fragrances in indoor air—An overview. *Environ. Int.* **2017**, *101*, 96–107. [CrossRef] [PubMed]
8. Basketter, D.; Kimber, I. Fragrance sensitisers: Is inhalation an allergy risk? *Regul. Toxicol. Pharmacol.* **2015**, *73*, 897–902. [CrossRef]
9. Nawaz, T.; Sengupta, S. Chapter 4—Contaminants of Emerging Concern: Occurrence, Fate, and Remediation. In *Advances in Water Purification Techniques*; Elsevier: Amsterdam, The Netherlands, 2018; ISBN 9780128147917.
10. Yadav, D.; Rangabhashiyam, S.; Verma, P.; Singh, P.; Devi, P.; Kumar, P.; Hussain, C.M.; Gaurav, G.K.; Kumar, K.S. Environmental and health impacts of contaminants of emerging concerns: Recent treatment challenges and approaches. *Chemosphere* **2021**, *272*, 129492. [CrossRef]
11. Enyoh, C.E.; Verla, A.W.; Qingyue, W.; Ohiagu, F.O.; Chowdhury, A.H.; Enyoh, E.C.; Chowdhury, T.; Verla, E.N.; Chinwendu, U.P. An overview of emerging pollutants in air: Method of analysis and potential public health concern from human environmental exposure. *Trends Environ. Anal. Chem.* **2020**, *28*, e00107. [CrossRef]
12. Christensson, J.B.; Hagvall, L.; Karlberg, A.T. Fragrance allergens, overview with a focus on recent developments and understanding of abiotic and biotic activation. *Cosmetics* **2016**, *3*, 19. [CrossRef]
13. Bickers, D.R.; Calow, P.; Greim, H.A.; Hanifin, J.M.; Rogers, A.E.; Saurat, J.H.; Sipes, I.G.; Smith, R.L.; Tagami, H. The safety assessment of fragrance materials. *Regul. Toxicol. Pharmacol.* **2003**, *37*, 218–273. [CrossRef]
14. Patel, S.; Homaei, A.; Sharifian, S. Need of the hour: To raise awareness on vicious fragrances and synthetic musks. *Environ. Dev. Sustain.* **2020**, *23* (Suppl. 3), 4764–4781. [CrossRef]
15. Dodson, R.E.; Nishioka, M.; Standley, L.J.; Perovich, L.J.; Brody, J.G.; Rudel, R.A. Endocrine disruptors and asthma-associated chemicals in consumer products. *Environ. Health Perspect.* **2012**, *120*, 935–943. [CrossRef] [PubMed]
16. Wieck, S.; Olsson, O.; Kümmerer, K.; Klaschka, U. Fragrance allergens in household detergents. *Regul. Toxicol. Pharmacol.* **2018**, *97*, 163–169. [CrossRef]
17. Chen, J.; Møller, K.H.; Wennberg, P.O.; Kjaergaard, H.G. Unimolecular Reactions following Indoor and Outdoor Limonene Ozonolysis. *J. Phys. Chem. A* **2021**, *125*, 669–680. [CrossRef]
18. Nørgaard, A.W.; Kofoed-Sørensen, V.; Mandin, C.; Ventura, G.; Mabilia, R.; Perreca, E.; Cattaneo, A.; Spinazzè, A.; Mihucz, V.G.; Szigeti, T.; et al. Ozone-initiated terpene reaction products in five European offices: Replacement of a floor cleaning agent. *Environ. Sci. Technol.* **2014**, *48*, 13331–13339. [CrossRef]
19. Lamas, J.P.; Sanchez-Prado, L.; Garcia-Jares, C.; Llompart, M. Determination of fragrance allergens in indoor air by active sampling followed by ultrasound-assisted solvent extraction and gas chromatography-mass spectrometry. *J. Chromatogr. A* **2010**, *1217*, 1882–1890. [CrossRef]
20. Lamas, J.P.; Sanchez-Prado, L.; Lores, M.; Garcia-Jares, C.; Llompart, M. Sorbent trapping solid-phase microextraction of fragrance allergens in indoor air. *J. Chromatogr. A* **2010**, *1217*, 5307–5316. [CrossRef]
21. Regueiro, J.; Garcia-Jares, C.; Llompart, M.; Lamas, J.P.; Cela, R. Development of a method based on sorbent trapping followed by solid-phase microextraction for the determination of synthetic musks in indoor air. *J. Chromatogr. A* **2009**, *1216*, 2805–2815. [CrossRef]
22. Ramírez, N.; Marcé, R.M.; Borrull, F. Development of a thermal desorption-gas chromatography-mass spectrometry method for determining personal care products in air. *J. Chromatogr. A* **2010**, *1217*, 4430–4438. [CrossRef]
23. Balci, E.; Genisoglu, M.; Sofuoglu, S.C.; Sofuoglu, A. Indoor air partitioning of Synthetic Musk Compounds: Gas, particulate matter, house dust, and window film. *Sci. Total Environ.* **2020**, *729*, 138798. [CrossRef] [PubMed]
24. Fontal, M.; van Drooge, B.L.; Grimalt, J.O. A rapid method for the analysis of methyl dihydrojasmonate and galaxolide in indoor and outdoor air particulate matter. *J. Chromatogr. A* **2016**, *1447*, 135–140. [CrossRef] [PubMed]
25. Van Drooge, B.L.; Rivas, I.; Querol, X.; Sunyer, J. Organic Air Quality Markers of Indoor and Outdoor PM 2. 5 Aerosols in Primary Schools from Barcelona. *Int. J. Environ. Res. Public Health* **2020**, *17*, 3685. [CrossRef] [PubMed]

26. Kruza, M.; McFiggans, G.; Waring, M.S.; Wells, J.R.; Carslaw, N. Indoor secondary organic aerosols: Towards an improved representation of their formation and composition in models. *Atmos. Environ.* **2020**, *240*, 117784. [CrossRef]
27. Angulo-Milhem, S.; Verriele, M.; Nicolas, M.; Thevenet, F. Indoor use of essential oils: Emission rates, exposure time and impact on air quality. *Atmos. Environ.* **2021**, *244*, 117863. [CrossRef]
28. Bokowa, A.; Diaz, C.; Koziel, J.A.; Mcginley, M.; Barclay, J.; Guillot, J.; Sneath, R.; Capelli, L.; Zorich, V. Summary and Evaluation of the Odour Regulations Worldwide. *Atmosphere* **2021**, *12*, 206. [CrossRef]
29. Delgado-Saborit, J.M.; Aquilina, N.; Baker, S.; Harrad, S.; Meddings, C.; Harrison, R.M. Determination of atmospheric particulate-phase polycyclic aromatic hydrocarbons from low volume air samples. *Anal. Methods* **2010**, *2*, 231–242. [CrossRef]
30. Hayward, S.J.; Lei, Y.D.; Wania, F. Sorption of a diverse set of organic chemical vapors onto XAD-2 resin: Measurement, prediction and implications for air sampling. *Atmos. Environ.* **2011**, *45*, 296–302. [CrossRef]
31. Balducci, C.; Cecinato, A.; Paolini, V.; Guerriero, E.; Perilli, M.; Romagnoli, P.; Tortorella, C.; Nacci, R.M.; Giove, A.; Febo, A. Volatilization and oxidative artifacts of PM bound PAHs at low volume sampling (2): Evaluation and comparison of mitigation strategies effects. *Chemosphere* **2017**, *189*, 330–339. [CrossRef]
32. Cerasa, M.; Guerriero, E.; Mosca, S. Evaluation of Extraction Procedure of PCDD/Fs, PCBs and Chlorobenzenes from Activated Carbon Fibers (ACFs). *Molecules* **2021**, *26*, 6407. [CrossRef]
33. Morcia, C.; Tumino, G.; Ghizzoni, R.; Terzi, V. Carvone (*Mentha spicata* L.) oils. In *Essential Oils in Food Preservation, Flavor and Safety*; Elsevier: Amsterdam, The Netherlands, 2016; ISBN 9780124166448.
34. Mercer, D.G.; Rodriguez-Amaya, D.B. Reactions and interactions of some food additives. In *Chemical Changes During Processing and Storage of Foods*; Elsevier: Amsterdam, The Netherlands, 2021; ISBN 9780128173800.
35. Kholibrina, C.R.; Aswandi, A. The Consumer Preferences for New Sumatran Camphor Essential Oil-based Products using a Conjoint Analysis Approach. *IOP Conf. Ser. Earth Environ. Sci.* **2021**, *715*, 012078. [CrossRef]
36. Cansian, R.L.; Astolfi, V.; Cardoso, R.I.; Paroul, N.; Roman, S.S.; Mielniczki-Pereira, A.A.; Pauletti, G.F.; Mossi, A.J. Atividade inseticida e repelente do óleo essencial de Cinnamomum camphora var. linaloolifera Y. Fujita (Ho-Sho) e Cinnamomum camphora (L.) J Presl. var. hosyo (Hon-Sho) sobre Sitophilus zeamais Mots. (Coleoptera, Curculionedae). *Rev. Bras. Plantas Med.* **2015**, *17*, 769–773. [CrossRef]
37. Zuccarini, P.; Soldani, G. Camphor: Benefits and risks of a widely used natural product. *Acta Biol. Szeged.* **2009**, *53*, 77–82. [CrossRef]
38. Cobo-Golpe, M.; Ramil, M.; Cela, R.; Rodríguez, I. Portable dehumidifiers condensed water: A novel matrix for the screening of semi-volatile compounds in indoor air. *Chemosphere* **2020**, *251*, 126346. [CrossRef] [PubMed]

Article

# Facile Preparation of Phenyboronic-Acid-Functionalized Fe₃O₄ Magnetic Nanoparticles for the Selective Adsorption of Ortho-Dihydroxy-Containing Compounds

Hongmei Zhou, Junhui Zhang, Aihong Duan, Bangjin Wang *, Shengming Xie * and Liming Yuan

Department of Chemistry, Yunnan Normal University, Kunming 650500, China
* Correspondence: wangbangjin711@163.com or wangbangjin@ynnu.edu.cn (B.W.); xieshengming_2006@163.com (S.X.)

**Abstract:** A new facile strategy was designed to prepare the phenyboronic acid-functionalized $Fe_3O_4$ magnetic nanoparticles ($Fe_3O_4$@PBA) via direct silanization and thiol-ene click chemistry for the selective adsorption of ortho-dihydroxy-containing compounds. The three kinds of $Fe_3O_4$@PBA nanoparticles obtained showed excellent adsorption capacity and selectivity for ortho-dihydroxy-containing compounds including adenosine and o-dihydroxybenzene. Among them, the $Fe_3O_4$@MPS@MPBA exhibited the highest adsorption capacity and selectivity for adenosine and o-dihydroxybenzene, followed by $Fe_3O_4$@MPTES@AAPBA and $Fe_3O_4$@MPTES@VPBA. A synthesis method of superparamagnetic and boronate affinity nanocomposites with mild reaction conditions and simple process has been developed, which also provides a novel way for the synthesis of other types of enrichment materials of ortho-dihydroxy-containing compounds.

**Keywords:** phenyboronic-acid-functionalized; $Fe_3O_4$ magnetic nanoparticles; ortho-dihydroxy-containing compounds

**Citation:** Zhou, H.; Zhang, J.; Duan, A.; Wang, B.; Xie, S.; Yuan, L. Facile Preparation of Phenyboronic-Acid-Functionalized Fe₃O₄ Magnetic Nanoparticles for the Selective Adsorption of Ortho-Dihydroxy-Containing Compounds. *Separations* **2023**, *10*, 4. https://doi.org/10.3390/separations 10010004

Academic Editor: Pavel Nikolaevich Nesterenko

Received: 29 October 2022
Revised: 27 November 2022
Accepted: 13 December 2022
Published: 21 December 2022

**Copyright:** © 2022 by the authors. Licensee MDPI, Basel, Switzerland. This article is an open access article distributed under the terms and conditions of the Creative Commons Attribution (CC BY) license (https:// creativecommons.org/licenses/by/ 4.0/).

## 1. Introduction

Due to the advantages of superparamagnetism, stability and low cytotoxicity [1–3], magnetic nanomaterials are widely used in various fields, such as pollutant detection and separation [4–7], drug delivery [8–10], magnetic resonance imaging [11–13] and biosensing [14–16]. Such a wide range of applications is attributed to the various functionalizations of magnetic nanomaterials, and phenylboric acid is one of the important functional modification groups. Phenylboric acid can selectively adsorb ortho-dihydroxy-containing compounds through the formation of reversible five- or six-membered cycloesters. The adsorption and desorption can be conveniently controlled by adjusting the pH value, which gives boric acid affinity materials incomparable advantages in the recognition and separation of biomolecules [17–25]. Accordingly, it has become a challenge to develop a facile modification strategy for the preparation of magnetic nanomaterials modified by phenylboric acid. Click chemistry has become a powerful means to solve the above challenge because of its high reaction efficiency, high reliability and good selectivity [26–29].

In this study, the direct silylation and thiol-ene click reaction were used for functionalized modification of $Fe_3O_4$ magnetic nanoparticles. The adsorption capacity and selectivity of the three kinds of prepared phenyboronic-acid-functionalized $Fe_3O_4$ magnetic nanoparticles ($Fe_3O_4$@PBA) toward ortho-dihydroxy-containing compounds were evaluated by adsorption experiments of adenosine and o-dihydroxybenzene.

## 2. Materials and Methods

### 2.1. Reagents and Material

Ferric trichloride hexahydrate ($FeCl_3·6H_2O$), ammonium hydrogen carbonate, sodium chloride, ethylene glycol, anhydrous sodium acetate, glacial acetic acid, methanol, ethanol and aqueous ammonia (25 wt%) were obtained from FengChuan Fine Chemical Research Institute (Tianjin, China). Adenosine, deoxyadenosine, azodiisobutyronitrile (AIBN), 3-mercaptopropyltrimethoxysilane (MPTES) and 4-vinylphenylboronic acid (VPBA) were obtained from Adamas-beta (Basel, Switzerland). 3-methacryloxypropyltrimethoxysilane (MPS) and 3-acrylamidophenylboronic acid (AAPBA) were purchased from Sigma-Aldrich (St. Louis, MO, USA). 4-Mercaptophenylboronic acid (MPBA), $o$-dihydroxybenzene, $m$-dihydroxybenzene, and $p$-dihydroxybenzene were obtained from Aladdin (Shanghai, China). Deionised water was used to prepare all buffer and analyte solutions.

### 2.2. Synthesis of $Fe_3O_4$@PBA Nanoparticles

The general scheme for the synthesis of magnetic nanoparticles modified by phenylboronic acid ($Fe_3O_4$@PBA) is illustrated in Figure 1. The $Fe_3O_4$ nanoparticles was synthesized by a previously reported solvothermal method [30]. The $Fe_3O_4$ nanoparticles were functionalized with the sulfydryl and vinyl groups through the direct silanization using MPTES and MPS [31]. Subsequently, the functionalization of the nanoparticles based on phenylboric acid was achieved by the thiol-ene click reaction using 4-vinylphenylboronic acid (VPBA), 3-acrylamidophenylboronic acid (AAPBA) and 4-mercaptophenylboronic acid (MPBA) [32].

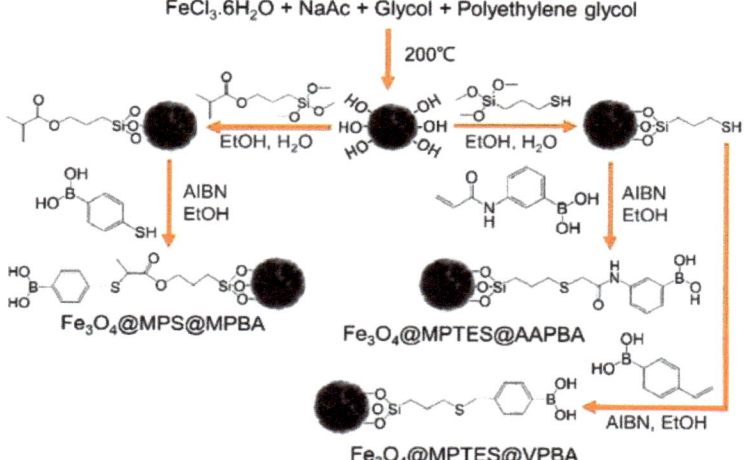

**Figure 1.** Schematic illustration of synthesis of phenylboronic-acid-modified magnetic nanoparticles.

#### 2.2.1. Synthesis of $Fe_3O_4$ Nanoparticles

A solution containing anhydrous sodium acetate (7.2 g), polyethylene glycol (2.0 g), $FeCl_3 \cdot 6H_2O$ (2.7 g) and ethylene glycol (80.0 mL) was stirred at 50 °C for 15 min. Then the solution was transferred to a PTFE reactor and heated at 200 °C for 8 h. The products were collected with magnets and washed repeatedly using deionized water and ethanol, respectively. Finally, the products were dried overnight in a vacuum at 50 °C.

#### 2.2.2. Synthesis of $Fe_3O_4$@MPTES

The $Fe_3O_4$ nanoparticles (200.0 mg), ethanol (25.0 mL) and deionized water (20.0 mL) were added into a 250 mL two-neck round-bottom flask, and the mixture was ultrasonically dispersed for 30 min. Then, 1.0 mL of glacial acetic acid was slowly added under mechanical agitation (350 rpm) to adjust the pH value of the solution to about 5, and stirring continued for 30 min. Subsequently, the direct silanization reaction was initiated by the dropwise addition of 5.0 mL MPTES, which was maintained in a water bath at 50 °C for 24 h. The products were separated with magnets, and washed with deionized water and ethanol for several times. Then, the products were dried overnight at 50 °C to obtain Fe$_3$O$_4$@MPTES nanoparticles.

2.2.3. Synthesis of Fe$_3$O$_4$@MPS

The Fe$_3$O$_4$ nanoparticles (200.0 mg), ethanol (40.0 mL), deionized water (10.0 mL) and ammonium hydroxide (1.5 mL) were added into a 250 mL two-neck round-bottom flask, and the mixture was ultrasonically dispersed for 30 min. Whereafter, 500 µL of MPS was dropwise added under mechanical agitation (350 rpm) to initiate the direct silanization reaction. After reacting at 50 °C for 24 h, the obtained Fe$_3$O$_4$@MPS was repeatedly washed using deionized water and ethanol and dried overnight at 50 °C.

2.2.4. Phenyboronic-Acid-Functionalized Fe$_3$O$_4$@MPS and Fe$_3$O$_4$@MPTES

200.0 mg of Fe$_3$O$_4$@MPS or Fe$_3$O$_4$@MPTES was ultrasonically dispersed in a solution containing 50 mL of ethanol and 200.0 mg of MPBA (400.0 mg of AAPBA or 200.0 mg of VPBA for Fe$_3$O$_4$@MPTES) for 30 min. Then the thiol-ene click reaction was initiated by the addition of AIBN (20.0) mg under nitrogen protection, mechanical stirring (400 rpm) and 50 °C for 24 h. The products were separated with magnets, and washed using deionized water and ethanol for several times. The vacuum drying process was maintained at 50 °C for 12 h. Finally, the obtained products were named as Fe$_3$O$_4$@MPTES@VPBA, Fe$_3$O$_4$@MPTES@AAPBA and Fe$_3$O$_4$@MPS@MPBA, respectively.

2.3. Binding Experiments

A static adsorption method was used to evaluate the adsorption capacity of the three kinds of Fe$_3$O$_4$@PBA nanoparticles: Briefly, Fe$_3$O$_4$@PBA (2.0 mg) was dispersed by ultrasound in test solution (0.2 mL) with different concentrations (0.1~1.0 mg/mL) which was prepared with an ammonium bicarbonate (50 mM, pH = 8.5, containing 500 mM NaCl) as buffer solution, and shaken at room temperature (1000 rpm) for 20~140 min. The concentration of the adsorbate in the test solution after adsorption was measured by UV–Vis spectrophotometry. Then the adsorption capacity ($Q_e$) was calculated according to the equation below:

$$Q_e = \frac{V(C_0 - C_e)}{m} \times 1000 \qquad (1)$$

where $C_0$ (mg/mL) is the initial concentration of the adsorbate in the test solution; $C_e$ (mg/mL) is the equilibrium concentration of the adsorbate; $V$ (mL) is the volume of the adsorbate solution; $m$ (mg) is the mass of Fe$_3$O$_4$@PBA. Contrast adsorption experiments were also performed with Fe$_3$O$_4$@MPTES and Fe$_3$O$_4$@MPS under the same conditions.

The Scatchard equation was employed to investigate the binding properties of the Fe$_3$O$_4$@PBA nanoparticles:

$$\frac{Q}{C} = \frac{(Q_{max} - Q)}{K_D} \qquad (2)$$

where $Q$ (mg/g) is the equilibrium adsorption capacity of the material to the adsorbate, $C$ (mg/mL) is the adsorbate concentration remaining in the test solution after adsorption equilibrium, $Q_{max}$ (mg/g) is the maximum apparent binding amount, $K_D$ is the equilibrium dissociation constant.

2.4. Selectivity Experiments

The selective adsorption capacity of Fe$_3$O$_4$@PBA nanoparticles can be investigated by adsorption experiments on mixed solutions of two or more adsorbates with similar structure but different in the presence of cis-diol. Two mixed solutions (1.0 mg/mL) with equal mass ratios of adsorbates were used in the selective adsorption experiments of Fe$_3$O$_4$@PBA nanoparticles, including adenosine and deoxyadenosine, and three dihydroxybenzene isomers. The procedure of adsorption experiments is the same as in Section 2.3, except that the Fe$_3$O$_4$@PBA nanoparticles were eluted by acetic acid buffer solution (0.05 mL, pH = 3.0) for 60 min after adsorption and rinsed. The eluent was analysed by HPLC. The

Chromatographic analysis was carried out using a P230II HPLC with a Shim-pack C18 column (150 mm × 4.6 mm, 5 μm), and the mobile phase was methanol-water (12:88, v/v) at the flow rate of 1.0 mL/min. The column temperature was maintained at 30 °C, and the wavelength of the UV detector was 260 nm for adenosine and deoxyadenosine (280 nm for dihydroxybenzene isomers).

## 3. Results and Discussion

### 3.1. Characterization of $Fe_3O_4$@PBA Nanoparticles

Figure 2 is the transmission electron microscope (TEM) images of $Fe_3O_4$@PBA nanoparticles. As shown in Figure 2, three kinds of nanoparticles formed a core-shell structure including a core composed of $Fe_3O_4$ and a modified layer about 5 nm thick.

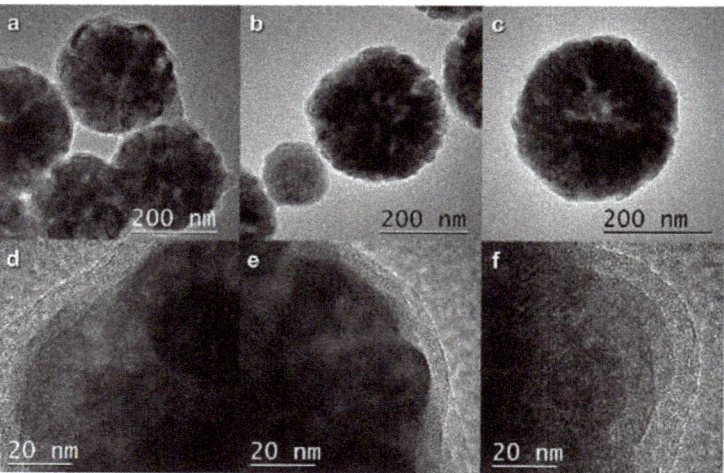

**Figure 2.** TEM images of the phenylboronic-acid-modified magnetic nanoparticles: $Fe_3O_4$@MPTES@VPBA (**a,d**), $Fe_3O_4$@MPTES@AAPBA (**b,e**), $Fe_3O_4$@MPS@MPBA (**c,f**).

The results of the scanning transmission electron microscopy (STEM) analysis of element distribution (Figure 3) shows that B, Si, N and S are uniformly distributed around the core composed of Fe and O, indicating that the modified layer composed of organosilane and phenylboric acid has been successfully introduced on the particle surface.

The above elements' distribution of $Fe_3O_4$@PBA nanoparticles was also confirmed by X-ray photoelectron spectroscopy (XPS). As shown in Figure 4, strong signal peaks at 530 eV and 284.8 eV indicate that there are abundant oxygen-containing groups and carbon elements on the surface of the nanoparticles, and the peaks at 710.5 eV, 399.6 eV, 285.0 eV, 191.4 eV, 163.2 eV, and 102.1 eV correspond to Fe2p, N1s, C1s, B1s, $S2p^3$ and $Si2p^2$, respectively. The signal peaks of S in Figure 4a,b shows that the hydrosulphonyl groups were introduced to the surface of $Fe_3O_4$ nanoparticles by the coating of MPTES, and the signal peak of N in Figure 4b and the signal peak of B in Figure 4a–c all indicate that phenylboric acid has been successfully modified on the particle surface.

The weight loss of different modified nanoparticles can provide more supporting evidence for the modification process under nitrogen atmosphere with a heating rate of 10 °C/min. As can be seen from Figure 5, the $Fe_3O_4$ nanoparticles had no significant weight loss during the heating process. However, with the modification of organosilane and the introduction of phenylboronic acid groups, the weight loss rate of the modified nanoparticles increased.

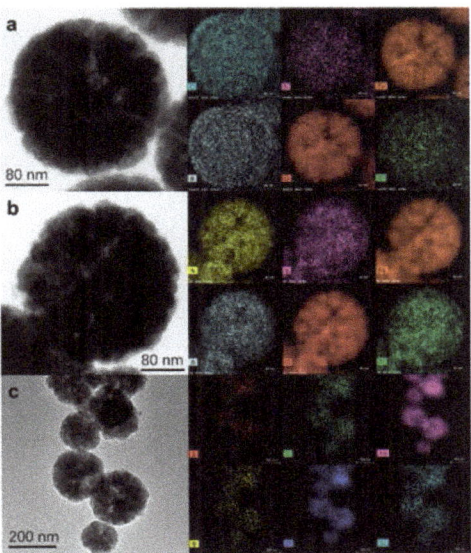

**Figure 3.** STEM elemental mapping images of the phenylboronic-acid-modified magnetic nanoparticles Fe$_3$O$_4$@MPTES@VPBA (**a**), Fe$_3$O$_4$@MPTES@AAPBA (**b**) and Fe$_3$O$_4$@MPS@MPBA (**c**).

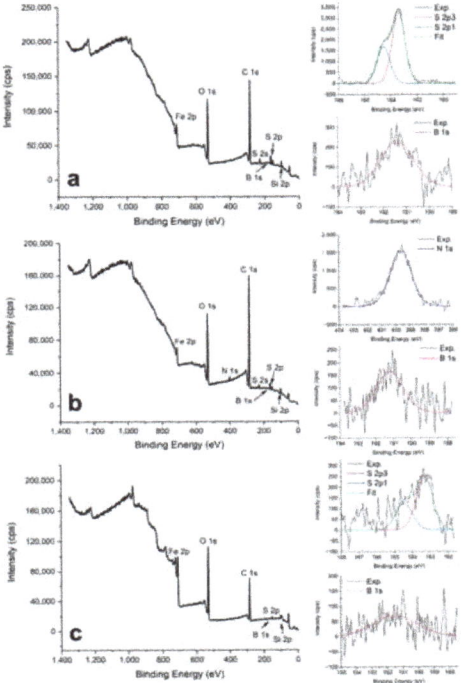

**Figure 4.** XPS spectra of the phenylboronic-acid-modified magnetic nanoparticles (**a**) Fe$_3$O$_4$@MPTES@VPBA, (**b**) Fe$_3$O$_4$@MPTES@AAPBA and (**c**) Fe$_3$O$_4$@MPS@MPBA.

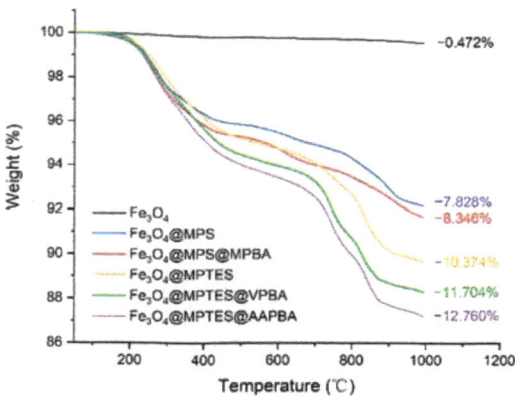

**Figure 5.** Thermogravimetric analysis curves of phenylboronic-acid-modified magnetic nanoparticles.

The room temperature magnetism of the prepared Fe$_3$O$_4$@PBA nanoparticles was analysed by using the vibration sample magnetometer (VSM). As can be seen from Figure 6, the room temperature saturation magnetic curves of the three kinds of Fe$_3$O$_4$@PBA nanoparticles have no obvious hysteresis loop and coercivity. In addition, nanoparticles dispersed in water can be reaggregated within 15 s under an applied magnetic field. The above analysis results and experimental phenomena shown that the prepared Fe$_3$O$_4$@PBA nanoparticles have excellent superparamagnetism.

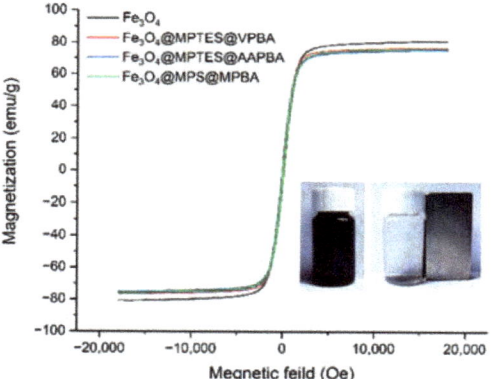

**Figure 6.** Magnetization curves of phenylboronic-acid-modified magnetic nanoparticles at room temperature.

### 3.2. Binding Properties of Fe$_3$O$_4$@PBA Nanoparticles

#### 3.2.1. Thermodynamics of Adsorption

The adsorption isotherms of the Fe$_3$O$_4$@MPTES@VPBA toward different adsorbates are shown in Figure 7. As can be seen from the adsorption isotherms, when the concentration of adsorbate is low, the adsorption capacity of the Fe$_3$O$_4$@MPTES@VPBA on the adsorbate increases synchronously with the concentration of the adsorbate. However, due to the high concentration of adsorbate, the boric acid binding site on the surface of the material tends to saturate, which makes the increasing trend of adsorption capacity slow down gradually. When the initial concentration of adsorbed substance was 1.0 mg/mL, the adsorption capacity basically reaches the maximum. Similarly, the isothermal adsorption curves of Fe$_3$O$_4$@MPTES@AAPBA and Fe$_3$O$_4$@MPS@MPBA shown the same trend of adsorption capacity as that of Fe$_3$O$_4$@MPTES@VPBA.

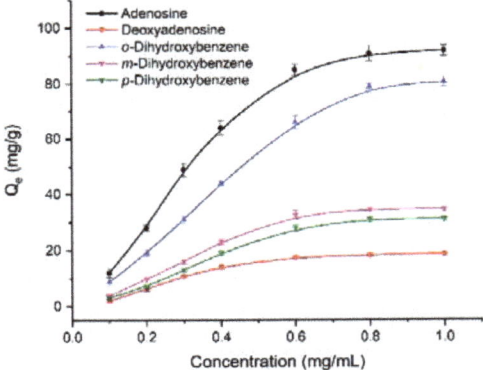

**Figure 7.** Adsorption isothermal curves of Fe$_3$O$_4$@MPTES@VPBA.

The adsorption isotherm of Fe$_3$O$_4$@MPS, Fe$_3$O$_4$@MPTES, and three kinds of Fe$_3$O$_4$@PBA nanoparticles for the adsorption of adenosine and o-dihydroxybenzene were analysed by Scatchard equation [33,34]. The fitting curves of Fe$_3$O$_4$@MPTES@VPBA and Fe$_3$O$_4$@MPTES are shown in Figure 8. As can be seen from Figure 8, due to the introduction of phenylboric acid high-affinity sites, the fitting curves of Fe$_3$O$_4$@MPTES@VPBA are composed of two straight lines representing high-affinity sites and low-affinity sites, respectively (Figure 8a,c). In contrast, the fitting curves of Fe$_3$O$_4$@MPTES show a linear relationship over the whole concentration range due to the absence of high-affinity sites (Figure 8b,d). Actually, the three kinds of Fe$_3$O$_4$@PBA nanoparticles have similar fitting analysis results. According to Table 1, there are two binding sites of high affinity and low affinity on the three kinds of Fe$_3$O$_4$@PBA nanoparticles. The equilibrium dissociation constant ($K_d$) of the high-affinity binding sites for the adsorption of adenosine and o-dihydroxybenzene was significantly lower than those of the low-affinity sites, indicating that the specific adsorption caused by boric acid affinity had a higher affinity than the non-specific adsorption caused by hydrogen bonding and electrostatic interaction.

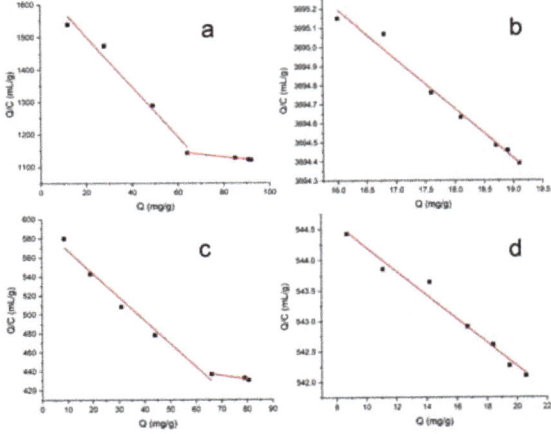

**Figure 8.** Scatchard fitting curve of Fe$_3$O$_4$@MPTES@VPBA (**a,c**) and Fe$_3$O$_4$@MPTES (**b,d**) for the adsorption of adenosine (**a,b**) and o-dihydroxybenzene (**c,d**).

Table 1. $K_D$ of different $Fe_3O_4$@PBA for the adsorption of adenosine and o-dihydroxybenzene.

| Adsorbate | Adsorbent | $K_D$ (mg/mL) | |
|---|---|---|---|
| | | High-Affinity Binding Sites | Low-Affinity Binding Sites |
| Adenosine | $Fe_3O_4$@MPTES@VPBA | 0.13 | 1.29 |
| | $Fe_3O_4$@MPTES@AAPBA | 0.05 | 1.95 |
| | $Fe_3O_4$@MPTES | — | 3.68 |
| | $Fe_3O_4$@MPS@MPBA | 0.05 | 1.33 |
| | $Fe_3O_4$@MPS | — | 3.45 |
| o-Dihydroxybenzene | $Fe_3O_4$@MPTES@VPBA | 0.41 | 2.50 |
| | $Fe_3O_4$@MPTES@AAPBA | 0.33 | 2.64 |
| | $Fe_3O_4$@MPTES | — | 5.15 |
| | $Fe_3O_4$MPS@MPBA | 0.13 | 2.73 |
| | $Fe_3O_4$@MPS | — | 4.17 |

Boric acid groups can specifically form ester rings with ortho-dihydroxy-containing compounds, resulting in selective adsorption. Unlike adenosine, deoxyadenosine does not have adjacent hydroxyl groups. The adsorption of deoxyadenosine by $Fe_3O_4$@PBA depends only on hydrogen bonding and electrostatic interaction, which results in the adsorption capacity of deoxyadenosine being significantly lower than that of adenosine.

Of the three isomers of dihydroxybenzene, o-Dihydroxybenzene is the only one that has adjacent hydroxyl groups, and it forms ester rings with boric acid much more easily than m-dihydroxybenzene and p-dihydroxybenzene. Therefore, the adsorption capacity and selectivity of $Fe_3O_4$@PBA for o-dihydroxybenzene are greater than that of m-dihydroxybenzene and p-dihydroxybenzene.

### 3.2.2. Kinetics of Adsorption

As can be seen from the adsorption kinetics curve of $Fe_3O_4$@MPTES@VPBA (Figure 9), at the initial stage of adsorption, a relatively stable complex was formed between the adsorbate and the boric acid binding site, which causes the adsorption capacity to rise rapidly. When most of the boric acid binding sites on the surface are occupied, it becomes difficult for the adsorbate to bind to the deeper binding sites, resulting in a slow increase in the amount of adsorption. With the continuous progress of the adsorption, all binding sites in the material are saturated, and the adsorption capacity tends to be stable. As shown in Figure 9, all the three kinds of prepared $Fe_3O_4$@PBA nanoparticles showed the same trend of adsorption capacity and reached their maximum adsorption capacity at 80~100 min.

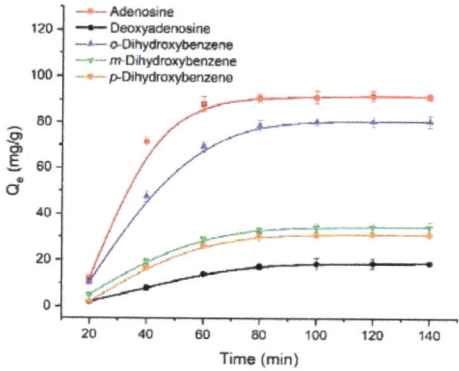

Figure 9. Kinetics of adsorption curves of $Fe_3O_4$@MPTES@VPBA.

## 3.3. Selectivity

A solution of adenosine or deoxyadenosine (1.0 mg/L) was adsorbed by the three prepared $Fe_3O_4$@PBA nanoparticles, respectively. The adsorption selectivity factor ($\alpha = Q_{adenosine}/Q_{deoxyadenosine}$) was calculated. Similarly, the adsorption selectivity factors of dihydroxybenzene isomers were also obtained by the same method. The calculation results are shown in Table 2.

**Table 2.** Absorption capacity ($Q_e$) and selectivity factor ($\alpha$) of different adsorbates by $Fe_3O_4$@PBA nanoparticles.

| Adsorbate | $Fe_3O_4$@MPTES@VPBA | | $Fe_3O_4$@MPTES@AAPBA | | $Fe_3O_4$@MPS@MPBA | |
|---|---|---|---|---|---|---|
| | $Q_e$ (mg/g) | $\alpha$ | $Q_e$ (mg/g) | $\alpha$ | $Q_e$ (mg/g) | $\alpha$ |
| Adenosine | 91.9 | - | 92.7 | - | 94.6 | - |
| Deoxyadenosine | 18.5 | 4.97 | 19.1 | 4.85 | 17.3 | 5.47 |
| o-Dihydroxybenzene | 80.6 | - | 79.3 | - | 83.1 | - |
| m-Dihydroxybenzene | 34.6 | 2.33 | 36.4 | 2.18 | 28.5 | 2.92 |
| p-Dihydroxy-benzene | 31.1 | 2.59 | 32.6 | 2.43 | 26.2 | 3.17 |

According to the Table 2, the adsorption capacity of the three kinds of $Fe_3O_4$@PBA nanoparticles toward adenosine was significantly higher than that of deoxyadenosine, and the adsorption selectivity factor reached more than 4.5. Furthermore, the adsorption selectivity factor for o-dihydroxybenzene ranged from 2.18 to 3.17, indicating that all three kinds of $Fe_3O_4$@PBA nanoparticles offered excellent adsorption selectivity toward adenosine and o-dihydroxybenzene. Particularly, the $Fe_3O_4$@MPS@MPBA exhibited the highest adsorption capacity and selectivity factor for adenosine and o-dihydroxybenzene.

The adsorption selectivity was further confirmed by HPLC analysis of the eluents after adsorption of the mixed solutions. Figure 10 presents the HPLC chromatograms of the eluents of the mixed solution (1.0 mg/mL) with equal mass ratio of adenosine and deoxyadenosine adsorbed by three kinds of $Fe_3O_4$@PBA nanoparticles, respectively. As can be seen from Figure 10, the chromatographic peak area of adenosine in the eluents was significantly higher than that of deoxyadenosine, which proved that three kinds of $Fe_3O_4$@PBA nanoparticles also had obvious adsorption selectivity for adenosine in the mixed solution of adenosine and deoxyadenosine.

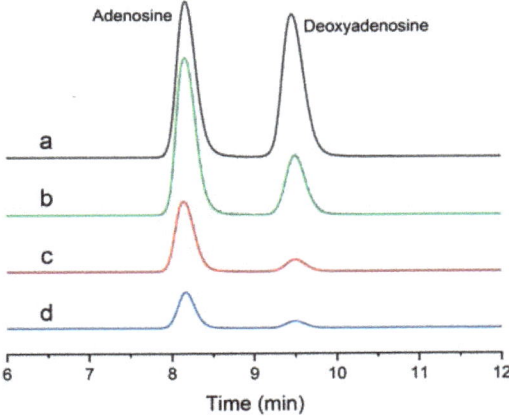

**Figure 10.** HPLC chromatograms of the mixed solution of adenosine and deoxyadenosine (**a**) and the chromatograms of the eluent for $Fe_3O_4$@MPS@MPBA (**b**), $Fe_3O_4$@MPTES@AAPBA (**c**) and $Fe_3O_4$@MPTES@VPBA (**d**) after being adsorbed in the mixture solution. Peak area ratio of adenosine to deoxyadenosine: 1:0.26 (**b**), 1:0.29 (**c**), 1:0.27 (**d**).

The HPLC chromatograms of the eluents of the mixed solution (1.0 mg/mL) with an equal mass ratio of o-dihydroxybenzene, m-dihydroxybenzene and p-dihydroxybenzene adsorbed by the three kinds of Fe$_3$O$_4$@PBA nanoparticles, respectively, are shown in Figure 11. Obviously, the peak area of o-dihydroxybenzene was the largest in all the eluents. It was also significantly different from that of m-dihydroxybenzene and p-dihydroxybenzene, further indicating that the three kinds of Fe$_3$O$_4$@PBA nanoparticles also showed the adsorption selectivity of o-dihydroxybenzene in the mixed solution of the dihydroxybenzene isomers.

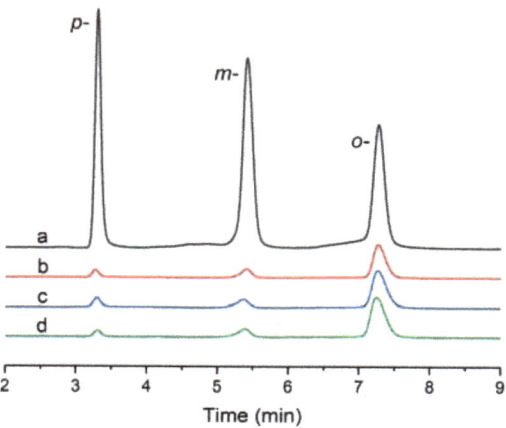

**Figure 11.** Chromatograms of mixed solution of o-dihydroxybenzene, m-dihydroxybenzene and p-dihydroxybenzene (**a**) and the chromatograms of the eluent for Fe$_3$O$_4$@MPS@MPBA (**b**), Fe$_3$O$_4$@MPTES@AAPBA (**c**) and Fe$_3$O$_4$@MPTES@VPBA (**d**) after being adsorbed in the mixture solution. Peak area ratio of o-dihydroxybenzene to m- and p-: 1:0.12: 0.06 (**b**), 1:0.26: 0.17 (**c**), 1:0.21:0.14 (**d**).

## 4. Conclusions

In this study, a new modification strategy for Fe$_3$O$_4$ magnetic nanoparticles with phenylboric acid was designed using the direct silanizing method and thiol-ene click reaction. Based on this new strategy, three kinds of phenylboric-acid-modified magnetic nanoparticles (Fe$_3$O$_4$@PBA) were prepared by a simple process under mild reaction conditions. The three kinds of Fe$_3$O$_4$@PBA nanoparticles obtained showed excellent adsorption capacity and selectivity for ortho-dihydroxy-containing compounds, including adenosine and o-dihydroxybenzene. The reactivity of acrylates was higher than that of acrylamide and styrene among the double-bond monomers that produce thiol-ene click chemistry, which resulted in a relatively larger number of phenylboronic acid binding sites on the surface of magnetic nanoparticles modified with 4-mercaptophenylboronic acid (MPBA). Therefore, Fe$_3$O$_4$@MPS@MPBA exhibited the highest adsorption capacity and selectivity factor for adenosine and o-diphenol among the three kinds of Fe$_3$O$_4$@PBA.

Furthermore, boric acid with low $pk_a$ value should be used in the modification of nanoparticles to reduce the bonding pH value of Fe$_3$O$_4$@PBA, which enables Fe$_3$O$_4$@PBA to selectively adsorb glycopeptides and glycoproteins in biological samples.

**Author Contributions:** Conceptualization, L.Y.; methodology, S.X.; validation, A.D.; investigation, H.Z.; data curation, J.Z.; writing—original draft preparation, H.Z.; writing—review and editing, B.W. All authors have read and agreed to the published version of the manuscript.

**Funding:** This research was funded by the Applied Basic Research Foundation of Yunnan Province, grant number 202201AT070029.

**Institutional Review Board Statement:** Not applicable.

**Informed Consent Statement:** Not applicable.

23. Zhang, W.; Liu, W.; Li, P.; Xiao, H.B.; Wang, H.; Tang, B. A fluorescence nanosensor for glycoproteins with activity based on the molecularly imprinted spatial structure of the target and boronate affinity. *Angew. Chem. Int. Ed. Engl.* **2014**, *53*, 12489–12493. [CrossRef] [PubMed]
24. Wei, J.R.; Ni, Y.L.; Zhang, W.; Zhang, Z.Q.; Zhang, J. Detection of glycoprotein through fluorescent boronic acid-based molecularly imprinted polymer. *Anal. Chim. Acta* **2017**, *960*, 110–116. [CrossRef]
25. Ye, J.; Chen, Y.; Liu, Z. A boronate affinity sandwich assay: An appealing alternative to immunoassays for the determination of glycoproteins. *Angew. Chem. Int. Ed. Engl.* **2014**, *53*, 10386–10389. [CrossRef]
26. Hoyle, C.E.; Bowman, C.N. Thiol-ene click chemistry. *Angew. Chem. Int. Ed. Engl.* **2010**, *49*, 1540–1573. [CrossRef]
27. Such, G.K.; Johnston, A.P.R.; Liang, K.; Caruso, F. Synthesis and functionalization of nanoengineered materials using click chemistry. *Prog. Polym. Sci.* **2012**, *37*, 985–1003. [CrossRef]
28. Hayase, G.; Kanamori, K.; Hasegawa, G.; Maeno, A.; Kaji, H.; Nakanishi, K. A superamphiphobic macroporous silicone monolith with marshmallow-like flexibility. *Angew. Chem. Int. Ed. Engl.* **2013**, *52*, 10788–10791. [CrossRef]
29. Li, P.; Wang, X.; Zhao, Y. Click chemistry as a versatile reaction for construction and modification of metal-organic frameworks. *Coord. Chem. Rev.* **2019**, *380*, 484–518. [CrossRef]
30. Deng, H.; Lin, L.X.; Qing, P.; Wang, X.; Chen, J.P.; Li, Y.D. Monodisperse magnetic single-crystal ferrite microspheres. *Angew. Chem. Int. Ed.* **2005**, *44*, 2782–2785. [CrossRef]
31. Bi, C.; Zhang, S.; Li, Y.; He, X.; Chen, L.; Zhang, Y. Boronic acid-functionalized iron oxide magnetic nanoparticles via distillation-precipitation polymerization and thiol-yne click chemistry for the enrichment of glycoproteins. *New J. Chem.* **2018**, *42*, 17331–17338. [CrossRef]
32. Zhang, S.; He, X.; Chen, L.; Zhang, Y. Boronic acid functionalized magnetic nanoparticles via thiol–ene click chemistry for selective enrichment of glycoproteins. *New J. Chem.* **2014**, *38*, 4212. [CrossRef]
33. Zhu, S.; Xia, M.; Chu, Y.; Khan, M.A.; Lei, W.; Wang, F.; Muhmood, T.; Wang, A. Adsorption and Desorption of Pb(II) on l-Lysine Modified Montmorillonite and the simulation of Interlayer Structure. *Appl. Clay Sci.* **2019**, *169*, 40–47. [CrossRef]
34. Zhu, S.; Khan, M.A.; Kameda, T.; Xu, H.; Wang, F.; Xia, M.; Yoshioka, T. New insights into the capture performance and mechanism of hazardous metals Cr(3+) and Cd(2+) onto an effective layered double hydroxide based material. *J. Hazard Mater.* **2022**, *426*, 128062. [CrossRef] [PubMed]

**Disclaimer/Publisher's Note:** The statements, opinions and data contained in all publications are solely those of the individual author(s) and contributor(s) and not of MDPI and/or the editor(s). MDPI and/or the editor(s) disclaim responsibility for any injury to people or property resulting from any ideas, methods, instructions or products referred to in the content.

**Data Availability Statement:** Not applicable.

**Conflicts of Interest:** The authors declare no conflict of interest. The funders had no role in the design of the study.

## References

1. Prucek, R.; Tucek, J.; Kilianova, M.; Panacek, A.; Kvitek, L.; Filip, J.; Kolar, M.; Tomankova, K.; Zboril, R. The targeted antibacterial and antifungal properties of magnetic nanocomposite of iron oxide and silver nanoparticles. *Biomaterials* **2011**, *32*, 4704–4713. [CrossRef] [PubMed]
2. Erathodiyil, N.; Ying, J.Y. Functionalization of inorganic nanoparticles for bioimaging applications. *Acc. Chem. Res.* **2011**, *44*, 925–935. [CrossRef] [PubMed]
3. Liu, G.; Gao, J.H.; Ai, H.; Chen, X.Y. Applications and potential toxicity of magnetic iron oxide nanoparticles. *Small* **2013**, *9*, 1533–1545. [CrossRef] [PubMed]
4. Babamiri, B.; Salimi, A.; Hallaj, R. Switchable electrochemiluminescence aptasensor coupled with resonance energy transfer for selective attomolar detection of $Hg^{2+}$ via CdTe@CdS/dendrimer probe and Au nanoparticle quencher. *Biosens. Bioelectron.* **2018**, *102*, 328–335. [CrossRef] [PubMed]
5. Moazzen, M.; Ahmadkhaniha, R.; Gorji, M.E.; Yunesian, M.; Rastkari, N. Magnetic solid-phase extraction based on magnetic multi-walled carbon nanotubes for the determination of polycyclic aromatic hydrocarbons in grilled meat samples. *Talanta* **2013**, *115*, 957–965. [CrossRef] [PubMed]
6. Facchi, D.P.; Cazetta, A.L.; Canesin, E.A.; Almeida, V.C.; Bonafé, E.G.; Kipper, M.J.; Martins, A.F. New magnetic chitosan/alginate/$Fe_3O_4$@$SiO_2$ hydrogel composites applied for removal of Pb(II) ions from aqueous systems. *Chem. Eng. J.* **2018**, *337*, 595–608. [CrossRef]
7. Fan, R.; Min, H.; Hong, X.; Yi, Q.; Liu, W.; Zhang, Q.; Luo, Z. Plant tannin immobilized $Fe_3O_4$@$SiO_2$ microspheres: A novel and green magnetic bio-sorbent with superior adsorption capacities for gold and palladium. *J. Hazard. Mater.* **2019**, *364*, 780–790. [CrossRef]
8. Cai, Y.R.; Pan, H.H.; Xu, X.R.; Hu, Q.H.; Li, L.; Tang, R.K. Ultrasonic controlled morphology transformation of hollow calcium phosphate nanospheres: A smart and biocompatible drug release system. *Chem. Mater.* **2007**, *19*, 3081–3083. [CrossRef]
9. Dobson, J. Gene therapy progress and prospects: Magnetic nanoparticle-based gene delivery. *Gene Ther.* **2006**, *13*, 283–287. [CrossRef]
10. Zhang, Z.; Niu, N.; Gao, X.; Han, F.; Chen, Z.; Li, S.; Li, J. A new drug carrier with oxygen generation function for modulating tumor hypoxia microenvironment in cancer chemotherapy. *Colloids Surf. B. Biointerfaces* **2019**, *173*, 335–345. [CrossRef] [PubMed]
11. Justin, R.; Tao, K.; Román, S.; Chen, D.; Xu, Y.; Geng, X.; Ross, I.M.; Grant, R.T.; Pearson, A.; Zhou, G.; et al. Photoluminescent and superparamagnetic reduced graphene oxide-iron oxide quantum dots for dual-modality imaging, drug delivery and photothermal therapy. *Carbon* **2016**, *97*, 54–70. [CrossRef]
12. Sonvico, F.; Dubernet, C.; Colombo, P.; Couvreur, P. Metallic colloid nanotechnology, applications in diagnosis and therapeutics. *Curr. Pharm. Des.* **2005**, *11*, 2095–2105. [CrossRef] [PubMed]
13. Xu, C.; Sun, S. Monodisperse magnetic nanoparticles for biomedical applications. *Polym. Int.* **2007**, *56*, 821–826. [CrossRef]
14. Villalonga, M.L.; Borisova, B.; Arenas, C.B.; Villalonga, A.; Arévalo-Villena, M.; Sánchez, A.; Pingarrón, J.M.; Briones-Pérez, A.; Villalonga, R. Disposable electrochemical biosensors for *Brettanomyces bruxellensis* and total yeast content in wine based on core-shell magnetic nanoparticles. *Sensor. Actuat. B Chem.* **2019**, *279*, 15–21. [CrossRef]
15. Gu, H.; Ho, P.L.; Tsang, K.W.; Wang, L.; Xu, B. Using biofunctional magnetic nanoparticles to capture vancomycin-resistant enterococci and other gram-positive bacteria at ultralow concentration. *J. Am. Chem. Soc.* **2003**, *125*, 15702–15703. [CrossRef] [PubMed]
16. Jiang, W.; Wu, L.; Duan, J.; Yin, H.; Ai, S. Ultrasensitive electrochemiluminescence immunosensor for 5-hydroxymethylcytosine detection based on $Fe_3O_4$@$SiO_2$ nanoparticles and PAMAM dendrimers. *Biosens. Bioelectron.* **2018**, *99*, 660–666. [CrossRef] [PubMed]
17. Liu, Z.; He, H. Synthesis and applications of boronate affinity materials: From class selectivity to biomimetic specificity. *Acc. Chem. Res.* **2017**, *50*, 2185–2193. [CrossRef]
18. Brooks, W.L.; Sumerlin, B.S. Synthesis and applications of boronic acid-containing polymers: From materials to medicine. *Chem. Rev.* **2016**, *116*, 1375–1397. [CrossRef]
19. Li, D.J.; Chen, Y.; Liu, Z. Boronate affinity materials for separation and molecular recognition: Structure, properties and applications. *Chem. Soc. Rev.* **2015**, *44*, 8097–8123. [CrossRef]
20. Nishiyabu, R.; Kubo, Y.; James, T.D.; Fossey, J.S. Boronic acid building blocks: Tools for sensing and separation. *Chem. Commun.* **2011**, *47*, 1106–1123. [CrossRef]
21. Xing, R.; Wang, S.; Bie, Z.; He, H.; Liu, Z. Preparation of molecularly imprinted polymers specific to glycoproteins, glycans and monosaccharides via boronate affinity controllable-oriented surface imprinting. *Nat. Protoc.* **2017**, *12*, 964–987. [CrossRef] [PubMed]
22. Luo, J.; Huang, J.; Cong, J.J.; Wei, W.; Liu, X.Y. Double recognition and selective extraction of glycoprotein based on the molecular imprinted graphene oxide and boronate affinity. *ACS Appl. Mater. Interfaces* **2017**, *9*, 7735–7744. [CrossRef] [PubMed]

MDPI  
St. Alban-Anlage 66  
4052 Basel  
Switzerland  
Tel. +41 61 683 77 34  
Fax +41 61 302 89 18  
www.mdpi.com

*Separations* Editorial Office  
E-mail: separations@mdpi.com  
www.mdpi.com/journal/separations